VOLUME FIVE HUNDRED AND SIXTY

METHODS IN ENZYMOLOGY

RNA Modification

METHODS IN ENZYMOLOGY

VOLUME FIVE HUNDRED AND SIXTY

METHODS IN ENZYMOLOGY

RNA Modification

Edited by

CHUAN HE
Howard Hughes Medical Institute
The University of Chicago, Chicago
Illinois, USA

ELSEVIER

AMSTERDAM • BOSTON • HEIDELBERG • LONDON
NEW YORK • OXFORD • PARIS • SAN DIEGO
SAN FRANCISCO • SINGAPORE • SYDNEY • TOKYO
Academic Press is an imprint of Elsevier

Academic Press is an imprint of Elsevier
225 Wyman Street, Waltham, MA 02451, USA
525 B Street, Suite 1800, San Diego, CA 92101-4495, USA
125 London Wall, London, EC2Y 5AS, UK
The Boulevard, Langford Lane, Kidlington, Oxford OX5 1GB, UK

First edition 2015

ISBN: 978-0-12-802192-7
ISSN: 0076-6879

For information on all Academic Press publications
visit our website at http://store.elsevier.com/

CONTENTS

CONTRIBUTORS

Ninette Amariglio
Cancer Research Center, Chaim Sheba Medical Center, Tel Hashomer, and Sackler School of Medicine, Tel Aviv University, Tel Aviv, Israel

Thomas J. Begley
College of Nanoscale Engineering and Science, State University of New York, Albany, New York, USA

Weiling Maggie Cai
Department of Microbiology, National University of Singapore, and Singapore-MIT Alliance for Research and Technology, Singapore

Bradley R. Cairns
HHMI, Department of Oncological Sciences, Huntsman Cancer Institute, University of Utah School of Medicine, Salt Lake City, Utah, USA

Thomas M. Carlile
Department of Biology, Massachusetts Institute of Technology, Cambridge, Massachusetts, USA

Kai Chen
Department of Chemistry, Institute for Biophysical Dynamics, Howard Hughes Medical Institute, The University of Chicago, Chicago, Illinois, USA

Yok Hian Chionh
Department of Microbiology, National University of Singapore, and Singapore-MIT Alliance for Research and Technology, Singapore

Alexander H. Dalpke
Department of Infectious Diseases, Medical Microbiology and Hygiene, University of Heidelberg, Heidelberg, Germany

Peter C. Dedon
Singapore-MIT Alliance for Research and Technology, Singapore; Department of Biological Engineering, and Center for Environmental Health Sciences, Massachusetts Institute of Technology, Cambridge, Massachusetts, USA

Dan Dominissini
Department of Chemistry, and Institute for Biophysical Dynamics, The University of Chicago, Chicago, Illinois, USA

Tatjana Eigenbrod
Department of Infectious Diseases, Medical Microbiology and Hygiene, University of Heidelberg, Heidelberg, Germany

Danica Galonić Fujimori
Department of Cellular and Molecular Pharmacology, and Department of Pharmaceutical Chemistry, University of California San Francisco, San Francisco, California, USA

Wendy V. Gilbert
Department of Biology, Massachusetts Institute of Technology, Cambridge, Massachusetts, USA

Chen Gu
Department of Biological Engineering, Massachusetts Institute of Technology, Cambridge, Massachusetts, USA

Michael P. Guy
Department of Biochemistry and Biophysics, Center for RNA Biology, University of Rochester School of Medicine, Rochester, New York, USA

Chuan He
Department of Chemistry, Department of Biochemistry and Molecular Biology, Institute for Biophysical Dynamics; Howard Hughes Medical Institute, The University of Chicago, Chicago, Illinois, USA

Mark Helm
Institute of Pharmacy and Biochemistry, Johannes Gutenberg-University of Mainz, Mainz, Germany

Fabian Hia
Singapore-MIT Alliance for Research and Technology, Singapore

Ya-Ming Hou
Department of Biochemistry and Molecular Biology, Thomas Jefferson University, Philadelphia, Pennsylvania USA

Chao Huang
Process Science Downstream, Bristol-Myers Squibb Company, Syracuse, New York, USA

Steffen Kaiser
Institute of Pharmacy and Biochemistry, Johannes Gutenberg-University of Mainz, Mainz, Germany

Byeong-il Kang
Department of Chemistry and Biotechnology, Graduate School of Engineering, University of Tokyo, Bunkyo-ku, Tokyo, Japan

Patrick Keller
Institute of Pharmacy and Biochemistry, Johannes Gutenberg-University of Mainz, Mainz, Germany

Stefanie Kellner
Department of Biological Engineering, Massachusetts Institute of Technology, Cambridge, Massachusetts, USA

Vahid Khoddami*
HHMI, Department of Oncological Sciences, Huntsman Cancer Institute, University of Utah School of Medicine, Salt Lake City, Utah, USA

*Current address: Department of Cell Biology, Howard Hughes Medical Institute, Harvard Medical School, 240 Longwood Avenue, Boston, MA 02115, USA

Xiaoyu Li
State Key Laboratory of Protein and Plant Gene Research, School of Life Sciences, Peking University, Beijing, China

Kok Seong Lim
Department of Biological Engineering, Massachusetts Institute of Technology, Cambridge, Massachusetts, USA

Jianzhao Liu
Department of Chemistry; Institute for Biophysical Dynamics, and Howard Hughes Medical Institute, The University of Chicago, Chicago, Illinois, USA

Nian Liu
Department of Chemistry, University of Chicago, Chicago, Illinois, USA

Guan-Zheng Luo
Department of Chemistry, Institute for Biophysical Dynamics, Howard Hughes Medical Institute, The University of Chicago, Chicago, Illinois, USA

Shiqing Ma
State Key Laboratory of Protein and Plant Gene Research, School of Life Sciences; Peking-Tsinghua Center for Life Sciences, and Academy for Advanced Interdisciplinary Studies, Peking University, Beijing, China

Isao Masuda
Department of Biochemistry and Molecular Biology, Thomas Jefferson University, Philadelphia, Pennsylvania USA

Megan E. McBee
Singapore-MIT Alliance for Research and Technology, Singapore, and Department of Biological Engineering, Massachusetts Institute of Technology, Cambridge, Massachusetts, USA

Kenjyo Miyauchi
Department of Chemistry and Biotechnology, Graduate School of Engineering, University of Tokyo, Bunkyo-ku, Tokyo, Japan

Sharon Moshitch-Moshkovitz
Cancer Research Center, Chaim Sheba Medical Center, Tel Hashomer, and Sackler School of Medicine, Tel Aviv University, Tel Aviv, Israel

Chee Sheng Ng
Singapore-MIT Alliance for Research and Technology, and School of Biological Sciences, Nanyang Technological Institute, Singapore

Shunpei Okada
Department of Chemistry and Biotechnology, Graduate School of Engineering, University of Tokyo, Bunkyo-ku, Tokyo, Japan

Tao Pan
Department of Biochemistry and Molecular Biology, and Institute of Biophysical Dynamics, University of Chicago, Chicago, Illinois, USA

Yan Ling Joy Pang
Department of Biological Engineering, Massachusetts Institute of Technology, Cambridge, Massachusetts, USA

Matthew J. Payea
Department of Biochemistry and Biophysics, Center for RNA Biology, University of Rochester School of Medicine, Rochester, New York, USA

Eric M. Phizicky
Department of Biochemistry and Biophysics, Center for RNA Biology, University of Rochester School of Medicine, Rochester, New York, USA

Erin G. Prestwich
Department of Biological Engineering, Massachusetts Institute of Technology, Cambridge, Massachusetts, USA

I. Ramesh Babu
Department of Biological Engineering, Massachusetts Institute of Technology, Cambridge, Massachusetts, USA

Gideon Rechavi
Cancer Research Center, Chaim Sheba Medical Center, Tel Hashomer, and Sackler School of Medicine, Tel Aviv University, Tel Aviv, Israel

Katharina Rimbach
Department of Infectious Diseases, Medical Microbiology and Hygiene, University of Heidelberg, Heidelberg, Germany

Maria F. Rojas-Duran
Department of Biology, Massachusetts Institute of Technology, Cambridge, Massachusetts, USA

Yuriko Sakaguchi
Department of Chemistry and Biotechnology, Graduate School of Engineering, University of Tokyo, Bunkyo-ku, Tokyo, Japan

Masayuki Sakurai
Department of Chemistry and Biotechnology, Graduate School of Engineering, University of Tokyo, Bunkyo-ku, Tokyo, Japan

Matthias Schaefer
Vienna Biocenter, Max F. Perutz Laboratories, Department of Biochemistry and Cell Biology, Universität Wien, Vienna, Austria

Vanja Stojković
Department of Cellular and Molecular Pharmacology, University of California San Francisco, San Francisco, California, USA

Tsutomu Suzuki
Department of Chemistry and Biotechnology, Graduate School of Engineering, University of Tokyo, Bunkyo-ku, Tokyo, Japan

Hiroki Ueda
Department of Chemistry and Biotechnology, Graduate School of Engineering, University of Tokyo, Bunkyo-ku, Tokyo, Japan

Guowei Wu
Department of Cellular and Molecular Medicine, University of California, San Diego, La Jolla, California, USA

Archana Yerra
HHMI, Department of Oncological Sciences, Huntsman Cancer Institute, University of Utah School of Medicine, Salt Lake City, Utah, USA

Chengqi Yi
State Key Laboratory of Protein and Plant Gene Research, School of Life Sciences; Peking-Tsinghua Center for Life Sciences, and Department of Chemical Biology and Synthetic and Functional Biomolecules Center, College of Chemistry and Molecular Engineering, Peking University, Beijing, China

Yi-Tao Yu
Center for RNA Biology, Department of Biochemistry and Biophysics, University of Rochester Medical Center, Rochester, New York, USA

Yanan Yue
Department of Chemistry; Institute for Biophysical Dynamics, and Howard Hughes Medical Institute, The University of Chicago, Chicago, Illinois, USA

PREFACE

Protein posttranslational modifications and DNA methylation are well-known chemical modifications on biomacromolecules that play critical functional roles. Reversible chemical modifications on DNA and histone proteins represent a new layer of epigenetic regulation of gene expression in eukaryotes. Posttranscriptional modifications are also ubiquitous in RNA. To date, over 100 different RNA modifications have been identified with a wide variety of chemical diversities including methylation, hydroxylation, acetylation, deamination, isomerization, selenylation, conjugation with amino acids and sugars, and addition of short- and long-chain lipids and fatty acids, to adding a new aromatic ring whose installation requires six genes in the human genome. The wide variety of RNA modifications appears to be a strategy used by nature to entail and facilitate a great diversity of structures and cellular functions for different RNA species.

Transfer RNA (tRNA) and ribosomal RNA (rRNA) are known to carry a diverse range of chemical modifications. The chemical nature and positions of many of these modifications have been well defined with RNA digestion followed by liquid chromatography separation coupled with mass spectrometry or thin-layer chromatography. With the availability of much more sensitive mass spectrometry technologies, researchers have begun to not only uncover new forms of modifications on tRNA and rRNA but also quantify dynamic changes at the global scale. Genetic studies as well as recent explosion in genome sequencing have uncovered mutations of tRNA-/rRNA-modifying enzymes that are associated with noticeable phenotypes from model organisms to humans. RNA modifications are also involved in human immune response and bacterial antibiotic resistance. This volume of *Methods in Enzymology* presents essential tools and methods to characterize and investigate these diverse RNA modifications and their biological functions.

Because of the low abundance of messenger RNA (mRNA) and long noncoding RNA (lncRNA), modifications on these RNA species have been difficult to investigate in the past. With the availability of highly sensitive mass spectrometry tools that can accurately quantify the modification level, and the development of "next-generation" high-throughput sequencing technology, we can not only accurately monitor dynamic changes of these less abundant mRNA/lncRNA modifications but also determine their transcriptome-wide distributions. These technology advances, coupled with the ability to knock down or overexpress corresponding enzymes and proteins involved in installing, recognizing, and erasing RNA modifications, open new doors to investigate functions of mRNA/lncRNA modifications. This volume also includes methods to study reversible N^6-methyladenosine methylation in eukaryotic mRNA/lncRNA and sequencing technologies that have been developed to map different mRNA/lncRNA modifications. Functional studies have revealed reversible/dynamic RNA modification as a new mode of posttranscriptional gene expression regulation mechanism.

I want to thank all authors and experts for their efforts and contributions to this volume of *Methods in Enzymology*. I hope this volume will facilitate future investigations of RNA modifications in biological regulations.

CHUAN HE

CHAPTER ONE

Methodology for the High-Throughput Identification and Characterization of tRNA Variants That Are Substrates for a tRNA Decay Pathway

Matthew J. Payea, Michael P. Guy, Eric M. Phizicky[1]
Department of Biochemistry and Biophysics, Center for RNA Biology, University of Rochester School of Medicine, Rochester, New York, USA
[1]Corresponding author: e-mail address: eric_phizicky@urmc.rochester.edu

Contents

Abstract

The rapid tRNA decay (RTD) pathway is a tRNA quality control pathway known to degrade several specific hypomodified or destabilized tRNAs in the yeast *Saccharomyces cerevisiae*. In this chapter, we describe seven methods for identifying RTD substrates, with a focus on two new approaches: a high-throughput approach that utilizes a suppressor tRNA library, fluorescence-activated cell sorting, and deep sequencing, and has greatly expanded the known range of RTD substrates; and a poison primer extension assay that allows for the measurement of levels of suppressor tRNA variants, even in the presence of highly similar endogenous tRNAs. We also discuss different applications of the use of the high-throughput and poison primer extension methodologies for different problems in tRNA biology.

Methods in Enzymology, Volume 560
ISSN 0076-6879
http://dx.doi.org/10.1016/bs.mie.2015.03.003

1. INTRODUCTION

tRNA folding and stability is crucial for efficient translation, and defects in either property can lead to reduced quantities of tRNA, resulting in growth defects in yeast and disease in humans (Hopper, 2013; Yarham, Elson, Blakely, McFarland, & Taylor, 2010). In the yeast *Saccharomyces cerevisiae*, there are two major cellular quality control pathways known to degrade defective tRNA species. The first pathway is the nuclear surveillance pathway, which acts on pre-tRNA in the nucleus through the use of the nuclear exosome and the TRAMP complex (Kadaba, Wang, & Anderson, 2006; Vanacova et al., 2005) by degrading pre-$tRNA_i^{Met}$ lacking the m^1A_{58} modification or with a misprocessed 3′ trailer (Ozanick et al., 2009) and a fraction of wild-type (WT) pre-tRNAs (Gudipati et al., 2012). The second pathway is the rapid tRNA decay (RTD) pathway, which degrades specific mature, hypomodified, or destabilized tRNA species through the activity of the 5′–3′ exonucleases Rat1 and Xrn1 (Alexandrov et al., 2006; Chernyakov, Whipple, Kotelawala, Grayhack, & Phizicky, 2008). RTD is elicited in mutants lacking any of several modifications in the body of the tRNA or through destabilizing mutations, and for all identified RTD substrates, *MET22* deletion fully restores tRNA levels and growth (Alexandrov et al., 2006; Chernyakov, Whipple, et al., 2008; Dewe, Whipple, Chernyakov, Jaramillo, & Phizicky, 2012; Guy et al., 2014; Kotelawala, Grayhack, & Phizicky, 2008; Whipple, Lane, Chernyakov, D'Silva, & Phizicky, 2011). Suppression of RTD in *met22Δ* strains is presumed to be due to inhibition of the exonucleases Rat1 and Xrn1 by the metabolite 3′-phosphoadenosine-5′-phosphate, which has increased levels when Met22 is inhibited (Dichtl, Stevens, & Tollervey, 1997; Murguia, Belles, & Serrano, 1996).

RTD is known to act on several specific tRNA species, which have been identified and studied using seven approaches (Fig. 1). The first approach was to use microarrays to compare the tRNA levels on a genome-wide scale in *trm8Δ trm4Δ* temperature-sensitive modification mutants (lacking m^7G_{46} and m^5C) and in related strains under semipermissive conditions. In this way, we identified the RTD substrate $tRNA^{Val(AAC)}$, since it had reduced tRNA levels in the *trm8Δ trm4Δ* mutant relative to WT or the corresponding single mutants (Alexandrov et al., 2006).

In the second approach, northern blots were used to examine both the rate and the specificity of tRNA degradation for RTD substrates in

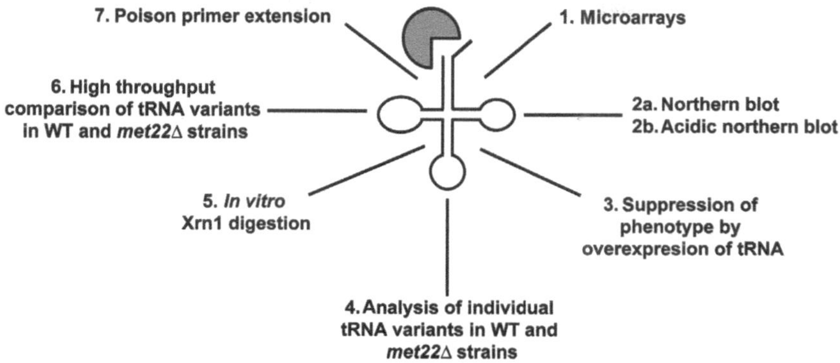

Figure 1 Different approaches used to identify and analyze RTD substrates.

temperature-sensitive modification mutants. In this approach, RNA isolated from cells at different time points after temperature shift was analyzed for levels of specific tRNAs. From this analysis, we found that 50% of the $tRNA^{Val(AAC)}$ was degraded in a *trm8Δ trm4Δ* mutant within 30 min of a shift from 28 to 37 °C, while the similarly hypomodified $tRNA_i^{Met}$, $tRNA^{Met}$, and $tRNA^{Phe}$ showed no decrease (Alexandrov et al., 2006; Chernyakov, Whipple, et al., 2008). Furthermore, the relative levels of charged and uncharged tRNA could be measured by performing the northern blot under acidic conditions, which showed that levels of charged $tRNA^{Val(AAC)}$ were reduced by 50% within 25 min of temperature shift in a *trm8Δ trm4Δ* mutant and that the uncharged $tRNA^{Val(AAC)}$ levels appeared unaffected (Alexandrov et al., 2006).

The third approach was through high-copy tRNA suppression, wherein a high-copy plasmid expressing a particular tRNA was introduced into a temperature-sensitive tRNA modification mutant. If the tRNA was an RTD substrate and the temperature sensitivity was the result of a single tRNA species being degraded, then overexpression of the tRNA would suppress the defect. Thus, we found that the temperature sensitivity of a *trm8Δ trm4Δ* mutant was suppressed by a high-copy plasmid expressing $tRNA^{Val(AAC)}$, indicating that temperature sensitivity was primarily due to the loss of $tRNA^{Val(AAC)}$, and that the missing modifications were important for tRNA stability (Alexandrov et al., 2006). Similarly, the RTD substrates of several other tRNA modification mutants have also been identified using this approach, including $tRNA^{Ser(CGA)}$ and $tRNA^{Ser(UGA)}$ in *tan1Δ trm44Δ* mutants (lacking ac^4C_{12} and Um_{44}) and in *trm1Δ trm4Δ* mutants (lacking $m^{2,2}G_{26}$ and m^5C) (Chernyakov, Whipple, et al., 2008; Dewe et al., 2012; Kotelawala et al., 2008).

Fourth, we used a genetic replacement approach to identify RTD determinants in the $tRNA^{Ser}$ family by substituting the single essential $tRNA^{Ser(CGA)}$ gene (*SUP61*) with different $tRNA^{Ser(CGA)}$ variants in the WT and the *met22Δ* strain, and then assaying for growth at different temperatures. Using this approach, we determined that the combined acceptor and T-stem stabilities were strong determinants for RTD susceptibility in the $tRNA^{Ser(CGA)}$ gene family (Whipple et al., 2011). This conclusion was further supported by a fifth approach to measure RTD, in which we showed *in vitro* that $tRNA^{Ser(CGA)}$ variants lacking ac^4C_{12} and Um_{44}, or with destabilizing mutations in the acceptor stem, were more prone to digestion by Xrn1 and more susceptible to 5′ phosphate removal by calf-intestinal phosphatase (Whipple et al., 2011).

In this review, we will discuss our recently developed sixth and seventh approaches for the study of RTD substrates, which have proven extremely valuable in broadening our understanding of the RTD pathway. The sixth approach uses a fluorescent reporter to comprehensively analyze libraries of thousands of tRNA variants in WT and *met22Δ* strains. Through this approach, we have identified 643 likely RTD substrate candidates, many in regions not expected to elicit RTD based on previous work (Guy et al., 2014). We will show data demonstrating that this approach can be used to study tRNA function under different conditions and will discuss other applications of the approach.

The seventh approach employs poison primer extension to measure the tRNA levels in a WT and *met22Δ* strain and is valuable in its ability to specifically measure a variant tRNA even in the presence of the WT tRNA whose sequence may differ by only a single residue. We provide a detailed methodology of this approach and discuss some of its other possible applications.

2. HIGH-THROUGHPUT IDENTIFICATION OF tRNA SUBSTRATES DEGRADED BY THE RTD PATHWAY

In this approach, a fluorescent reporter that is sensitive to the levels of functional tRNA is used to identify tRNA variants that are subject to RTD because there is less fluorescence in a WT strain (in which RTD is functional) than in a *met22Δ* strain (in which RTD is inhibited). For this analysis, we used the previously developed RNA-ID fluorescent reporter, which contains the inducible $P_{GAL1,10}$ bidirectional promoter expressing red fluorescent protein (*RFP*) in one direction and GFP_{oc} (green fluorescent protein

(GFP) with a UAA nonsense codon) in the other direction (Dean & Grayhack, 2012). Expression of GFP_{oc} relative to RFP is 0.5% of that of the corresponding *GFP* reporter without a nonsense codon, and is increased to 94% if the strain has an integrated $SUP4_{oc}$ gene (encoding $tRNA^{Tyr}$ in which the GUA anticodon is mutated to UUA by a G34U mutation, Fig. 2A), which efficiently suppresses UAA nonsense codons (Guy et al., 2014). We tested the ability of our reporter to distinguish between substrates and nonsubstrates of the RTD pathway by examining a previously identified temperature-sensitive RTD substrate, $SUP4_{oc}$-*G62C* (Whipple et al., 2011), for a significant increase in GFP fluorescence when expressed in a *met22Δ* strain compared to a WT strain. We found that there was approximately a twofold increase in GFP fluorescence in the *met22Δ* strain at 28 °C, indicating that the levels of functional $SUP4_{oc}$ tRNA were increased by a measurable level upon inhibition of the RTD pathway (Guy et al., 2014).

The use of the RNA-ID reporter for the identification of RTD substrates offered two distinct advantages over previously described genetic approaches to define substrates. First, a growth or temperature-sensitive phenotype would no longer be required to assay RTD, and second, a library of thousands of tRNA variants could be tested in parallel.

We used this approach to analyze a comprehensive library of approximately 220,000 independent $SUP4_{oc}$ variants integrated into both WT and *met22Δ* strains. We sorted the libraries into four bins based on GFP fluorescence using fluorescent activated cell sorting (FACS), followed by deep sequencing of the extracted genomic DNA from each bin. Each variant was assigned a functional score (GFP^{SEQ}) based on its frequency of distribution in the four bins and the measured median GFP fluorescence of each bin, relative to that of $SUP4_{oc}$. This analysis resulted in GFP^{SEQ} scores for 9563 single and double mutant variants in the WT library, of which 4263 were also scored in the *met22Δ* library (Guy et al., 2014). The GFP^{SEQ} scores for each variant were then used to calculate the fold increase in a *met22Δ* strain compared to a WT strain (GFP^{SEQ} (*met22Δ*)/GFP^{SEQ} (WT)), called the GFP^{SEQ} RTD ratio, which was used to identify variants as putative RTD substrates (Fig. 2B). Based on the comparison of GFP fluorescence for $SUP4_{oc}$-*G62C* and the overall trend of RTD ratios for the library, an RTD ratio cutoff of 2 was chosen to identify putative RTD substrates. For single mutants, this resulted in the identification of 38 putative RTD substrates among the 213 possible single mutant variants (Guy et al., 2014).

To validate the results from our deep-sequencing analysis, we further analyzed 60 $SUP4_{oc}$ variants with a range of RTD ratios from 0.9 to 24.4

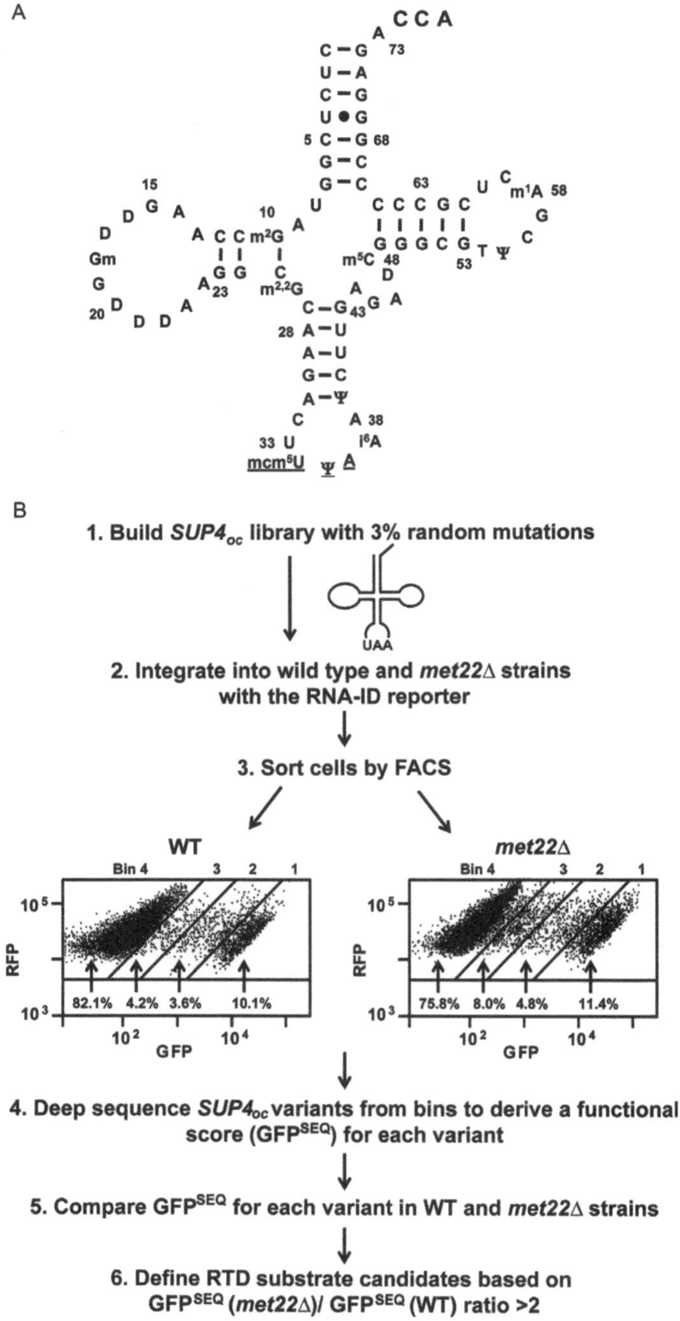

Figure 2 High-throughput approach for identification of *SUP4*$_{oc}$ RTD substrates. (A) Schematic representation of the sequence of *SUP4*$_{oc}$, a G34U mutant of a tRNA^{Tyr} gene. The anticodon residues are underlined. (B) General approach for identification of *SUP4*$_{oc}$ RTD substrates using yeast libraries of *SUP4*$_{oc}$ variants in WT and *met22*Δ RNA-ID reporter strains.

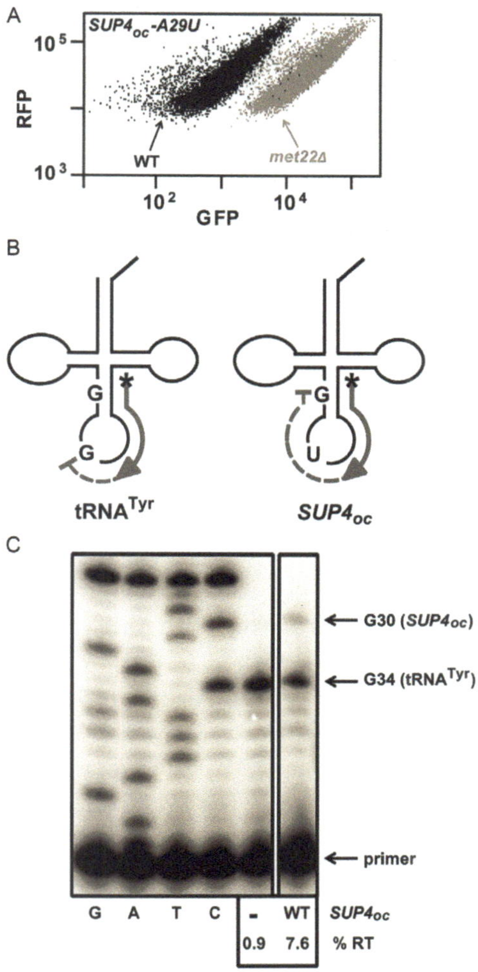

Figure 3 Methods for validating $SUP4_{oc}$ RTD candidates identified through a high-throughput approach. (A) Scatterplot of flow cytometry of the reconstructed $SUP4_{oc}$-*A29U* variant in WT (RTD on, median GFP/RFP $= 0.045 \pm 0.003$) and *met22*Δ cells (RTD off, median GFP/RFP $= 0.47 \pm 0.06$) with the integrated RNA-ID reporter. (B) Schematic representation of the poison primer extension assay with ddCTP used for determining levels of $SUP4_{oc}$ variants. (C) The poison primer extension assay measures the levels of $SUP4_{oc}$ in bulk RNA. Bulk RNA from the indicated strains was analyzed by primer extension in the presence of ddCTP. A sequencing ladder is on the left. (See the color plate.)

by reconstructing and retesting the variants using flow cytometry (Fig. 3A). Our results showed that 19 of 21 putative RTD substrates identified by a GFP^{SEQ} RTD ratio of 2 or more also had an RTD ratio of 2 or more when measured by flow cytometry, and that 26 of 30 strains with a GFP^{SEQ} RTD ratio less than 2 also had an RTD ratio of less than 2 when measured by flow cytometry. These experiments validated the ability of our high-throughput and deep-sequencing analysis to accurately predict *SUP4*$_{oc}$ variants as RTD substrates.

3. MEASUREMENT OF tRNA LEVELS OF RTD SUBSTRATE *SUP4*$_{oc}$ VARIANTS IN THE PRESENCE OF WT tRNA^{Tyr}

To determine if the increased GFP^{SEQ} of a *SUP4*$_{oc}$ variant in the *met22Δ* strain (relative to the WT strain) was due to increased tRNA, we sought to compare the levels of *SUP4*$_{oc}$ variants in each strain. However, this analysis was difficult because *SUP4*$_{oc}$ differs from WT tRNA^{Tyr} by only one nucleotide (the G34U mutation). Thus, it was not practical to either selectively purify *SUP4*$_{oc}$ away from the endogenous tRNA^{Tyr} or selectively probe for *SUP4*$_{oc}$ using a northern blot.

To solve this problem, we used a poison primer extension analysis (Driscoll, Wynne, Wallis, & Scott, 1989), which is performed similarly to conventional primer extension, except that one nucleotide triphosphate (NTP) in the reaction is replaced by a 2′,3′-dideoxy NTP that, once incorporated into the cDNA, prevents any further extension by the reverse transcriptase. This allows for one primer to extend two different tRNA species, in this case producing a signal unique to *SUP4*$_{oc}$ variants even in the presence of contaminating WT tRNA^{Tyr} (Guy et al., 2014).

Since *SUP4*$_{oc}$ bears a G34U mutation in the anticodon that is four bases 3′ of the next guanosine residue (G_{30}), we reverse transcribed RNA from a primer complementary to both WT tRNA^{Tyr} and *SUP4*$_{oc}$ in the presence of 2′,3′-dideoxy CTP (ddCTP), which stops the extension when the reverse transcriptase inserts ddCTP opposite a guanosine residue. Thus, the primer extension product produced from *SUP4*$_{oc}$ variants stops after G_{30}, whereas the product from WT tRNA^{Tyr} stops after G_{34} (Fig. 3B). As expected, in the absence of a *SUP4*$_{oc}$ variant, the ddCTP poison primer extension produces a stop primarily at G_{34}, corresponding to the WT tRNA^{Tyr}, whereas in the presence of *SUP4*$_{oc}$, there is an expected increase at G_{30} (Fig. 3C). Since both signals are easily distinguishable on a gel, this method also conveniently provides for an internal control for normalization of the *SUP4*$_{oc}$ variant relative to WT tRNA^{Tyr} (Fig. 3C).

We initially performed our poison primer extension analysis on tRNA that was purified using a biotinylated DNA probe complementary to residues 76–55 of tRNA^{Tyr}, which purified both WT tRNA^{Tyr} and *SUP4*$_{oc}$ variants. Although a poison primer extension assay of the purified tRNA discriminated between $tRNA^{Tyr}$ and $SUP4_{oc}$ with high resolution, the same analysis was equally discriminating with bulk low-molecular-weight RNA (Fig. 4A). We have used bulk RNA as our source for all further primer extensions, because it is simpler to make and it is not subject to the possible

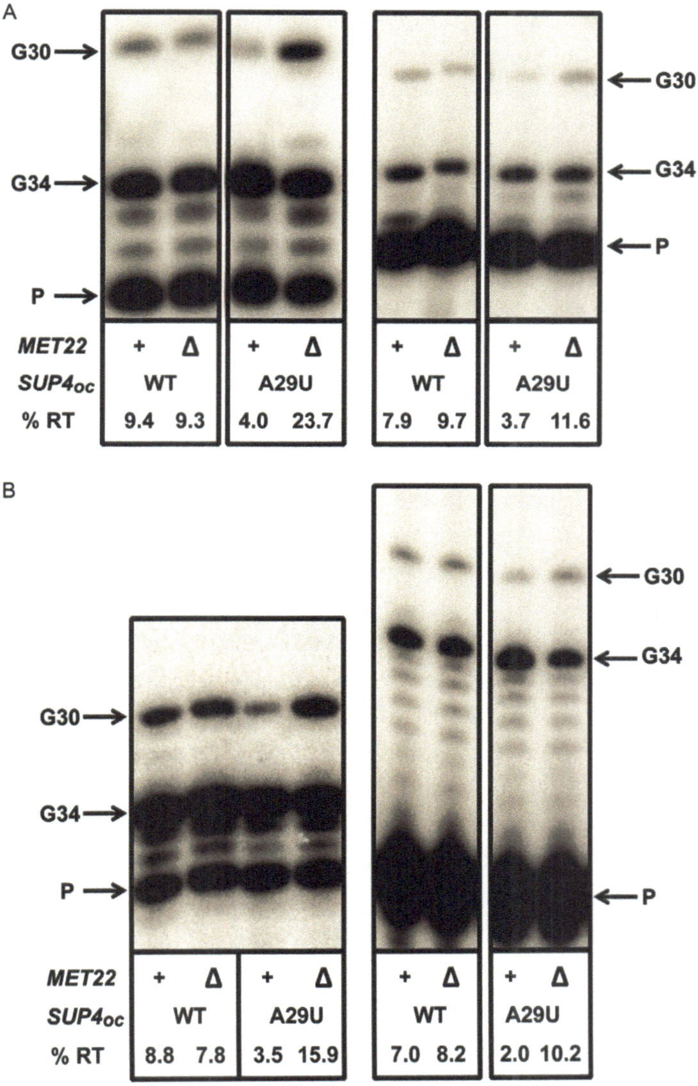

Figure 4 Analysis of levels of the *$SUP4_{oc}$-A29U* variant by poison primer extension. (A) Comparison of poison primer extension assay using either bulk RNA (left panel) or purified $tRNA^{Tyr}$ (right panel). (B) Comparison of poison primer extension assay using different primers. Primer P5 is complementary to residues 57–37 of $tRNA^{Tyr}$, and primer P7 is complementary to residues 62–43 of $tRNA^{Tyr}$. P, primer.

anomalies that could occur during purification of different *SUP4*$_{oc}$ variants together with endogenous $tRNA^{Tyr}$.

3.1 Preparation of Bulk Low-Molecular-Weight RNA

The bulk low-molecular-weight RNA was prepared by the hot phenol method as previously described (Chernyakov, Baker, Grayhack, & Phizicky, 2008), but modified slightly for the extraction from a smaller (2 OD-mL) pellet, harvested from 2 mL cells grown in rich (1% yeast extract, 2% bacto peptone (YP) containing glucose or raffinose and galactose) media to $OD_{600} = \sim 1.0$. The pellet is suspended in one volume of 200 μL of RNA extraction buffer (0.1 *M* NaOAc, pH 5.2, 20 m*M* EDTA, 1% SDS), and then mixed with an equivalent volume of phenol and lysed by the above method (Chernyakov, Baker, et al., 2008). The yield of RNA from 2 OD-mL cells is approximately 1.5–5.0 μg, based on A^{260}.

3.2 Preparation of Labeled DNA Primer

The selection of a primer for the poison primer extensions with ddCTP of *SUP4*$_{oc}$ variants was determined using two considerations. First, the primer had to anneal at a location in the tRNA where the first guanosine residue encountered during extension would be at residue 34, because any earlier encountered guanosine residue would result in only one signal for both *SUP4*$_{oc}$ and WT $tRNA^{Tyr}$. Second, as with all primer extensions, the primer had to have a high enough T_m for efficient binding to the tRNA (70–75 °C), but not too high so as to prevent nonspecific hybridization. Our first primer annealed to residues 57–37 (P5) of $tRNA^{Tyr}$ and *SUP4*$_{oc}$ variants. We also developed an alternative primer that annealed to residues 62–43 (P7) of $tRNA^{Tyr}$, still satisfying the above considerations, but allowing measurement of tRNA levels for variants with mutations at residues 42–38. For most of the examined variants, the two primers were interchangeable and although there was some variability in the measured read-through of variants in different experiments (possibly attributed to differing RNA amounts or annealing efficiency), the ratios of read-through between WT and *met22Δ* strains were largely unchanged for those variants (Fig. 4A, left panel, Fig. 4B). To label the primer for poison primer extension analysis, we did the following:

1. Incubate 2 μL of 30 μ*M* primer (~60 pmol) suspended in tris-EDTA (TE) 8 (10 m*M* Tris–HCl, pH 8.0, 1 m*M* EDTA, pH 8.0) with 60 pmol [γ-^{32}P] ATP (ICN, 6000 Ci/mmol), and T4 polynucleotide kinase in

the provided buffer (New England Biolabs, Cat# M0201L) for 1 h at 37 °C. The equimolar ratio of primer and ATP ensures high specific activity primers, which increases the sensitivity of primer extension.

2. Purify labeled primer from unincorporated nucleotides by passaging through a Micro BioSpin 6 column (BioRad, Cat# 732-6221) by spinning at 1000 × *g* for 4 min.
3. Discard the column and then mix the eluted sample with equal volume loading dye (100% formamide, 10 m*M* EDTA, pH 8.0, 1 mg/mL Bromophenol Blue, 1 mg/mL xylene cyanol) and then heat at 95 °C for 3 min.
4. Load sample onto a 15% 29:1 bis:acrylamide, 7 *M* urea, tris-borate-EDTA (TBE) pH 8.3 (0.88 *M* Tris base, 0.025 *M* Na_2EDTA, 0.9 *M* boric acid) sequencing gel and resolve the primer from contaminants by electrophoresis at 60 W in TBE buffer. The gel should be warm to the touch under these conditions.

   This step is important because we found that the primers used for our experiments (ordered from Integrated DNA Technologies) contained small amounts of contaminating species that were shorter than the designed primer. Although this would be a benign issue for northern blotting, it was a significant concern for the poison primer extension assay, because the contaminating primers were also extended and therefore produced multiple bands of varying lengths for a single stop.
5. Expose the gel to radioisotope film (Carestream Biomax MS Film, Cat# 829 4985) for 30 s in a dark room to locate the purified primer band. Cut band out of gel and mechanically break the gel by forcing it through a 5-mL syringe using the plunger into a 14-mL round-bottom tube (Falcon #2059).
6. Suspend the crushed gel in 500 μL of extraction buffer (0.5 *M* NH_4OAC, 0.10% SDS, 5 m*M* EDTA, pH 8.0) and incubate while shaking at 37 °C for 3–12 h.
7. After incubation, spin the tubes in a clinical centrifuge at low speed for 1 min. Remove the extraction buffer from the gel pieces by eluting through a Quik-Sep column (PerkinElmer Life Sciences, Cat# QS-PM) spun at 2000 rpm for 10 min in a clinical centrifuge at room temperature.
8. Add the eluted sample to a phase lock gel heavy tube and mix with 400 μL of a 25:24:1 mixture of phenol:chloroform:isoamyl alcohol (equilibrated with 0.5 *M* Tris–HCl, pH 7.5), and spin at a maximum speed of 5 min at room temperature in a microfuge.

9. Add the aqueous phase to a new tube containing 2 volumes of 100% ethanol with 5 μL of 5 mg/mL glycogen and freeze on dry ice for 30 min, spin for 30 min at a maximum speed in a microfuge at 4 °C, and then discard the supernatant. A pellet should be visible at this point (from the glycogen carrier), and the discarded supernatant should contain only a relatively small amount of radioactivity compared to the pellet.
10. Wash with 100 μL cold 70% ethanol without disturbing the pellet and then remove the supernatant.
11. Resuspend the pellet in 120 μL of water (~0.5 pmol primer/μL). The yield for 60 pmol of primer labeled with equimolar [γ-^{32}P] ATP (ICN, 6000 Ci/mmol) is variable among probes, but is approximately 2.5×10^7 CPM as measured by a scintillation counter.

3.3 Poison Primer Extension

1. Add 200 ng of bulk low-molecular-weight RNA (~2.4×10^{-4} mol tRNA, assuming the prep is 30% tRNA, corresponding to ~7 μmol tRNA^{Tyr}) to ~1.0 pmol of [γ-^{32}P] ATP-labeled primer from a stock of ~0.5 pmol/μL (as described in Section 3.2) and dilute with water to a final reaction volume of 5 μL.
2. Incubate the reaction at 95 °C for 3 min using a heat block to melt the tRNA, and then slow cool to 50 °C (about 1 h) and incubate for another 30 min, to allow for specific annealing of the labeled primer to the tRNA target.
3. Remove the annealed reaction tube from the heat block and place directly on ice for 2 min, followed by quick spin of the tube to collect liquid that may have condensed on the lid.
4. Add 5 μL of extension mix (1 m*M* dATP, 1 m*M* dGTP, 1 m*M* dTTP, 1 m*M* ddCTP, and 2 units Avian Myeloblastosis Virus (AMV) reverse transcriptase with the provided reaction buffer (Promega, Cat# PR-M9004)) to the annealed reaction on ice and then incubate for 1 h at 50 °C, prior to storage of reaction mixtures at −20 °C.

   We compared Superscript III (Invitrogen) and AMV and found that AMV had the most fidelity for incorporation of ddCTP and therefore the least background.

3.4 Analyzing Poison Primer Extension Using Polyacrylamide Gel Electrophoresis

1. Prerun a large 7 *M* urea 15% 29:1 bis:polyacrylamide sequencing gel at 60 W for 35–60 min, so that it is warm to the touch. Clear precipitated

urea from the lanes before and after prerunning the gel by blowing out each lane using a syringe filled with running buffer.

2. Thaw poison primer extension reactions and add 5 μL (half total reaction volume) to a 1.5 mL Eppendorf tube containing 5 μL loading dye (100% formamide, 10 m*M* EDTA, pH 8.0, 1 mg/mL Bromophenol Blue, 1 mg/mL xylene cyanol) and then heat at 95 °C for 3 min, followed by a low-speed spin.
3. Load samples onto the gel and resolve products by electrophoresis for 2.5–3 h at 60 W.
4. Remove gel from electrophoresis apparatus, remove glass plates, and then dry gel for 1.5 h at 83 °C using a gel dryer (BioRad Model 583 Gel Dryer) before exposing to a phosphorimager.

4. CONCLUSIONS AND ADDITIONAL APPLICATIONS OF THESE APPROACHES

The high-throughput method for identification of RTD substrates described here has greatly increased the number and the type of known substrates for this cellular pathway and has fueled experiments in our lab to understand the mechanisms of substrate recognition and RTD. Understanding these features of RTD in yeast may also shed light on specificity and mechanisms that occur during RTD in metazoans (Watanabe et al., 2013).

The high-throughput approach we described has a number of potential applications. We could use the same libraries to investigate temperature sensitivity of variants or to investigate RTD at different temperatures. As shown in Fig. 5A, the GFP/RFP value for $SUP4_{oc}$ is reduced 2.18-fold after growth at 37 °C (from GFP/RFP of 1.09 ± 0.02 at 28 °C to 0.50 ± 0.03 at 37 °C), whereas the GFP/RFP value for $SUP4_{oc}$-*A29U* GFP is reduced 3.75-fold (from 0.045 ± 0.003 to 0.012 ± 0.002), indicating that the $SUP4_{oc}$-*A29U* is slightly temperature sensitive. Media effects on suppression can also be investigated since they appear to differentially affect suppression of some variants. As shown in Fig. 5B, the GFP/RFP of $SUP4_{oc}$ is reduced by 1.45-fold in minimal media containing galactose relative to YP media containing galactose (1.07 ± 0.02 in YP galactose and 0.74 ± 0.04 in S-His galactose), whereas the $SUP4_{oc}$-*A29U* variant is reduced 2.64-fold (from 0.066 ± 0.01 to 0.025 ± 0.001). Similarly, differential effects of UAA nonsense codon context can be examined. In setting up our high-throughput analysis, we used a GFP_{oc} construct in a "poor" context, since this results in poor termination and correspondingly good suppression (Bonetti, Fu,

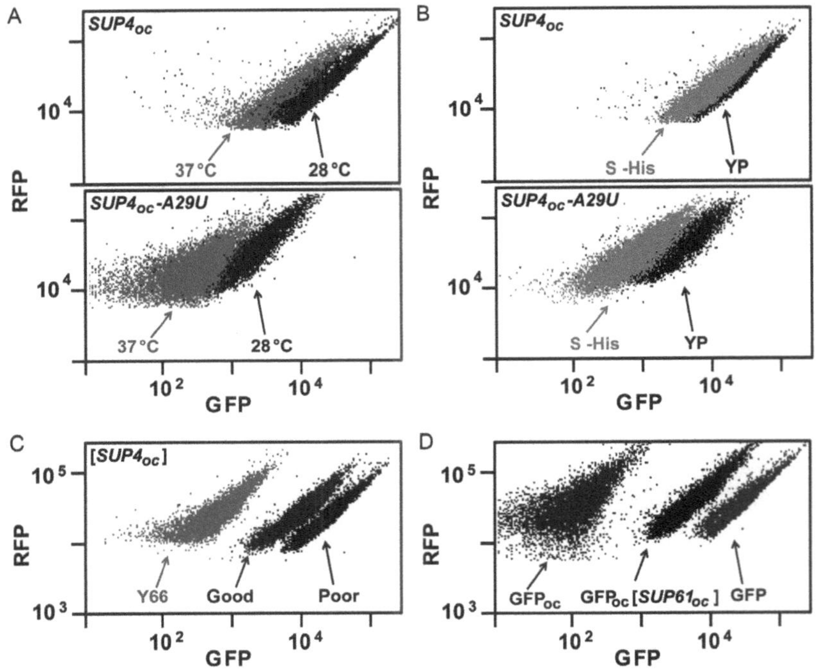

Figure 5 Effect of different variables on the activity of suppressor tRNAs as measured by flow cytometry. (A) Flow cytometry analysis comparing suppression by the *SUP4$_{oc}$-A29U* variant at different temperatures. *SUP4$_{oc}$* has a median GFP/RFP of 1.09 ± 0.02 at 28 °C and 0.50 ± 0.03 at 37 °C, and the *SUP4$_{oc}$* A29U variant has a median GFP/RFP of 0.045 ± 0.003 at 28 °C and 0.012 ± 0.002 at 37 °C. (B) Flow cytometry analysis comparing suppression by the *SUP4$_{oc}$* A29U variant in different media. *SUP4$_{oc}$* has a median GFP/RFP of 0.74 ± 0.04 in S-His galactose medium and 1.07 ± 0.02 in YP galactose medium, and the *SUP4$_{oc}$* A29U variant has a median GFP/RFP of 0.025 ± 0.001 in S-His galactose medium and 0.066 ± 0.01 in YP galactose medium. (C) Flow cytometry analysis comparing the effect of stop codon context on suppression by *SUP4$_{oc}$*. *SUP4$_{oc}$* has a median GFP/RFP of 1.04 ± 0.01 for a stop codon in a poor context, 0.36 ± 0.01 for a stop codon in a good context, and 0.032 ± 0.002 for the stop codon when inserted at amino acid 66 of GFP. (D) Flow cytometry analysis demonstrating that *SUP61$_{oc}$* suppresses *GFP$_{oc}$*. *SUP61$_{oc}$* has a median GFP/RFP of 0.22 ± 0.002 compared to 0.007 in the absence of a suppressor tRNA.

Moon, & Bedwell, 1995; Dean & Grayhack, 2012), giving us near-complete suppression (94% of that with a GFP reporter with no stop codon) and therefore a large dynamic range (235-fold) for scoring suppression frequency (Guy et al., 2014). Since suppression by *SUP4$_{oc}$* is reduced to 35% in a good-context *GFP$_{oc}$* reporter (Fig. 5C), one could use this reporter to examine a library of variants to identify those with differential suppression efficiency in this context, or with improved suppression efficiency relative to WT. Additionally, we could use a *GFP-Y66*$_{oc}$ construct to define determinants of TyrRS among our variants, since codon 66 of GFP is required to be tyrosine in GFP, and other variants do not fluoresce, or fluoresce with different excitation or emission maxima (Heim, Prasher, & Tsien, 1994). A *GFP-Y66*$_{oc}$ construct has only ~3.1% of the GFP/RFP signal (Fig. 5C), which should be enough to define determinants, but if not, the signal might be improved by changing the context of the stop codon or increasing copy number. We also have further applied this approach to the identification of RTD substrates in another tRNA species: tRNA$^{Ser(CGA)}$ (encoded by the *SUP61* gene) which when mutated to a UUA anticodon is a relatively efficient suppressor of our *GFP*$_{oc}$ construct (Fig. 5D).

In addition to the applications described above, variations of this high-throughput approach could be used to identify tRNA variants with altered function in any mutant background that impacts tRNA processing or function. This could include, for example, screening for tRNA variants that function better in reporter strains lacking different modification genes, in strains with mutations that affect tRNA function during translation, in strains with mutations that affect intracellular trafficking, or in strains with mutations that affect other quality control pathways, such as the nuclear surveillance pathway. With appropriate adaptation of this methodology, one might also extend this approach to other RNA molecules for which a selection can be applied, based on either growth or an appropriate reporter.

The poison primer extension methodology for measuring tRNA levels can be extended to analysis of relative levels of different isodecoders, tRNAs with the same anticodon but different tRNA bodies, which are commonly found in yeast and most other organisms (Goodenbour & Pan, 2006). A recent report shows that a specific isodecoder has a profound effect on the central nervous system in mice with a mutation in a factor that affects translation (Ishimura et al., 2014). In addition, poison primer extension methodology could be used to study the *in vitro* degradation of tRNA variants in complex RNA mixtures without the need for a previous purification step.

ACKNOWLEDGMENTS

We thank David Young, Stan Fields, Yohsiko Kon, Xiaoju Zhang, David Mathews, Kimberly Dean, and Elizabeth Grayhack for their important contributions in developing the high-throughput approach described in this chapter. This work was supported by NIH grant GM052347 to E.M.P., and M.J.P. was partially supported by NIH Training Grant in Cellular, Biochemical, and Molecular Sciences 5T32 GM068411.

REFERENCES

Alexandrov, A., Chernyakov, I., Gu, W., Hiley, S. L., Hughes, T. R., Grayhack, E. J., et al. (2006). Rapid tRNA decay can result from lack of nonessential modifications. *Molecular Cell, 21*(1), 87–96.

Bonetti, B., Fu, L., Moon, J., & Bedwell, D. M. (1995). The efficiency of translation termination is determined by a synergistic interplay between upstream and downstream sequences in Saccharomyces cerevisiae. *Journal of Molecular Biology, 251*(3), 334–345.

Chernyakov, I., Baker, M. A., Grayhack, E. J., & Phizicky, E. M. (2008). Chapter 11. Identification and analysis of tRNAs that are degraded in Saccharomyces cerevisiae due to lack of modifications. *Methods in Enzymology, 449*, 221–237.

Chernyakov, I., Whipple, J. M., Kotelawala, L., Grayhack, E. J., & Phizicky, E. M. (2008). Degradation of several hypomodified mature tRNA species in Saccharomyces cerevisiae is mediated by Met22 and the 5'–3' exonucleases Rat1 and Xrn1. *Genes & Development, 22*(10), 1369–1380.

Dean, K. M., & Grayhack, E. J. (2012). RNA-ID, a highly sensitive and robust method to identify cis-regulatory sequences using superfolder GFP and a fluorescence-based assay. *RNA, 18*(12), 2335–2344.

Dewe, J. M., Whipple, J. M., Chernyakov, I., Jaramillo, L. N., & Phizicky, E. M. (2012). The yeast rapid tRNA decay pathway competes with elongation factor 1A for substrate tRNAs and acts on tRNAs lacking one or more of several modifications. *RNA, 18*(10), 1886–1896.

Dichtl, B., Stevens, A., & Tollervey, D. (1997). Lithium toxicity in yeast is due to the inhibition of RNA processing enzymes. *The EMBO Journal, 16*(23), 7184–7195.

Driscoll, D. M., Wynne, J. K., Wallis, S. C., & Scott, J. (1989). An in vitro system for the editing of apolipoprotein B mRNA. *Cell, 58*(3), 519–525.

Goodenbour, J. M., & Pan, T. (2006). Diversity of tRNA genes in eukaryotes. *Nucleic Acids Research, 34*(21), 6137–6146.

Gudipati, R. K., Xu, Z., Lebreton, A., Seraphin, B., Steinmetz, L. M., Jacquier, A., et al. (2012). Extensive degradation of RNA precursors by the exosome in wild-type cells. *Molecular Cell, 48*(3), 409–421.

Guy, M. P., Young, D. L., Payea, M. J., Zhang, X., Kon, Y., Dean, K. M., et al. (2014). Identification of the determinants of tRNA function and susceptibility to rapid tRNA decay by high-throughput in vivo analysis. *Genes & Development, 28*(15), 1721–1732.

Heim, R., Prasher, D. C., & Tsien, R. Y. (1994). Wavelength mutations and posttranslational autoxidation of green fluorescent protein. *Proceedings of the National Academy of Sciences of the United States of America, 91*(26), 12501–12504.

Hopper, A. K. (2013). Transfer RNA post-transcriptional processing, turnover, and subcellular dynamics in the yeast Saccharomyces cerevisiae. *Genetics, 194*(1), 43–67.

Ishimura, R., Nagy, G., Dotu, I., Zhou, H., Yang, X. L., Schimmel, P., et al. (2014). Ribosome stalling induced by mutation of a CNS-specific tRNA causes neurodegeneration. *Science, 345*(6195), 455–459.

Kadaba, S., Wang, X., & Anderson, J. T. (2006). Nuclear RNA surveillance in Saccharomyces cerevisiae: Trf4p-dependent polyadenylation of nascent hypomethylated tRNA and an aberrant form of 5S rRNA. *RNA, 12*(3), 508–521.

Kotelawala, L., Grayhack, E. J., & Phizicky, E. M. (2008). Identification of yeast tRNA Um44 2'-O-methyltransferase (Trm44) and demonstration of a Trm44 role in sustaining levels of specific tRNASer species. *RNA, 14*, 158–169.

Murguia, J. R., Belles, J. M., & Serrano, R. (1996). The yeast HAL2 nucleotidase is an in vivo target of salt toxicity. *The Journal of Biological Chemistry, 271*(46), 29029–29033.

Ozanick, S. G., Wang, X., Costanzo, M., Brost, R. L., Boone, C., & Anderson, J. T. (2009). Rex1p deficiency leads to accumulation of precursor initiator tRNAMet and

polyadenylation of substrate RNAs in Saccharomyces cerevisiae. *Nucleic Acids Research, 37*(1), 298–308.
Vanacova, S., Wolf, J., Martin, G., Blank, D., Dettwiler, S., Friedlein, A., et al. (2005). A new yeast poly(A) polymerase complex involved in RNA quality control. *PLoS Biology, 3*(6), e189.
Watanabe, K., Miyagawa, R., Tomikawa, C., Mizuno, R., Takahashi, A., Hori, H., et al. (2013). Degradation of initiator tRNAMet by Xrn1/2 via its accumulation in the nucleus of heat-treated HeLa cells. *Nucleic Acids Research, 41*(8), 4671–4685.
Whipple, J. M., Lane, E. A., Chernyakov, I., D'Silva, S., & Phizicky, E. M. (2011). The yeast rapid tRNA decay pathway primarily monitors the structural integrity of the acceptor and T-stems of mature tRNA. *Genes & Development, 25*(11), 1173–1184.
Yarham, J. W., Elson, J. L., Blakely, E. L., McFarland, R., & Taylor, R. W. (2010). Mitochondrial tRNA mutations and disease. *Wiley Interdisciplinary Reviews. RNA, 1*(2), 304–324.

CHAPTER TWO

Nucleoside Analysis by Hydrophilic Interaction Liquid Chromatography Coupled with Mass Spectrometry

Yuriko Sakaguchi, Kenjyo Miyauchi, Byeong-il Kang, Tsutomu Suzuki[1]
Department of Chemistry and Biotechnology, Graduate School of Engineering, University of Tokyo, Bunkyo-ku, Tokyo, Japan
[1]Corresponding author: e-mail address: ts@chembio.t.u-tokyo.ac.jp

Contents

Abstract

RNA molecules contain a wide variety of chemical modifications that cannot be deduced from the genomic sequence. RNA modifications confer a chemical diversity to simple RNA molecules, enabling a greater variety of biological functions. To detect RNA modifications, highly sensitive analytical tools are required. Liquid chromatography/mass spectrometry (LC/MS) has been playing a vital role in analyzing minor modified nucleosides in RNA specimens from various sources. Reverse-phase chromatography (RPC) has been used for LC/MS for a long time because RPC is compatible with electrospray ionization (ESI) MS. However, RPC is not always suitable for detecting hydrophilic or polar nucleosides. We here describe a different mode of LC/MS for detecting RNA modifications using hydrophilic interaction liquid chromatography (HILIC). HILIC/ESI-MS is a valuable alternative for profiling modified nucleosides.

Methods in Enzymology, Volume 560
ISSN 0076-6879
http://dx.doi.org/10.1016/bs.mie.2015.03.015

1. INTRODUCTION

To date, over 100 different RNA modifications have been identified in RNA molecules across all domains of life. They have a wide variety of chemical diversity including methylation, hydroxylation, acetylation, deamination, isomerization, selenylation, reduction, cyclization, and conjugation with amino acids and sugars (Machnicka et al., 2013). In the past few years, our laboratory and others have reported several novel RNA modifications (Dumelin, Chen, Leconte, Chen, & Liu, 2012; Ikeuchi et al., 2010; Mandal et al., 2014; Miyauchi, Kimura, & Suzuki, 2013). The chemical space of RNA modification will likely expand with the development of improved analytical technologies. In contrast to the limited variation of DNA modifications, the wide variety of RNA modifications appears to be a strategy by which RNA molecules acquire a greater variety of cellular functions.

The modifications that occur in RNA molecules stabilize tertiary structures, modulate affinity for RNA-binding proteins, regulate decoding of genetic codes, determine the subcellular localization and lifetime of RNAs, and antagonize the innate immune system. However, the exact functions and biogenesis of many of these modifications still remain to be investigated. Moreover, a lack of RNA modification causes functional defects in RNA, leading to pathological consequences. Indeed, we previously reported that a lack of tRNA modification is a primary cause of mitochondrial encephalomyopathy (Suzuki, Nagao, & Suzuki, 2011a, 2011b). This is the first demonstration of human disease caused by an RNA modification disorder. Large-scale exome analyses have identified a number of genes encoding tRNA/rRNA-modifying enzymes that are associated with human diseases (Torres, Batlle, & Ribas de Pouplana, 2014).

To investigate the functional roles played by RNA modifications, it is essential to use highly sensitive analytical methods to detect and profile modified nucleosides in total RNA as well as individual RNAs. Reverse-phase chromatography (RPC) coupled with electrospray ionization (ESI) mass spectrometry (MS) has been used to analyze RNA modifications for several decades (Gaston & Limbach, 2014; Pomerantz & McCloskey, 1990; Su et al., 2014; Suzuki, Ikeuchi, Noma, Suzuki, & Sakaguchi, 2007). Our group has employed this technique combined with reverse genetics for the identification of novel genes responsible for the biogenesis of RNA modifications (Ikeuchi, Kitahara, & Suzuki, 2008; Ikeuchi, Shigi, Kato, Nishimura, & Suzuki, 2006; Kimura, Ikeuchi, Kitahara, Sakaguchi, &

Suzuki, 2012; Kimura & Suzuki, 2010; Kimura et al., 2014; Noma, Kirino, Ikeuchi, & Suzuki, 2006; Noma, Sakaguchi, & Suzuki, 2009; Soma et al., 2003; Suzuki et al., 2007). However, RPC/ESI-MS is not always suitable for detecting hydrophilic or polar compounds. In addition, the detection sensitivity for proton adducts of uridine and its derivatives is poor due to their low pK_a values (Banks, Shen, Whitehouse, & Fenn, 1994). Uridines tend to elute in high aqueous solvent in RPC as a result of their weak retention by octadecyl-silyl groups, which reduce their ionization efficiency in ESI.

Hydrophilic interaction liquid chromatography (HILIC) has been used for separation of polar compounds in complex mixtures (Alpert, 1990; Hemstrom & Irgum, 2006). Indeed, there are several reports of HILIC applied for analyzing nucleobases, nucleosides, and nucleotides (Marrubini, Mendoza, & Massolini, 2010). As an alternative to RPC/ESI-MS, HILIC can be applied to LC/MS analyses of modified nucleosides, especially for uridine derivatives to improve their detection sensitivity.

2. NUCLEOSIDE PREPARATION FOR LC/MS ANALYSIS

To analyze modified nucleosides in RNA, total RNA (or isolated tRNA) obtained from various organisms was digested with nuclease P1 (Wako Pure Chemical Industries) and bacterial alkaline phosphatase (BAP) derived from *Escherichia coli* strain C75 (BAP C75, Takara) (Suzuki, Suzuki, Wada, Saigo, & Watanabe, 2002; Suzuki et al., 2007). Nuclease P1, supplied as a lyophilized powder, was dissolved in ddH_2O. Both enzymes were dialyzed into ddH_2O by the drop dialysis method (Marusyk & Sergeant, 1980) and stored at −30 °C.

In the conventional method (Crain, 1990), total RNA is digested by nuclease P1 under acidic conditions (pH 5.3) and then dephosphorylated by BAP under basic conditions (pH 8.2), according to the optimal pH of each enzyme. However, in this condition, the cyclic form of N^6-threonylcarbamoyladenosine (ct^{6}A) can be converted to t^{6}A as a result of hydrolysis of the oxazolone ring in ct^{6}A under basic conditions (Miyauchi et al., 2013). Thus, we revised the protocol for preparing total nucleosides under acidic conditions as follows. The reaction solution (typically 40 μL) containing 1 μg/μL total RNA (40 μL) or individual RNA, 20 m*M* trimethylamine-acetate (pH 5.3), 0.1 units of nuclease P1, and 0.08–0.16 units of BAP C75 was incubated at 37 °C for 1 h. Although the activity of BAP decreases under acidic conditions, we confirmed that all nucleotides were completely dephosphorylated in this condition. Thus, the prepared nucleosides were ready for LC/MS analysis. The preparation can also be stored at −30 °C. It should be noted that some modified nucleosides tend to produce dinucleotides when this procedure is used (e.g., yWpA for wybutosine (Noma et al., 2006) and NmpN for 2′-*O*-methylated nucleosides).

3. HILIC/ESI-MS FOR TOTAL NUCLEOSIDES

A high-resolution mass spectrometer LTQ-Orbitrap (Thermo Fisher Scientific) equipped with an ESI source and HP1100 liquid chromatography system (Agilent Technologies) was used for analyzing the total nucleosides. For HILIC/ESI-MS, we used a ZIC-cHILIC column (3 μm particle size, 2.1 × 150 mm, Merck Millipore) and a guard column with the same resin (2.1 × 20 mm, Merck Millipore). The mobile phase is consisted of 5 m*M* ammonium acetate (pH 5.3) (solvent A) and acetonitrile (ACN) (solvent B), and the nucleosides (digested as described above) were injected and chromatographed with a flow rate of 100 μL/min in a multistep linear gradient; 90–40% B from 0 to 30 min, 40% B for 5 min, and then initialized to 90% B. The chromatographic eluent was directly introduced into the ESI source. Proton adducts of nucleosides were scanned in a positive polarity mode over a *m*/*z* range of 110–700 (or 110–900 for detecting dinucleotides) throughout the separation. Mass spectrometer parameters were tuned with uridine as a reference compound according to the manufacturer's instructions.

4. HILIC/ESI-MS VERSUS RPC/ESI-MS FOR NUCLEOSIDE ANALYSIS

First, the nucleoside profiles obtained by HILIC/ESI-MS were compared to those obtained by RPC/ESI-MS (Fig. 1). As a standard, 1 μL of a 100 μ*M* nucleoside solution (100 pmol each of A, U, G, and C) was mixed with 10 μL of ACN (11 μL in total with 91% ACN) and injected into the HILIC/ESI-MS system with the conditions as described in the previous section. For RPC/ESI-MS, 100 pmol of each of the four nucleosides in 11 μL dH_2O was injected into an ODS column in the conditions described previously (Suzuki et al., 2007). The same mass spectrometer (LTQ-Orbitrap) and its parameters were used for both systems. As shown in the UV trace (Fig. 1), the elution order of the four nucleosides was completely different between the two systems. In HILIC/ESI-MS, the most hydrophobic

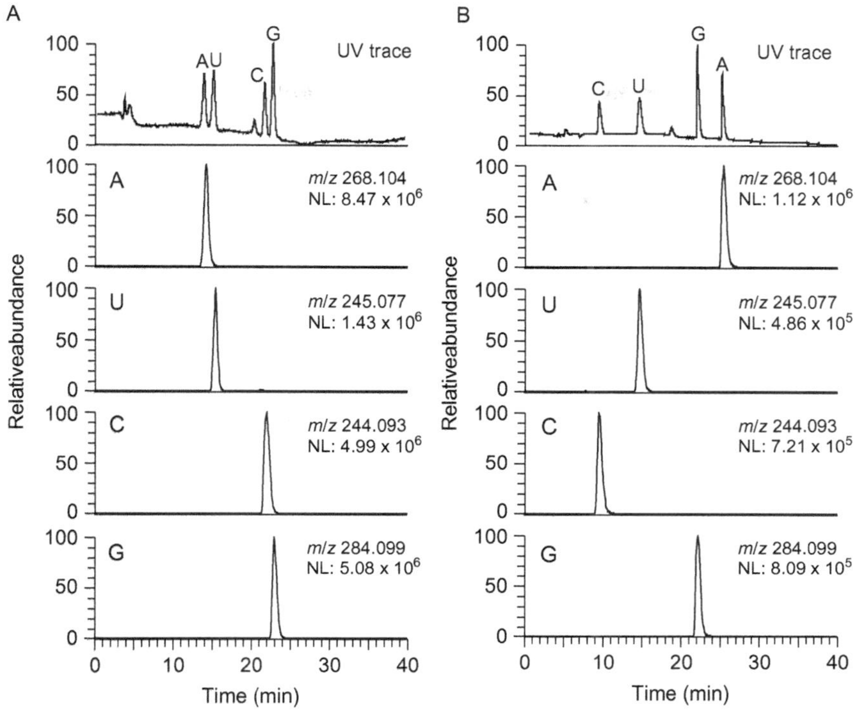

Figure 1 HILIC/ESI-MS versus RPC/ESI-MS for nucleoside analysis. HILIC/ESI-MS (A) and RPC/ESI-MS (B) were used for analyzing 100 pmol each of four nucleosides. Top panels show relative abundance in the UV trace at 254 nm. Peaks for A, U, G, and C are indicated in the chromatograms. Lower panels show mass chromatograms for monoisotopic ions of A (*m/z* 268.104), U (*m/z* 245.077), C (*m/z* 244.093), and G (*m/z* 284.099), respectively. NL, normalized largest value.

nucleoside, adenosine (A), was eluted earliest, while guanosine (G) was eluted last among the four nucleosides. According to the mass chromatograms (Fig. 1), the detection sensitivities for all the nucleosides in HILIC/ESI-MS were three to sevenfold greater than those in RPC/ESI-MS. This is because all nucleosides were eluted at a high concentration of ACN in HILIC as compared to that of RPC, facilitating desolvation and ionization by ESI.

5. PROFILING OF MODIFIED NUCLEOSIDES IN HILIC/ESI-MS

To demonstrate the performance of HILIC/ESI-MS for nucleoside analysis, total nucleosides from *E. coli* (A19) total RNA were analyzed. *E. coli* total RNA (12.5 μg) was digested into nucleosides as described above, followed by adjusting the sample volume to 25 μL in 90% ACN. Then, 23 μL of the sample was injected into the HILIC/ESI-MS system. In the UV trace and base peak chromatogram (Fig. 2), 34 species of modified RNA nucleosides were clearly assigned by the accurate determination of the molecular mass of their proton adducts (MH^+) (Fig. 3) and their

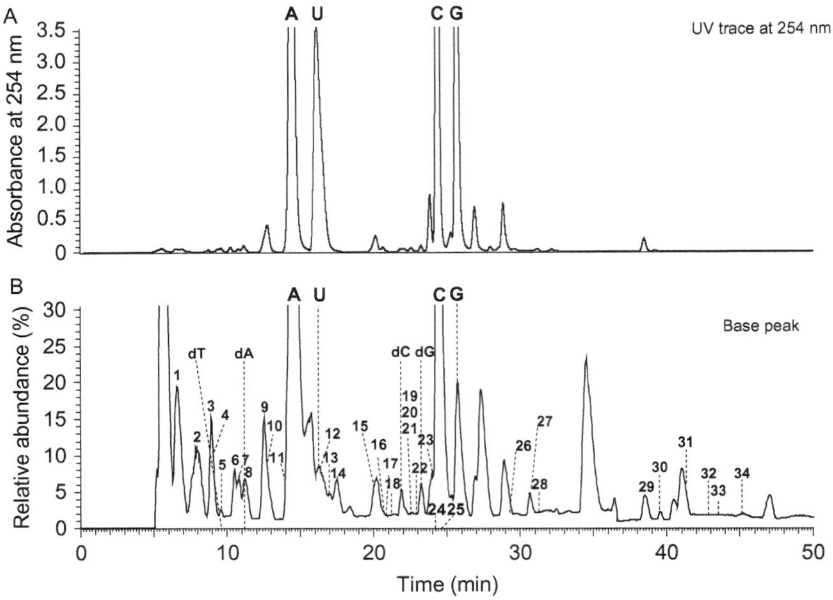

Figure 2 HILIC/ESI-MS nucleoside analysis for *E. coli* total RNA. UV trace at 254 nm (A) and base peak mass chromatogram (B) for detecting modified nucleosides in *E. coli* (A19) total RNA. The modified nucleosides detected in this analysis are numbered: 1, 2-methylthio-*N*^{6}-isopentenyladenosine (ms^{2}i^{6}A); 2, *N*^{6}-isopentenyladenosine (i^{6}A); 3, *N*^{6},*N*^{6}-dimethyladenosine (m^{6,6}A); 4, 2′-*O*-methyladenosine (Am); 5, 2′-*O*-methyluridine (Um); 6, 4-thiouridine (s^{4}U); 7, *N*^{6}-methyladenosine (m^{6}A); 8, uridine 5-oxyacetic acid methyl ester (mcmo^{5}U); 9, 2-methyladenosine (m^{2}A); 10, 5-methyluridine (m^{5}U); 11, *N*^{4}-acetylcytidine (ac^{4}C); 12, 1-methylguanosine (m^{1}G); 13, 2′-*O*-methylcytidine (Cm); 14, dihydrouridine (D); 15, 2′-*O*-methylguanosine (Gm); 16, 2-thiocytidine (s^{2}C); 17, cyclic *N*^{6}-threonylcarbamoyladenosine (ct^{6}A); 18, 5-hydroxycytidine (ho^{5}C); 19, 5-methylcytidine (m^{5}C); 20, inosine (I); 21, *N*^{6}-methyl-*N*^{6}-threonylcarbamoyladenosine (m^{6}t^{6}A); 22, *N*^{2}-methylguanosine (m^{2}G); 23, pseudouridine (Ψ); 24, *N*^{6}-threonylcarbamoyladenosine (t^{6}A); 25, uridine 5-oxyacetic acid (cmo^{5}U); 26, 5-carboxymethylaminomethyl-2′-*O*-methyluridine (cmnm^{5}Um); 27, 5-carboxymethylaminomethyl-2-thiouridine (cmnm^{5}s^{2}U); 28, 3-(3-amino-3-carboxypropyl)uridine (acp^{3}U); 29, 7-methylguanosine (m^{7}G); 30, 5-methylaminomethyl-2-thiouridine (mnm^{5}s^{2}U); 31, epoxyqueuosine (oQ); 32, 5-methylaminomethyluridine (mnm^{5}U); 33, queuosine (Q); 34, lysidine (L). A series of deoxyribonucleosides (dT, dA, dC, and dG) are also indicated.

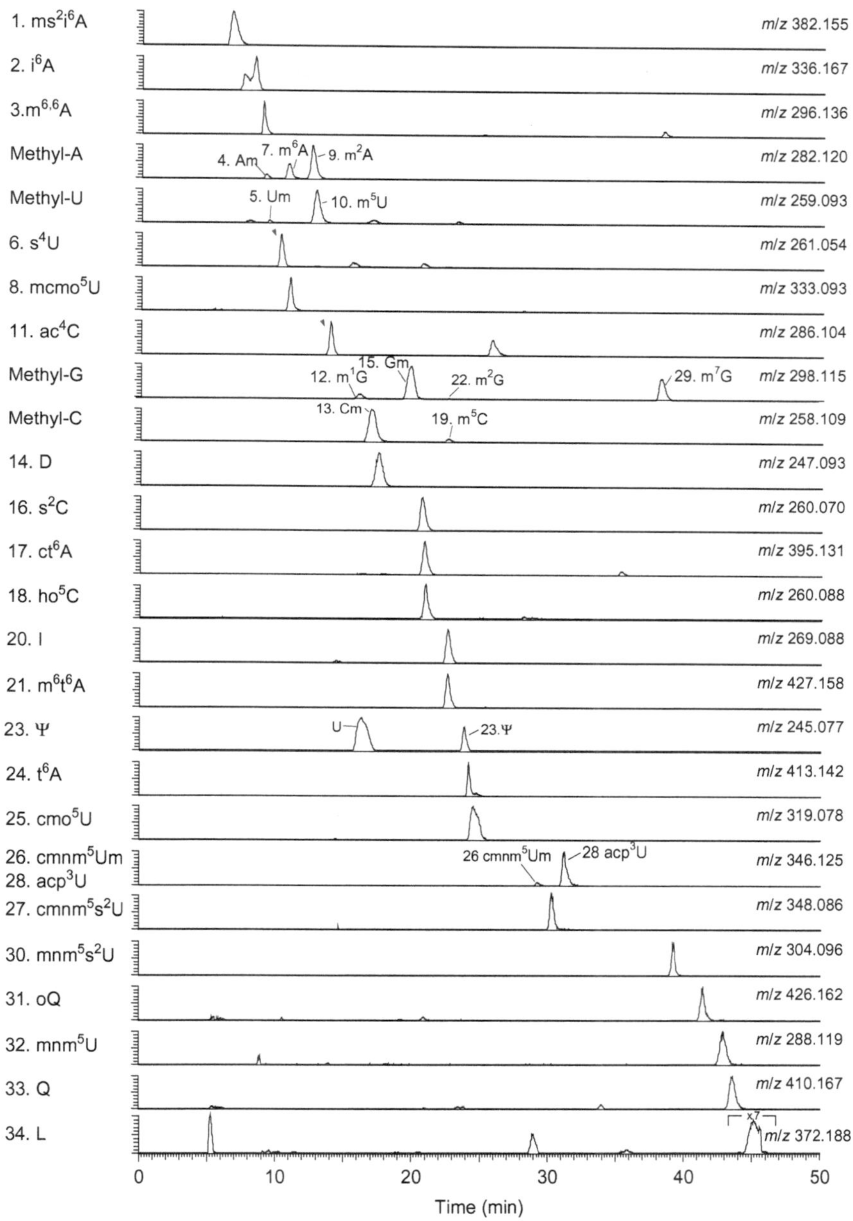

Figure 3 Mass chromatograms of HILIC/ESI-MS for modified nucleosides detected in *E. coli* total RNA. Each panel represents a mass chromatogram for a monoisotopic ion of a modified nucleoside with a specific *m/z* value as indicated on the left-handed side. The number of each nucleoside is the same as in Fig. 2.

base-related ions (BH_2^+), which spontaneously dissociated during ESI. Each modified nucleoside was numbered according to its elution order. Most of these nucleosides originate from tRNAs, while N^6,N^6-dimethyladenosine ($m^{6,6}A$), 5-hydroxycytidine (ho^5C), 5-methylcytidine (m^5C), and N^2-methylguanosine (m^2G) originate from rRNAs. Small amounts of deoxyribonucleoside (dT, dA, dC, and dG) contaminants in the total RNA were also detected as indicated by their retention times. Mass chromatograms for the monoisotopic ions of modified nucleosides (MH^+) are shown in Fig. 3. Peak signals for each modified nucleoside were unequivocally observed. In the mass chromatogram for monomethylated adenosine (*m/z* 282.12), three isomers (No. 4, 7, and 9) were clearly detected at distinct retention times. In contrast to RPC/ESI-MS (Suzuki et al., 2007), proton adducts of uridine derivatives were robustly detected with high sensitivity due to their strong retention in HILIC and high concentration of ACN in the eluates.

6. DISCUSSION

We here describe a detailed protocol for nucleoside analysis by HILIC/ESI-MS. As a practical alternative to RPC/ESI-MS, HILIC/ESI-MS is useful for profiling modified nucleosides from total RNA and individual RNAs. Chromatography with an ACN-rich solvent, which facilitates desolvation and ionization of modified nucleosides, is a substantial advantage of HILIC/ESI-MS. In addition, the use of the same solvent system for both HILIC and RPC is of practical merit when using both LC systems with one mass spectrometer. Moreover, strong retention of uridine derivatives is another advantage of HILIC/ESI-MS over RPC/ESI-MS. For further sensitivity, a HILIC capillary column (e.g., of 100 μm diameter) coupled with nanospray ionization could be applied for nucleoside analysis, as we performed previously with RPC/ESI-MS (Ohara, Sakaguchi, Suzuki, Ueda, & Miyauchi, 2007; Saito, Sakaguchi, Suzuki, Siomi, & Siomi, 2007; Zhou et al., 2010). To date, there are a number of commercially available columns and resins designed for HILIC. Further studies, examining different types of HILIC are required to obtain better performance in nucleoside analysis.

ACKNOWLEDGMENTS

We thank members of the Suzuki laboratory for fruitful discussions regarding this study. This work was supported by Grants-in-Aid for Scientific Research on Priority Areas from the Ministry of Education, Science, Sports, and Culture of Japan (to T. S.).

REFERENCES

Alpert, A. J. (1990). Hydrophilic-interaction chromatography for the separation of peptides, nucleic acids and other polar compounds. *Journal of Chromatography, 499*, 177–196.

Banks, J. F., Jr., Shen, S., Whitehouse, C. M., & Fenn, J. B. (1994). Ultrasonically assisted electrospray ionization for LC/MS determination of nucleosides from a transfer RNA digest. *Analytical Chemistry, 66*, 406–414.

Crain, P. F. (1990). Preparation and enzymatic hydrolysis of DNA and RNA for mass spectrometry. *Methods in Enzymology, 193*, 782–790.

Dumelin, C. E., Chen, Y., Leconte, A. M., Chen, Y. G., & Liu, D. R. (2012). Discovery and biological characterization of geranylated RNA in bacteria. *Nature Chemical Biology, 8*, 913–919.

Gaston, K. W., & Limbach, P. A. (2014). The identification and characterization of non-coding and coding RNAs and their modified nucleosides by mass spectrometry. *RNA Biology, 11*, 1568–1585.

Hemstrom, P., & Irgum, K. (2006). Hydrophilic interaction chromatography. *Journal of Separation Science, 29*, 1784–1821.

Ikeuchi, Y., Kimura, S., Numata, T., Nakamura, D., Yokogawa, T., Ogata, T., et al. (2010). Agmatine-conjugated cytidine in a tRNA anticodon is essential for AUA decoding in archaea. *Nature Chemical Biology, 6*, 277–282.

Ikeuchi, Y., Kitahara, K., & Suzuki, T. (2008). The RNA acetyltransferase driven by ATP hydrolysis synthesizes N4-acetylcytidine of tRNA anticodon. *The EMBO Journal, 27*, 2194–2203.

Ikeuchi, Y., Shigi, N., Kato, J., Nishimura, A., & Suzuki, T. (2006). Mechanistic insights into sulfur relay by multiple sulfur mediators involved in thiouridine biosynthesis at tRNA wobble positions. *Molecular Cell, 21*, 97–108.

Kimura, S., Ikeuchi, Y., Kitahara, K., Sakaguchi, Y., & Suzuki, T. (2012). Base methylations in the double-stranded RNA by a fused methyltransferase bearing unwinding activity. *Nucleic Acids Research, 40*, 4071–4085.

Kimura, S., Miyauchi, K., Ikeuchi, Y., Thiaville, P. C., Crecy-Lagard, V., & Suzuki, T. (2014). Discovery of the beta-barrel-type RNA methyltransferase responsible for N6-methylation of N6-threonylcarbamoyladenosine in tRNAs. *Nucleic Acids Research, 42*, 9350–9365.

Kimura, S., & Suzuki, T. (2010). Fine-tuning of the ribosomal decoding center by conserved methyl-modifications in the Escherichia coli 16S rRNA. *Nucleic Acids Research, 38*, 1341–1352.

Machnicka, M. A., Milanowska, K., Osman Oglou, O., Purta, E., Kurkowska, M., Olchowik, A., et al. (2013). MODOMICS: A database of RNA modification pathways—2013 update. *Nucleic Acids Research, 41*, D262–D267.

Mandal, D., Kohrer, C., Su, D., Babu, I. R., Chan, C. T., Liu, Y., et al. (2014). Identification and codon reading properties of 5-cyanomethyl uridine, a new modified nucleoside found in the anticodon wobble position of mutant haloarchaeal isoleucine tRNAs. *RNA, 20*, 177–188.

Marrubini, G., Mendoza, B. E., & Massolini, G. (2010). Separation of purine and pyrimidine bases and nucleosides by hydrophilic interaction chromatography. *Journal of Separation Science, 33*, 803–816.

Marusyk, R., & Sergeant, A. (1980). A simple method for dialysis of small-volume samples. *Analytical Biochemistry, 105*, 403–404.

Miyauchi, K., Kimura, S., & Suzuki, T. (2013). A cyclic form of N6-threonylcarbamoyladenosine as a widely distributed tRNA hypermodification. *Nature Chemical Biology, 9*, 105–111.

Noma, A., Kirino, Y., Ikeuchi, Y., & Suzuki, T. (2006). Biosynthesis of wybutosine, a hyper-modified nucleoside in eukaryotic phenylalanine tRNA. *The EMBO Journal, 25*, 2142–2154.

Noma, A., Sakaguchi, Y., & Suzuki, T. (2009). Mechanistic characterization of the sulfur-relay system for eukaryotic 2-thiouridine biogenesis at tRNA wobble positions. *Nucleic Acids Research*, *37*, 1335–1352.

Ohara, T., Sakaguchi, Y., Suzuki, T., Ueda, H., & Miyauchi, K. (2007). The 3′ termini of mouse Piwi-interacting RNAs are 2′-O-methylated. *Nature Structural & Molecular Biology*, *14*, 349–350.

Pomerantz, S. C., & McCloskey, J. A. (1990). Analysis of RNA hydrolyzates by liquid chromatography-mass spectrometry. *Methods in Enzymology*, *193*, 796–824.

Saito, K., Sakaguchi, Y., Suzuki, T., Siomi, H., & Siomi, M. C. (2007). Pimet, the Drosophila homolog of HEN1, mediates 2′-O-methylation of Piwi-interacting RNAs at their 3′ ends. *Genes & Development*, *21*, 1603–1608.

Soma, A., Ikeuchi, Y., Kanemasa, S., Kobayashi, K., Ogasawara, N., Ote, T., et al. (2003). An RNA-modifying enzyme that governs both the codon and amino acid specificities of isoleucine tRNA. *Molecular Cell*, *12*, 689–698.

Su, D., Chan, C. T., Gu, C., Lim, K. S., Chionh, Y. H., McBee, M. E., et al. (2014). Quantitative analysis of ribonucleoside modifications in tRNA by HPLC-coupled mass spectrometry. *Nature Protocols*, *9*, 828–841.

Suzuki, T., Ikeuchi, Y., Noma, A., Suzuki, T., & Sakaguchi, Y. (2007). Mass spectrometric identification and characterization of RNA-modifying enzymes. *Methods in Enzymology*, *425*, 211–229.

Suzuki, T., Nagao, A., & Suzuki, T. (2011a). Human mitochondrial diseases caused by lack of taurine modification in mitochondrial tRNAs. *Wiley Interdisciplinary Reviews RNA*, *2*, 376–386.

Suzuki, T., Nagao, A., & Suzuki, T. (2011b). Human mitochondrial tRNAs: Biogenesis, function, structural aspects, and diseases. *Annual Review of Genetics*, *45*, 299–329.

Suzuki, T., Suzuki, T., Wada, T., Saigo, K., & Watanabe, K. (2002). Taurine as a constituent of mitochondrial tRNAs: New insights into the functions of taurine and human mitochondrial diseases. *The EMBO Journal*, *21*, 6581–6589.

Torres, A. G., Batlle, E., & Ribas de Pouplana, L. (2014). Role of tRNA modifications in human diseases. *Trends in Molecular Medicine*, *20*, 306–314.

Zhou, X., Zuo, Z., Zhou, F., Zhao, W., Sakaguchi, Y., Suzuki, T., et al. (2010). Profiling sex-specific piRNAs in zebrafish. *Genetics*, *186*, 1175–1185.

CHAPTER THREE

A Platform for Discovery and Quantification of Modified Ribonucleosides in RNA: Application to Stress-Induced Reprogramming of tRNA Modifications

Weiling Maggie Cai[*,†], Yok Hian Chionh[*,†], Fabian Hia[†], Chen Gu[‡], Stefanie Kellner[‡], Megan E. McBee[†,‡], Chee Sheng Ng[†,§], Yan Ling Joy Pang[‡], Erin G. Prestwich[‡,1], Kok Seong Lim[‡,2], I. Ramesh Babu[‡,3], Thomas J. Begley[¶], Peter C. Dedon[†,‡,||,4]

*Department of Microbiology, National University of Singapore, Singapore
†Singapore-MIT Alliance for Research and Technology, Singapore
‡Department of Biological Engineering, Massachusetts Institute of Technology, Cambridge, Massachusetts, USA
§School of Biological Sciences, Nanyang Technological Institute, Singapore
¶College of Nanoscale Engineering and Science, State University of New York, Albany, New York, USA
||Center for Environmental Health Sciences, Massachusetts Institute of Technology, Cambridge, Massachusetts, USA
[4]Corresponding author: e-mail address: pcdedon@mit.edu

Contents

[1] Current address: Clark and Elbing LLP, Boston, Massachusetts, USA.
[2] Current address: TechAtlas, Boston, Massachusetts, USA.
[3] Current address: Alnylum Pharmaceuticals, Cambridge, Massachusetts, USA.

Methods in Enzymology, Volume 560
ISSN 0076-6879
http://dx.doi.org/10.1016/bs.mie.2015.03.004

Abstract

Here we describe an analytical platform for systems-level quantitative analysis of modified ribonucleosides in any RNA species, with a focus on stress-induced reprogramming of tRNA as part of a system of translational control of cell stress response. This chapter emphasizes strategies and caveats for each of the seven steps of the platform workflow: (1) RNA isolation, (2) RNA purification, (3) RNA hydrolysis to individual ribonucleosides, (4) chromatographic resolution of ribonucleosides, (5) identification of the full set of modified ribonucleosides, (6) mass spectrometric quantification of ribonucleosides, (6) interrogation of ribonucleoside datasets, and (7) mapping the location of stress-sensitive modifications in individual tRNA molecules. We have focused on the critical determinants of analytical sensitivity, specificity, precision, and accuracy in an effort to ensure the most biologically meaningful data on mechanisms of translational control of cell stress response. The methods described here should find wide use in virtually any analysis involving RNA modifications.

1. INTRODUCTION

In addition to the canonical adenosine, guanosine, cytidine, and uridine, nearly every form of RNA in prokaryotes and eukaryotes contains posttranscriptionally modified ribonucleosides. There are currently more than 120 different enzymatically derived chemical derivatives of the canonical ribonucleosides, with about 25–40 of these modifications present in any organism (Agris, Vendeix, & Graham, 2007; Bjork et al., 1999; Crain, Rozenski, & McCloskey, 1999; Czerwoniec et al., 2009; Urbonavicius, Qian, Durand, Hagervall, & Bjork, 2001; Yarian et al., 2002). Among the different RNA species, tRNA is heavily decorated with modified ribonucleosides. Individual modifications have been found to influence tRNA structure, folding, and stability, such as stabilization of base stacking by 2′-*O*-methylation (Agris, 2004; Davis, 1995; Motorin & Helm, 2010), enhancement of flexibility by dihydrouridine (D) (Dalluge, Hashizume, Sopchik, McCloskey, & Davis, 1996). tRNA modifications also affect the recognition and subsequent amino acid charging of tRNA species by tRNA synthases (Nameki, 1995; Nameki, Asahara, Shimizu, Okada, & Himeno, 1995; Putz, Florentz, Benseler, & Giege, 1994; Rudinger, Florentz, & Giege, 1994).

Modifications located within the anticodon loop and particularly those at the wobble position have been demonstrated to be important in translational fidelity and efficiency (Agris et al., 2007; Bjork et al., 1999; Urbonavicius et al., 2001; Yarian et al., 2002), preventing frameshift mutations (Bjork et al., 1999; Bregeon, Colot, Radman, & Taddei, 2001), and cell growth (Emilsson, Naslund, & Kurland, 1992). More recently, systems-level analyses have revealed a transcendent property of the dozens of tRNA modifications in cellular stress responses (Begley et al., 2007; Chan et al., 2010, 2012; Patil et al., 2012; Rezgui et al., 2013). The involvement of tRNA modifications in the cell stress response was first reported by Begley et al., with alkylation stress leading to upregulation of mcm^5U at the wobble position of tRNA species reading AGA and GAA codons that are enriched in genes critical to the DNA damage response (Begley et al., 2007). This model was expanded at a systems-level in studies of the response of budding yeast to exposure to different stresses, in which each chemical stressor produced highly predictive, signature changes in the relative quantities of modified ribonucleosides (Chan et al., 2010, 2015), with this reprogramming of tRNA modifications leading to selective translation of codon-biased stress response genes (Chan et al., 2012, 2015).

These new systems-level models of translational control of cell phenotype and stress response were enabled by the development and application of both chromatographic methods for purifying different types of RNA and chromatography-coupled mass spectrometric methods for quantifying modified ribonucleosides in tRNA and other RNA species. While approaches to ribonucleoside analysis such as thin-layer chromatography and high-performance liquid chromatography (HPLC) with UV–VIS spectrometric detection provide a cost-effective means to characterize subsets of RNA modifications and modified oligonucleotides (Grosjean, Droogmans, Roovers, & Keith, 2007; Köhrer & Rajbhandary, 2008), the methods are at best semiquantitative, lack the sensitivity needed for the detection of low-abundance modifications, and do not allow quantification of all tRNA modifications at once. The recent emergence of chromatography-coupled mass spectrometric(LC–MS) methods has facilitated significant advancements in discovering, identifying, and quantifying modified 2-deoxyribonucleosides and ribonucleosides in all organisms. Based on this foundation, we developed a sensitive and robust LC–MS platform for analyzing the full set of modified ribonucleosides in any RNA species, with first application to purified tRNA from *Saccharomyces cerevisiae* (Su et al., 2014). This systems-level approach provided the first evidence of stress-induced changes in the relative abundance of tRNA modifications, with the emergence of stress-specific signature patterns, or reprogramming, of the modified ribonucleosides (Chan et al., 2012, 2015). Subsequent mining of this dataset in conjunction with genomic and proteomic analyses revealed a new system of stress response involving reprogramming of tRNA wobble modifications coordinated with selective translation of codon-biased mRNAs required for the stress response (Chan et al., 2012).

Here we describe this LC–MS approach for the discovery and quantification of modified ribonucleosides from purified RNA in biological systems, with an emphasis on the strategies and caveats for each of the six steps of the platform: (1) RNA isolation and purification, (2) hydrolysis of RNA into individual ribonucleosides, (3) liquid chromatographic resolution of ribonucleosides, (4) identification of modified ribonucleosides by high-mass accuracy spectrometric methods, (5) quantification of ribonucleosides by tandem quadrupole mass spectrometry (Su et al., 2014), and (6) interrogation of the stress-reprogrammed tRNA modification profiles and mapping the location of altered modifications in individual tRNA molecules.

2. METHODS

2.1 RNA Isolation

The importance of RNA modifications in biological systems is underscored by recent discoveries of modified ribonucleosides in virtually all forms of RNA, both coding and noncoding (Carlile et al., 2014; He, 2010; Kirino & Mourelatos, 2007; Ohara et al., 2007; Zheng et al., 2013). This importance is underscored by the advent of RNA sequencing and bioinformatics, which has led to the discovery of new noncoding RNA (ncRNA) species and the implication of ncRNA in a variety of host–pathogen interactions, genetic disorders, and adaptive responses to stress (Arnvig & Young, 2012; Chan et al., 2010; Rederstorff & Hüttenhofer, 2010; Zhang et al., 2012). The diversity of ncRNA species containing modified ribonucleosides necessitates an unbiased method for isolating and purifying the individual ncRNAs. In this section, we discuss the steps and caveats involved in RNA isolation and the methods employed for assessing the quality and quantity of isolated RNA.

RNA isolation generally consists of several steps: (1) cell lysis and homogenization, (2) quenching of biochemical processes, (3) nucleic acid partitioning, (4) RNA retrieval and crude purification, and (5) assessing

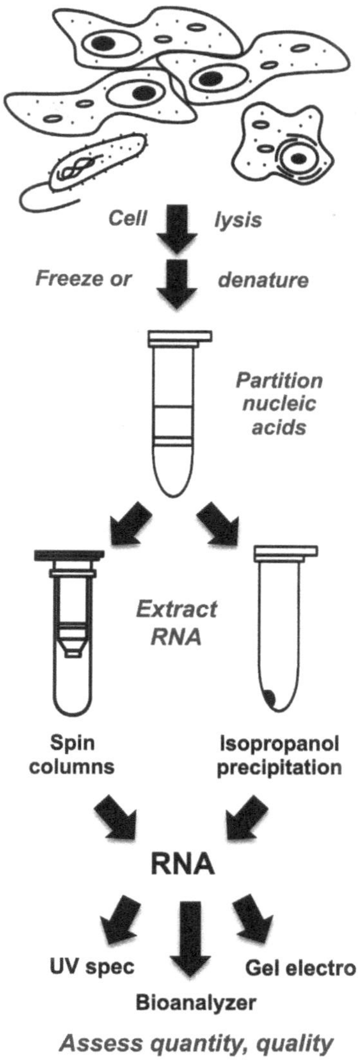

Figure 1 Flowchart of the general RNA extraction process.

the quality of the extracted RNA (Fig. 1). We consider each of these steps to address specific problems, but some or all steps can be consolidated depending upon need and the use of commercial kits.

Step 1: Cell lysis and homogenization

The first step requires effective cell lysis following homogenization for the complete release of nucleic acids. Several methods include chemical treatments such as TRIzol or detergents that disrupt cells to release cellular contents.

In addition to chemical methods, enzymatic means may be employed, such as treatment with lysozymes or enzymatic spheroplasting to weaken the cell walls for homogenization (Akhtar, Sarkar, Mishra, & Sarkar, 2011). For cells that are refractory to these treatments, mechanical disruption such as reciprocal bead-beating (Hia et al., 2015) or mechanical shearing with a French pressure cell (Kinger, Verma, & Tyagi, 1993) may be employed.

Step 2: Quenching of biochemical processes

During cell lysis, previously compartmentalized biomolecules are released and subjected to the milieu of enzymatic activities that may compromise RNA integrity. Therefore, solvents that solubilize cell contents should be denaturing (e.g., phenol ± chloroform) or contain chaotropic agents such as guanidinium thiocyanate or urea. In most instances, cell lysis/homogenization and quenching of biochemical processes are performed in the same step.

Step 3: Nucleic acid partitioning

Many commercial preparations use phenol-chloroform-based extractions to isolate nucleic acids. Phase separation is usually achieved when their constituents are centrifuged to separate aqueous and organic phases. In nucleic acid isolation, the partitioning of DNA from RNA depends on pH (Brawerman, Mendecki, & Lee, 1972; Perry, La Torre, Kelley, & Greenberg, 1972). At an alkaline pH, DNA and RNA are retained in the aqueous phase. However, as pH decreases, DNA increasingly migrates from the aqueous phase to the organic phase and interphase. Therefore, phenol equilibrated to a pH of ~8 is used for the extraction of DNA while acid phenol of pH 4.8 is used to isolate RNA. The critical role of pH in the stability of some modified ribonucleosides is discussed shortly.

Step 4: RNA retrieval and crude purification. With RNA partitioned in a solvent, the next step is to retrieve the RNA with high yield and minimal artifacts

RNA contained in the aqueous phase can be extracted using a variety of methods. One of the commonly used methods is through precipitation with isopropanol. However, as the final step involves drying the RNA pellet, caution must be exercised as excessive drying may result in poor resolubilization of RNA, which may affect subsequent purification and downstream analysis. Additionally, it was reported that isopropanol-based precipitation results in a retrieval bias of small RNA (Kim, Yeo, Kim, Ha, & Kim, 2012). As an alternative to precipitation, silica-based microcentrifuge spin-columns provide rapid purification of RNA. In general, ethanol is added to the aqueous phase and the resulting mixture is applied to the columns, with the RNA-containing column washed to remove contaminants and the RNA subsequently eluted. In a recent study, a two-column RNA extraction method based on these principles was employed to purify total ncRNA from mycobacteria (Hia et al., 2015). Using a mixture of 35% ethanol, large RNA species are recovered first. The resulting flow through is adjusted to 70% ethanol and applied to a second column, resulting in the recovery of smaller RNA species (Fig. 2). This approach facilitates an unbiased extraction of the full spectrum of ncRNA species.

When deciding on an appropriate RNA isolation strategy, consideration should be given to possible artifact generation during extraction or downstream processing, such as hydrolysis of nucleoside modifications under alkaline conditions. For example, cyclic N^6-threonylcarbamoyladenosine (ct^6A), a nucleotide base modification, is hydrolyzed to N^6-threonylcarbamoyladenosine (t^6A) under slightly alkaline conditions (Miyauchi, Kimura, & Suzuki, 2013). Similarly, under mild alkaline conditions, glutamyl-queuosine, which occurs at the wobble position of tRNAAsp, is hydrolyzed to a queuosine base (Salazar, Ambrogelly, Crain, McCloskey, & Söll, 2004).

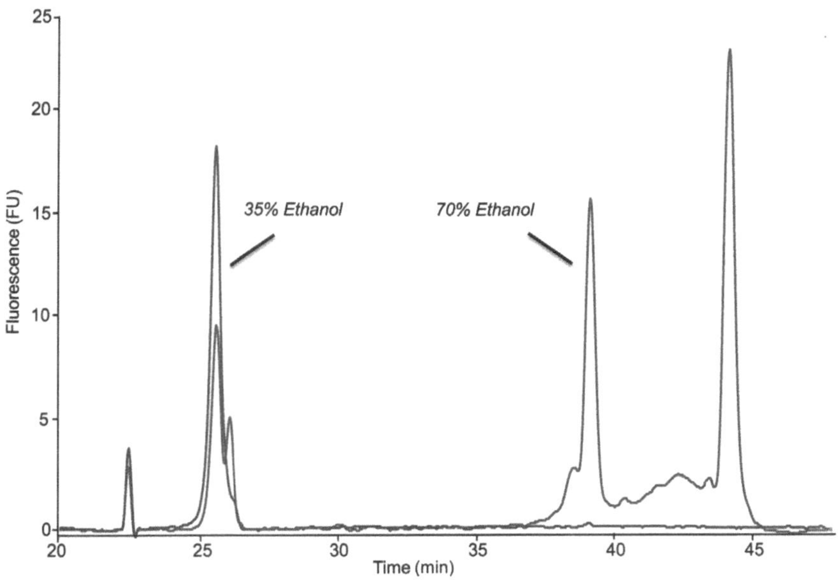

Figure 2 Representative 2100 Bioanalyzer electropherogram for isolated BCG RNA using the two-column approach. Large RNA species are recovered by elution from the column with 35% ethanol while small RNA species are recovered from the flow through using 70% ethanol.

Step 5: Assessing the quality of extracted RNA

Several methods are employed to assess the quality of extracted RNA. The easiest but most problematic approach is to estimate RNA concentration and purity by UV absorbance. The main drawbacks of spectroscopic approaches are the inability to evaluate RNA integrity, the strong potential for artifacts caused by contaminating phenol, DNA, and protein, and a lack of sensitivity for low concentration RNA samples. The issue of RNA integrity can be addressed by electrophoresis on denaturing gels, which are relatively cost-effective but time consuming in the casting and running the gels. While densitometric analysis of RNA band intensities provides semiquantitative information, at best, the major disadvantage of gel electrophoresis is the requirement for at least nanogram quantities of DNA for adequate visualization. Perhaps the most sensitive and accurate method for RNA quantification involves the use of fluorescent dyes with simple corrections for RNA secondary structure influences in different RNA species (Chionh et al., 2013).

Of all the available methods, perhaps the most efficient approach for assessing both RNA quantity and integrity involves the microfluidic chip-based platforms offered by Agilent Technologies (Bioanalyzer) and BioRad (Experion), with options for RNA, DNA, and proteins using specific chips. Using fluorescent dye-based electrophoresis, the 2100 Bioanalyzer detects and resolves up to picogram quantities of total RNA or small RNA depending on the chip used (Kuschel & Ausserer, 2002; Tissot, 2008). As a measure of RNA integrity, the instrument software assigns a RNA integrity number from 1 to 10 (with 10 indicating full integrity) based on the electrophoretic trace of the sample (Schroeder et al., 2006). The Bioanalyzer tracings in recent publications show the full range of ncRNA species with excellent resolution (Chionh et al., 2013; Hia et al., 2015). The chips and reagents for RNA analysis are highly cost-effective for analysis of multiple samples, relative to other rigorous techniques, though the specialized instrument for processing and analyzing the chips can be expensive for small research laboratories.

In conclusion, RNA isolation is a flexible process that can be customized according to specific laboratory requirements and costs. However, designing an RNA isolation strategy should incorporate considerations such as the toughness of the cell, pH of the extraction solvent, and an unbiased method of RNA retrieval—issues that are fundamentally important to producing reliable and biologically meaningful results.

2.2 RNA Purification

RNA modifications occur in nearly all forms of RNA (Wang & He, 2014), so the ability to purify a specific ncRNA species is critical to accurate analysis of species-specific modified ribonucleosides. While liquid chromatography (LC) (Easton, Shibata, & Lukavsky, 2010; Kim, McKenna, Viani Puglisi, & Puglisi, 2007), affinity purification (Batey & Kieft, 2007), and gel electrophoresis (Cunningham, Kittikamron, & Lu, 1996) techniques have been developed for purifying various RNA species, the systems-level study of RNA modifications requires the ability to fractionate all species of ncRNA from a single sample of total RNA, which poses a challenge to most RNA purification methods. RNA separation techniques had steadily improved over the last half-century, with each breakthrough leading to improved resolution, sensitivity, reproducibility, and ease-of-use. Density gradient centrifugation was replaced by polyacrylamide slab gel electrophoresis (Lewicki & Sinskey, 1970; Varricchio & Ernst, 1975), which in turn is currently being replaced by capillary electrophoresis (CE) and LC (Biba et al., 2013; Katsivela & Hofle, 1995; Skeidsvoll & Ueland, 1996; Warren & Vella, 1995). Here we review strengths and weaknesses of the latter two approaches and then focus on the most efficient approach to complete resolution of ncRNA species.

The choice of RNA purification method is largely dependent on the length, and thus molecular weight, of the RNA of interest. Denaturing pulsed-field CE offers unparalleled resolution for large RNA molecules longer than 2000 nucleotides (nt) while anion exchange (AX) and ion-pair reverse phase (IP RP) chromatography are generally used for smaller RNA species (~10–600 nt) (Azarani & Hecker, 2001; Easton et al., 2010; Koubek, Lin, Chen, Cheng, & Huang, 2013; Li et al., 2012; Li, Dou, Ni, Sumitomo, & Yamaguchi, 2010; McCarthy, Gilar, & Gebler, 2009; Shanagar, 2005; Waghmare, Pousinis, Hornby, & Dickman, 2009). Size exclusion chromatography (SEC) bridges this gap in separating RNA species ranging from ~20 to 2500 nt in length (Chionh et al., 2013; Kim et al., 2007). Affinity chromatography, on the other hand, is suitable for the targeted isolation of tagged RNA species and sequences, but efficiency falls off for longer RNA species (Aviv & Leder, 1972; Kieft & Batey, 2004; Luo, Eldho, Sintim, & Dayie, 2011; Srisawat & Engelke, 2001; Srisawat, Goldstein, & Engelke, 2001; Walker, Scott, Srisawat, & Engelke, 2008; Ward, Gunaratne, & Garcia-Blanco, 2014). AXC, IP RPC, and SEC are suitable for native tRNA isolation from crude cellular total RNA, with

IP RPC further fractionating total tRNA into multiple tRNA peaks (Chionh et al., 2013; Waghmare et al., 2009), though this level of resolution is not likely to be useful do to the lack of characterization of the individual fractions. Furthermore, the triethylammonium ions used in IP RPC for the neutralization of the polyanionic RNA molecules should be removed prior to LC–MS to prevent ionization suppression and contamination of the MS. SEC, conversely, is tractable and readily multiplexed to other online LC systems (such as IP RPC) for orthogonal separations and to various MS systems for detection (Fujii et al., 2014), making it ideal for the downstream analysis of modified ribonucleosides.

Building on the utility of SEC, we developed a multidimensional SEC-based method for nondenaturing purification of ncRNA by analytical or semipreparative scale HPLC (Chionh et al., 2013). The system demonstrates sufficient separation of major species of ncRNA of varying sizes from miRNA to 28S rRNA. The method uses Bio SEC3 300 Å or Bio SEC5 1000 Å columns (Agilent Technologies) on an Agilent 1200 HPLC system to achieve one-dimensional separation using an isocratic elution with 100 m*M* ammonium acetate (pH 7.0) at a temperature of 60 °C to minimize cross-hybridization entanglement of RNA species. The Bio SEC3 300 Å column sufficiently resolves the small ncRNA (tRNA; 5S, 5.8S rRNA) but not the large ncRNA (16S, 18S, 23S, 26S rRNA), while the Bio SEC5 1000 Å column resolves large ncRNA but not small ncRNA. By positioning both SEC3 300 Å and SEC5 1000 Å columns in tandem, coupled to a valve switching device, a two-dimensional SEC HPLC configuration is achieved that allows good resolution of the entire range of major ncRNA species in a single run (Chionh et al., 2013). Total RNA is first injected onto the SEC5 column to resolve large ncRNA from the small ncRNA. The eluting small ncRNA fraction is then directed to the SEC3 column to resolve each small ncRNA species. The large ncRNA bypasses the SEC3 column and is eluted for collection of individual fractions. It should be noted that, while the 2D-SEC HPLC platform provides an efficient means to purify all ncRNA, individual 1D SEC HPLC systems can be employed separately to achieve the same results albeit with additional steps in processing.

A major problem facing all of these RNA purifications strategies involves contamination of individual ncRNA species with fragments arising from degradation of larger species or from smaller species by cross-hybridization during the chromatography. One can minimize degradation with careful and gentles RNA isolation, and one can address contamination using

RNA-seq or Northern blots to identify and quantify contaminating species. However, contamination is inevitable and must be considered in any analysis of modified ribonucleosides presumed to arise in a specific ncRNA species. Ultimately, LC–MS-based modification mapping strategies RNAs (Castleberry & Limbach, 2010; Chan et al., 2012; Hossain & Limbach, 2007) can be used to define the association of RNA modifications with specific RNA species.

2.3 RNA Hydrolysis

With purified RNA in hand, the next step is to hydrolyze the RNA into oligonucleotide fragments or individual ribonucleotides, with the latter dephosphorylated to ribonucleoside form for LC–MS analysis. The oligonucleotides are used for localizing and quantifying modified ribonucleosides in specific tRNA species (Castleberry & Limbach, 2010; Chan et al., 2012; Hossain & Limbach, 2007), while the ribonucleosides can be identified and quantified by LC–MS as discussed later in this chapter. We focus here on the hydrolysis of RNA into ribonucleosides for analysis of stress-induced changes and patterns in translational response mechanisms.

There are several methods for hydrolyzing RNA into ribonucleosides. The classical approach of dephosphorylating ribonucleotides released by alkaline hydrolysis, which exploits the nucleophilicity of the 2′-hydroxyl group, causes degradation of many labile modifications (Taniguchi et al., 2013), while it is inhibited by ribonucleosides with 2′-*O*-methylation, most abundantly found in eukaryotic rRNA (Birkedal et al., 2015). A more rigorous approach involves enzymatic hydrolysis of RNA under conditions of physiological pH. Traditionally, a two-step method involving three enzymes has been used (Chan et al., 2010; Crain, 1990). RNA is incubated with nuclease P1 at acidic pH (5.3) for a few hours to overnight at 37 °C, followed by adjustment of the pH to 8.5 and addition of snake venom phosphodiesterase I (PDE I) to complete the hydrolysis and calf intestine phosphatase (CIP) to dephosphorylate the ribonucleotides. PDE I is used in combination with nuclease P1 because certain modified ribonucleosides are resistant to cleavage by the latter. In some cases, PDE I has been used alone overnight to achieve complete hydrolysis (Klootwijk & Planta, 1973), but the completeness of the hydrolysis has not been established.

This two-step, three-enzyme method is time consuming and exposes RNA to both acidic and alkaline pH values for extended periods, which increases the risk of degrading sensitive ribonucleosides, as discussed shortly.

To address these issues, we developed an enzymatic approach that completely hydrolyzes and dephosphorylates RNA in a single 3-h step (Su et al., 2014). Instead of nuclease P1, a highly robust and efficient endonuclease, benzonase, is used simultaneously with PDE I and CIP to hydrolyze the RNA and dephosphorylate the ribonucleotides. All three enzymes work at pH 7.8 and the high efficiency of benzonase allows complete hydrolysis within in a few hours. One note of caution is that benzonase requires Mg^{2+} that can alter the chromatographic behavior of nucleophilic and negatively charged ribonucleosides during reversed-phase HPLC. We have observed this behavior with 5-carboxymethyluridine (cm^5U), presumably through the coordination of its carboxylic acid group with Mg^{+2} (unpublished observation).

At this point, it is important to discuss the risks of adventitious chemical alterations of modified ribonucleosides during RNA isolation, purification, and hydrolysis. A widely known example is the claim that most of the t^6A observed in *E. coli* tRNA arises from the adventitious ring-opening of its cyclic form, ct^6A, during RNA hydrolysis (Taniguchi et al., 2013). By performing the digestion at neutral pH and shortening reaction time, Miyauchi et al. were able to see a significant increase in the level of ct^6A. They also reported that the primary amines of the commonly used buffer Tris (tris (hydroxymethyl)aminomethane) act as nucleophiles to cleave the ring structure of ct^6A and form a Tris-t^6A adduct (Miyauchi et al., 2013). We have also observed increased detection of ct^6A and reduced levels of t^6A during benzonase digestion of RNA (Su et al., 2014) when ammonium acetate is used as the buffer in place of Tris during hydrolysis of *Mycobacterium smegmatis* tRNA (unpublished data). Other adventitious reactions can be avoided with additives during the isolation and hydrolysis. For instance, CIP sold by some commercial vendors possesses significant adenosine deaminase activity, which, if not properly controlled, can lead to biologically irrelevant levels of inosine (I) (Dong, Wang, Deen, & Dedon, 2003). To avoid this reaction, we frequently include coformycin, an adenosine deaminase inhibitor, and tetrahydrouridine, a cytidine deaminase inhibitor, in hydrolysis reactions (Su et al., 2014). Another set of additives that we frequently add during hydrolysis includes antioxidants such as butylated hydroxytoluene, a free radical scavenger, and deferoxamine, a potent iron chelator that inhibits iron-based Fenton chemistry. The presence of these antioxidants could explain our ability to detect 5-hydroxycytidine (oh^5C), an extremely redox-sensitive ribonucleoside species, in the 23S

E. coli rRNA (unpublished data). There are conflicting reports of this unstable RNA modification using similar LC-MS methods (Havelund et al., 2011).

2.4 HPLC Resolution of Modified Ribonucleosides

Historically, reverse phase (RP) HPLC with or without ion pairing has used for separation of nucleosides (Buck, Connick, & Ames, 1983; Gehrke & Kuo, 1989; Gehrke, Kuo, McCune, Gerhardt, & Agris, 1982; Gehrke, Kuo, & Zumwalt, 1980; Pomerantz & McCloskey, 1990; Werner, 1993). The use of ion-pairing agents is not recommended for in-line MS analysis, so we will not address this method here. When designing or choosing a method for separation of modified ribonucleosides for LC–MS analysis, it is important to consider the major factors affecting retention and separation, including the nature of the stationary and mobile phases, and the pH. Polar nucleosides are poorly retained on standard C18 RP columns, so manufacturers have expanded the choice of columns by polar functionalization of the C18 solid phase allowing for increased retention of polar compounds. Each type of functionalization yields different retention and elution times, thus columns should be tested for particular applications. Unlike traditional C18 columns, the polar functionalized C18 columns are compatible with 100% water, thus 100% water or buffer is used as mobile phase A and 100% methanol or acetonitrile as mobile phase B. Additives such as 0.1% formic acid or 10 m*M* ammonium acetate (pH 4.5) added to both mobile phases increases the ionization of analytes, particularly those eluting in 100% water. Note that use of ammonium acetate results in ammonium adducts of many ribonucleosides, which should be considered during mass spectrometry acquisition steps. Using 100% mobile phase A for 3–6 min allows sufficient retention and separation of the highly polar ribonucleosides, such as pseudouridine, modified uridines—which are common at the wobble position of tRNA (Agris, 1991; Chan et al., 2010)—as well as uridine and cytidine, while multiple gradients to achieve 6% mobile phase B improves separation and decreases run time (Buck et al., 1983; Chan et al., 2010; Pomerantz & McCloskey, 1990; Su et al., 2014).

There are some predictable chromatographic behaviors of nucleosides that assist in developing LC methods and discovering new nucleosides. As noted earlier, RP columns have poor retention of polar compounds, so pyrimidine ribonucleosides generally elute prior to purine ribonucleosides due to their increased polarity. If 2-deoxyribonucleosides are included in the sample, they elute after their equivalent ribonucleoside with the exception of thymidine that elutes with purines. Modification of the nucleosides on the sugar or base alters the elution behavior. The abundant 2′-*O*-methylation of ribose decreases polarity, thus causing these ribonucleosides to elute after their unmethylated counterparts (Buck et al., 1983; Chan et al., 2010; Farand & Gosselin, 2009; Pomerantz & McCloskey, 1990; Su et al., 2014). The low pH of mobile phases used in positive-ion mode MS causes protonation of basic groups on nucleosides, which results in increased polarity and reduced retention time (Gehrke et al., 1980). Therefore, additives and mobile phase pH will affect polarity and retention of ribonucleosides according to their acid–base properties.

As uridine and its modified derivatives often elute near the solvent front, alternatives to RP HPLC, such as hydrophilic interaction chromatography (HILIC), are useful to enhance retention and resolution of polar compounds (Alpert, 1990; Zhao et al., 2013). HILIC retains polar compounds through interaction with the hydrophilic solid phase enhanced by a high-percentage organic mobile phase. Similar to RP HPLC, the percent organic solvent has a strong influence on retention, with less significant pH effects (Alpert, 1990; Guo & Gaiki, 2005). This is an important consideration when using multiple reaction monitoring (MRM) for quantification as enhanced retention and resolution will decrease the compounds screened and increase the scan time for MS analysis. HILIC is also useful when in-line precolumns and an high-organic mobile phase are used, as has been applied to quantification of urinary nucleosides (Tuytten et al., 2008). Ribonucleoside separation for LC–MS analysis does not require baseline separation of modified nucleosides, thus shorter runtimes are feasible with UPLC systems and sub-2 μm particle size columns (Farand & Gosselin, 2009; Zhou et al., 2014). Regardless of separation method, the MS instrument sensitivity and coelution of modified nucleosides should be evaluated for specific applications as this will be a limiting factor.

2.5 Ribonucleoside Identification and Structural Characterization

A major challenge in quantitative analysis of tRNA modifications is limited information about the full set of modified ribonucleosides in an organism. Apart from well-studied species such as humans, yeast, and *E. coli*, there are no other organisms for which the complete set is known. This dilemma is resolved to an extent by the high degree of conservation of a core set of modified ribonucleosides, such as singly methylated nucleobases, in every organism. However, systems-level analyses of modification reprogramming in tRNA and other RNA species requires identification of all modifications, both known and novel. Though >120 RNA modifications have now been reported (Machnicka et al., 2013), two-thirds were discovered before 1980 (Cantara et al., 2011), which represent the most abundant modified ribonucleosides. Less abundant and unstable species are only recently being discovered as more sensitive technology is brought to bear on the problem and attention is paid artifact-free RNA extraction and isolation (Fu et al., 2013; Giessing et al., 2009; Miyauchi et al., 2013). High-mass accuracy and tandem quadrupole mass spectrometric methods have emerged in the past decade as the standard tools for identifying and characterizing modified ribonucleosides, which has greatly aided RNA modification discovery. For example, mass spectrometric methods have been used to discover a variety of new tRNA modifications in the past several years, such as 5-cyanomethyl uridine (cnm^5U), ct^6A, geranylated 2-thiouridine (ges^2U), geranylated 5-methylaminomethyl-2-thiouridine (mnm^5ges^2U), and geranylated 5-carboxymethylaminomethyl-2-thiouridine ($cmnm^5ges^2U$), and N^6,N^6-dimethyladenosine ($m^{6,6}A$). Two of these (cnm^5U and ct^6A) were discovered by hypothesis-driven, targeted LC–MS/MS of unknown modified tRNA bases from isolated tRNA isoacceptors (Mandal et al., 2014; Miyauchi et al., 2013), while the other four were identified by data-driven chemical screens for ribonucleoside variants (Chan et al., 2011; Chen, Kowtoniuk, Agarwal, Shen, & Liu, 2009; Kellner, Neumann, et al., 2014; Su et al., 2014). Subsequently, the chemical structures of these modified ribonucleosides were validated by the comparison of LC retention times, tandem mass spectrometry (MS^n) collision-induced dissociation (CID) fragments and ^1H- and ^{13}C-NMR comparisons with those produced by synthetic standards (Table 1).

Our laboratory has used a variety of approaches to defining the complete set of modified ribonucleosides in eukaryotes and prokaryotes, including a two-pronged chemical fingerprint-based screening strategy. The first uses neutral loss analysis to scan for molecules containing ribose or a 2′-*O*-methylribose among molecules with *m*/*z* ranging from 200 to 700. Digested ribonucleosides are resolved on a Hypersil gold aQ HPLC column (3 μm particle size; 150 × 2.1 mm; Thermo) on a Agilent 1290 uHPLC system using a two-component mobile phase (Solvent A: 100% ultrapure water with 0.1% formic acid v/v; Solvent B: 100% LC–MS-grade acetonitrile with 0.1% formic acid v/v) at flow rates of 0.3 ml/min under gradient conditions defined previously (Su et al., 2014). An Agilent 6460 triple quadrupole mass spectrometer with an electrospray ionization (ESI) Jetstream ionization

Table 1 Methods Used in the Discovery and Characterization of RNA Modifications (2009–2014)

Identity	Symbol	Structure	RNA Species	Discovery	Validation	Quantification
5-Cyanomethyluridine	cnm^{5}U	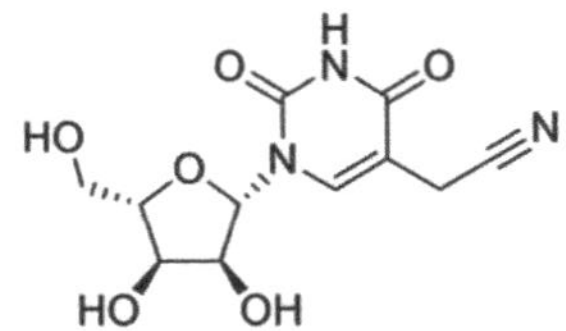	tRNA	Targeted analysis of mutant phenotype	Std, pMS^3	MRM
N^6-hydroxymethyladenosine	hm^{6}A	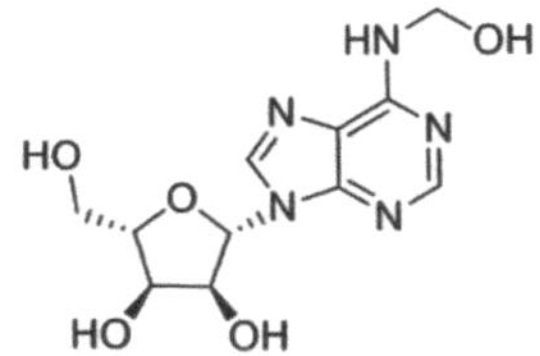	mRNA	Targeted analysis of modified synthetic ssRNA	Std, MS^2	MRM
N^6-formyladenosine	f^{6}A	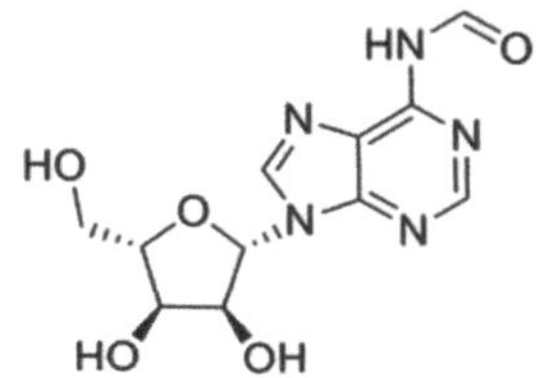	mRNA	Targeted analysis of modified synthetic ssRNA	Std, MS^2	MRM
Cyclic N^6-threonylcarbamoyladenosine	t^{6}A	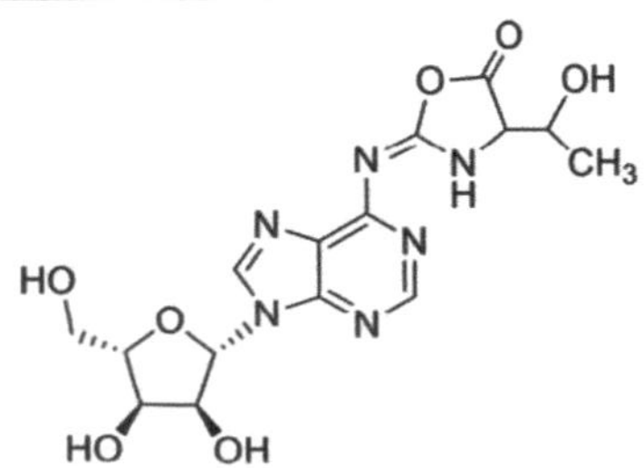	tRNA	Targeted analysis of natural product	Std, MS^n, ^{18}O labeling, NMR	UV absorbance

2-Geranylthiouridines[a]	ge^5s^2U	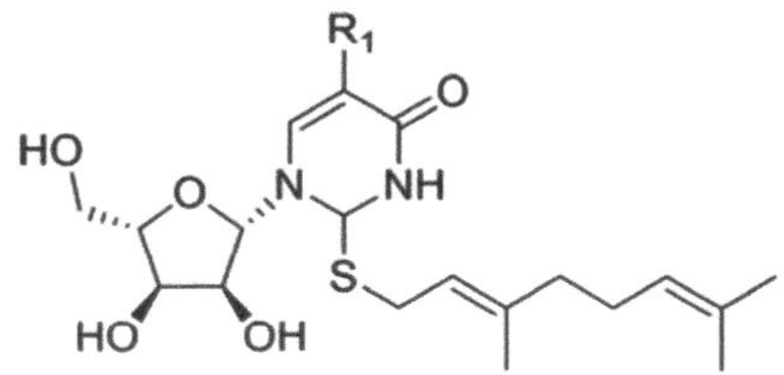	tRNA	LC–MS/MS screen of nucleotides	Std, MS^2, ^{13}C labeling	External calibration
N^6, N^6-dimethyladenosine	$m^{6,6}A$	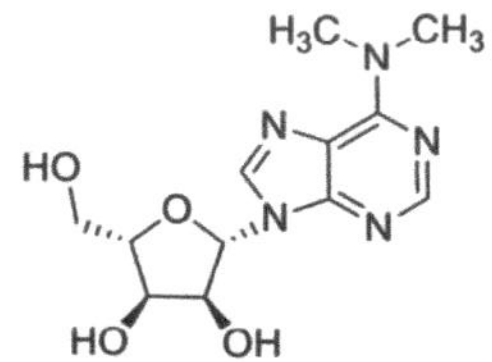	tRNA	LC–MS/MS screen of nucleotides	Std, pMS^3	MRM
2,8-Dimethyladenosine	$m^{2,8}A$	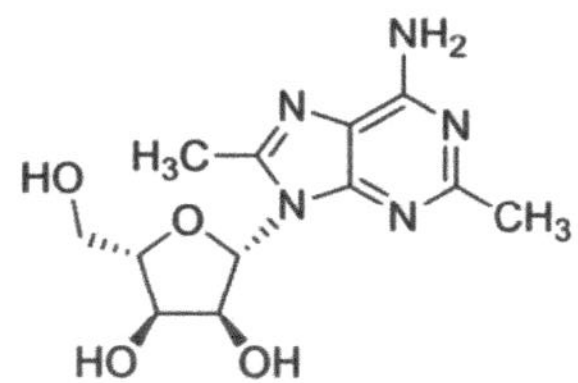	rRNA	Targeted analysis of mutant phenotype	Std, MS^n	Not performed
N-ribosylnicotinamide	r-NA	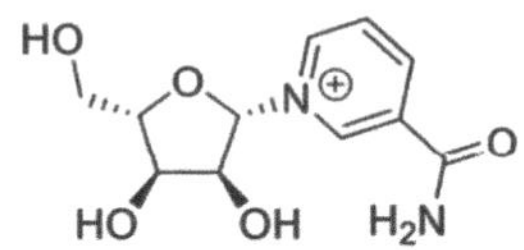	5′cap of sRNAs (>200 nt)	LC–MS/MS screen of nucleotides	Std, MS^2, ^{13}C labeling	External calibration, $^{13}C/^{15}N$-labeled isotope dilution

Continued

Table 1 Methods Used in the Discovery and Characterization of RNA Modifications (2009–2014)—cont'd

Identity	Symbol	Structure	RNA Species	Discovery	Validation	Quantification
Co-enzyme A[b]	coA	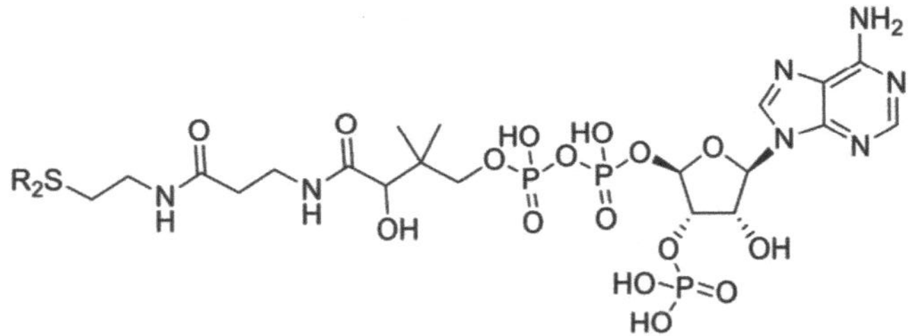	5′cap of sRNAs (>200 nt)	LC–MS/MS screen of nucleotides	Std, MS^2, ^{18}O labeling	External calibration

[a]R_1 can be hydrogen, methylaminomethyl-, or carboxymethylaminomethyl-groups (Dumelin, Chen, Leconte, Chen, & Liu, 2012).
[b]R_2 can be hydrogen, succinyl-, acetyl-, or methylmalonyl-groups (Kowtoniuk, Shen, Heemstra, Agarwal, & Liu, 2009).
Std, synthetic standards; pMS3, pseudo-MS3 by controlled in-source fragmentation; MS^n, tandem-in-time fragmentation on ion-trap mass spectrometers.

source operated in a positive-ion mode is used to acquire tandem-in-space information of the compounds found in the digestion mixture. In neutral loss mode, quadrupoles 1 and 3 are scanned together, but with constant mass offsets matching those of ribose (−132.0 *m*/*z*) and a 2′-*O*-methylribose (−146.0 *m*/*z*). This allows the selective recognition of all molecular ions that, by fragmentation in quadrupole 2, lead to the loss of an uncharged (hence, neutral) ribose or 2′-*O*-methylribose fragment—the defining molecular signatures of modified ribonucleosides. The presence or absence of known RNA modifications could be determined through targeted data extraction for ribose or 2′-*O*-methylribose containing precursor ions whose molecular masses corresponds to those cataloged in chemical databases such as Modomics (http://modomics.genesilico.pl/; Machnicka et al., 2013), The RNA Modification Database (http://mods.rna.albany.edu/; Cantara et al., 2011), and ChemSpider (http://www.chemspider.com/; Little, Williams, Pshenichnov, & Tkachenko, 2012). However, without chemical standards to define retention time, this approach would fail to differentiate positional isomers. Both 3- and 5-methylcytidine, for instance, would produce the same precursor ions of 258.1 *m*/*z*, albeit at two different times along the total ion chromatogram. Hence, libraries of chemical standards are prepared and analyzed using the same analytical system with the subsequent comparison of retention times and fragmentation patterns allowing for the positive identification of known modified ribonucleosides. Another limitation of neutral loss analysis is the failure to detect pseudouridine and its hypermodified derivatives: 1-methylpseudouridine, $m^1\Psi$; 3-methylpseudouridine, $m^3\Psi$; 2′-*O*-methylpseudouridine, Ψm; 3-(3-amino-3-carboxypropyl)pseudouridine, $acp^3\Psi$; and 1-methyl-3-(3-amino-3-carboxypropyl)pseudouridine, $m^1acp^3\Psi$. In these modifications, the stable C–C glycosidic bond does not readily fragment to produce neutral loss of a ribose residue (Dudley et al., 2005).

This limitation is directly addressed by use of a complementary automated MS/MS structural assignment screen in which compounds are identified by combining accurate mass, isotopic distribution, and fragmentation patterns. A quadrupole time-of-flight mass spectrometer (Agilent 6510 or 6520) was used to perform mass scanning and CID fragmentation and analyzed using both molecular feature extraction (MFE) and molecular structural correlation (MSC) (Agilent MassHunter software suite, B05.00; Hill & Mortishire-Smith, 2005; Little, Cleven, & Brown, 2011; Little et al., 2012). Elemental compositions of unknown compounds could be determined using the nitrogen rule (McLafferty, 1993; Sparkman, 2007),

Bristow's consideration for rings and double bonds based on the number of C, H, and N atoms that a molecule contains (Bristow, 2006), and Kind and Fiehn's second isotopic signal intensity approach to structural assignment (Kind & Fiehn, 2006). More information on structural arrangements can be gleamed from the CID fragmentation patterns. Knowing breakages will occur at the weakest points of the ion allows one to determine which molecules contain a *N*-glycosidic linkage that fragments before other bonds in the nucleobase when sufficient energy is provided. For example, pseudouridine can be identified by the presence of its signature m/z 125 fragment from CID fragmentation, which corresponds to the nucleobase with a ribose C1 methylene group attached (Dudley et al., 2005). Comparison of the mass spectra against mock-digest controls, for example with heat-killed nucleases (Dumelin et al., 2012), and controls lacking RNA further narrow down the number of chemical entities to be considered. Together, integration of datasets from both neutral loss analysis and MFE-MSC screens provides a list of modified ribonucleoside candidates that can be confirmed by a variety of methods, including analysis of commercially available standard compounds, high-mass accuracy tandem MS (Lim, Chen, Sensenhauser, Cook, & Subrahmanyam, 2007), ion trap MS^n mass spectrometry, controlled in-source fragmentation in space (pseudo-MS^3) (Chan et al., 2011; Mandal et al., 2014), homonuclear ^1H-2D-NMR (COSY, TOCSY, and NOESY), or heteronuclear ^1H-^{13}C 2D-NMR (HMQC) (Reynolds & Enriquez, 2002), and by their characteristic chemical derivatization reactions (Kellner, Burhenne, & Helm, 2010).

2.6 Mass Spectrometric Quantification of Modified Ribonucleosides

Having identified the full set of modified ribonucleosides in the model cell type or organism, it is now possible to assess changes in the relative quantities of each ribonucleoside as a function of stress or environmental conditions. This systems-level analysis provides the initial insights into the translational control of the cell stress response in terms of mechanistic links between the reprogrammed tRNAs and selective translation of codon-biased mRNAs (Begley et al., 2007; Chan et al., 2010, 2012, 2015). While descriptive analysis of ribonucleosides using MS was established two decades ago (Pomerantz & McCloskey, 1990; Suzuki, Ikeuchi, Noma, & Sakaguchi, 2007), quantitative MS analysis of ribonucleosides presents significantly greater challenges (Russell & Limbach, 2013; Su et al., 2014). Both relative and absolute quantification are best achieved using HPLC-coupled tandem or triple quadrupole (QQQ) mass spectrometry (LC–MS/MS) with ESI, given the increased sensitivity over other instrument types. In a mixture of ribonucleosides progressively eluting from the HPLC column in LC–MS/MS, individual ribonucleosides are ionized by ESI and then propelled into the first quadrupole (Q1) for initial detection. With the QQQ programmed to operate in MRM mode, which is used to instruct the system to acquire data for each ribonucleoside and its predictable MS behavior, the ion of interest (parent ion) is preselected in Q1 and then fragmented in the collision cell (Q2), followed by the selection of specific fragment ions (product ions) in the third quadrupole (Q3) (Fig. 3). The LC-coupled QQQ system enables the use of a dynamic MRM (DMRM) program, which allows the instrument to acquire MRM data only during the elution time window of each targeted ribonucleoside resolved on the HPLC column. The DMRM setting can minimize the waste of MS duty cycle, reduce concurrent ion transitions, and optimize dwell time and peak symmetry, which leads to further improvement of sensitivity and accuracy. In all types of quantitative analysis, however, it is critical to optimize the MS parameters for each ribonucleoside using either manual approaches or MS-specific software (Su et al., 2014).

While quantification of relative changes (i.e., fold-change) in ribonucleoside levels among samples is significantly easier (and more common) than absolute quantification (mass or moles), both approaches require control over a variety of parameters to achieve high precision (i.e., small variation in repeat or replicate analyses) and, in the case of absolute quantification, accuracy. Absolute quantification of ribonucleosides involves correlating the signal intensity with the analyte amount using a response factor (Fig. 4), though variation in the response factor can lead to poor precision during relative quantification. The response factor in mass spectrometry depends on at least three parameters, namely (i) instrumental parameters, (ii) sample parameters, and (iii) physicochemical parameters of the analyte/solvent system. A point in case is the recent report by the Limbach group, in which the reproducibility of modified nucleoside quantification by LC-UV/MS in four replicative measurements was analyzed. For MS detection, the average percent relative standard deviation (RSD) in peak areas measured for the ions was 5.9%. RSD variability in MS peak areas in this study ranged from 1.0% to 12.4% (Russell & Limbach, 2013). Instrumental parameters take account of the instrumental settings and performance of the source, the analyzer (Q1, collision cell, Q3), and the detector. To minimize the problem on quantitative precision and accuracy caused by variation in

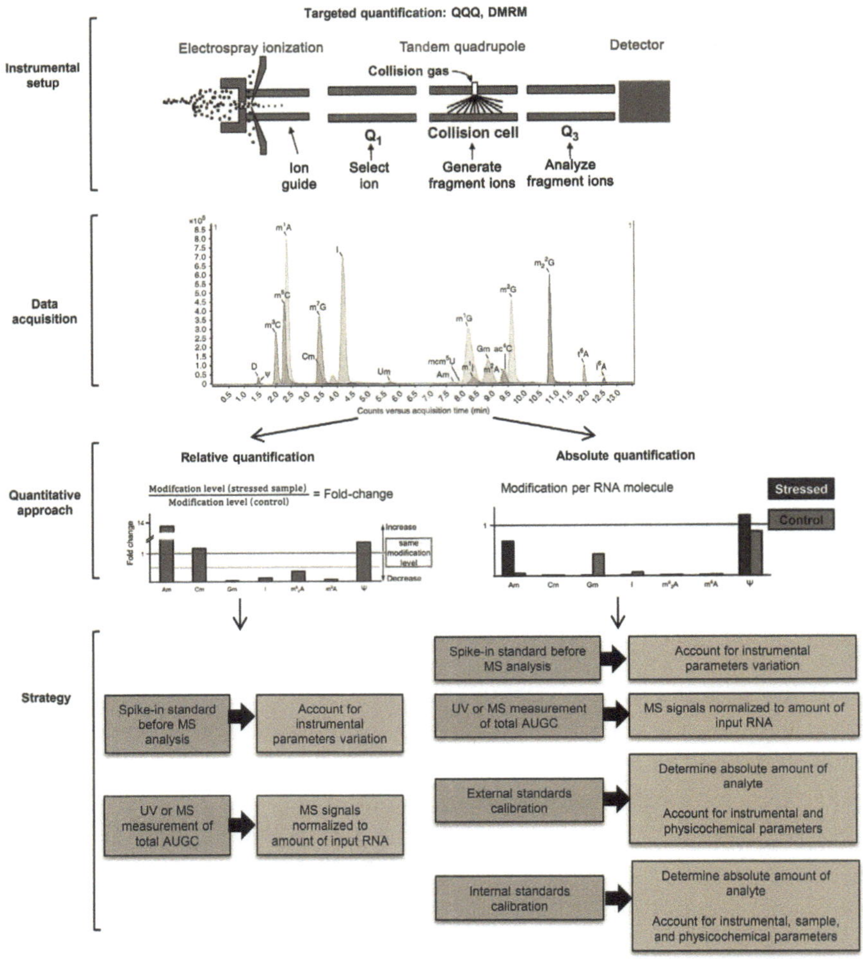

Figure 3 Conceptual workflow for targeted quantification of modified ribonucleosides by LC–MS/MS. Instrument setup: The MS system is programmed with a work list containing estimated retention times and CID transition information for each ribonucleoside, with dynamic MRM (DMRM) mode for data acquisition. Data acquisition: The signal intensity for each ribonucleoside is acquired in DMRM mode as the ribonucleoside elutes from the HPLC, with the peak area calculated from these data. Quantitative approach: Relative quantification provides fold-change information for stress-induced changes in modification levels (left), while absolute quantification indicates the abundance of the ribonucleoside in measurements per nucleotide or per tRNA molecule (right). *Strategy*: Steps to achieve precision and accuracy in both relative (left) and absolute (right) quantification are illustrated. Refer to main text for detailed description. (See the color plate.)

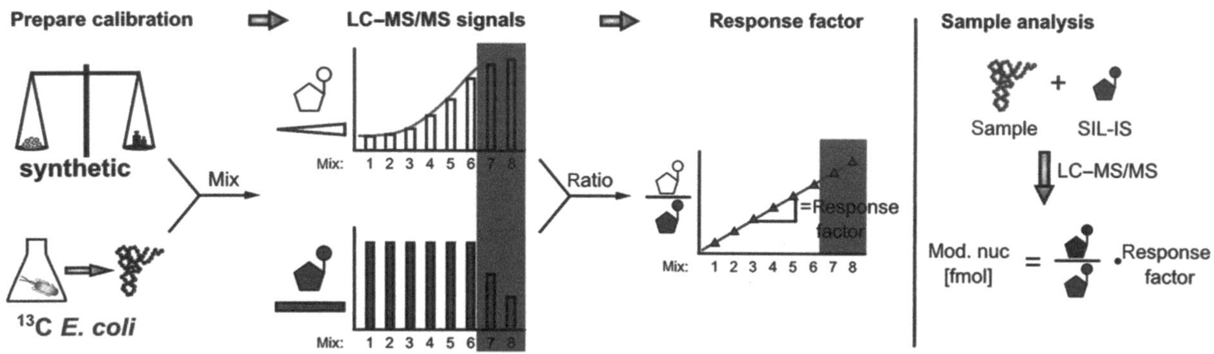

Figure 4 Calibration measurements to determine the mass spectrometric response factor for a ribonucleoside in absolute quantification approaches. The synthetically available and biosynthetically produced SIL-IS are mixed in different ratios (SIL-IS concentration is kept constant). The calibration mixes are then measured and the signals plotted. The increasing amounts of the unlabeled ribonucleoside result in a sigmoid signal plot. The gray area indicates reduced signals due to saturation effects. The signal ratios of unlabeled and labeled analyte are plotted and the slope determined. The slope represents the response factor of the analyte. Note the extended dynamic range in the gray area (saturation range). *Adapted and modified from Gu, Begley, and Dedon (2014), Kellner, Ochel, et al. (2014), and Su et al. (2014).*

instrument parameters over time, a synthetic standard such as $^{15}N_5$-2-deoxyadenosine ($^{15}N_5$-dA) with known concentration can be spiked into the processed analyte immediately before LC–MS/MS analysis and then quantified by the DMRM program together with the targeted transitions (Figs. 3 and 4). The MS signals of [$^{15}N_5$]-dA from different analyses are eventually used to account for hour-to-hour and day-to-day variations in instrumental parameters (Su et al., 2014). The sample parameters include salt load and any unknown impurities in the samples that might affect the MS detection due to ion suppression. The physicochemical parameters take account of the buffer pH, as well as the elution and ionization efficiency of the analyte in the solvent system. The inherent complex nature of response factors in MS highlights the importance of developing a robust and reproducible method to determine relative and absolute abundance of modified ribonucleosides in an experiment.

The relative quantification approach is applied to compare the fold-changes in modification levels between stressed samples and untreated controls (Fig. 3). To ensure accurate quantitative comparison of changes in modified ribonucleosides across different samples, the MS signals of individual modified ribonucleosides should be normalized to the input RNA concentration in each fully processed sample at the time of LC–MS analysis (Fig. 3). This precaution is needed to prevent imprecision and inaccuracy in quantification by accounting for any loss of RNA during sample processing before LC–MS analysis or variations in sample loading. To accurately assess the concentration of input RNA during LC–MS analysis, the canonical ribonucleosides (i.e., cytidine, uridine, guanosine, and adenosine) can be quantified by MS in a second run on the instrument with known dilution of the original sample. Alternatively, an in-line UV absorbance detector can be used with an external calibration curve to quantify the canonical ribonucleosides and normalize the ribonucleoside signal intensities.

To achieve absolute quantification of modified ribonucleosides, external calibration or isotope-dilution techniques can be applied, with the latter providing the greatest possible precision and accuracy (Figs. 3 and 4). An external calibration curve with unlabeled standards can be plotted by injecting serial dilutions of synthetic standard solutions of defined concentrations into LC–MS system, with the absolute amount of each modified ribonucleoside determined from the calibration curve at the linear dynamic range. Ideally, the standard should be identical to the analyte for maximal accuracy, but a structurally homologous ribonucleoside can also be used in the calibration curve of an identical standard, which is not available, with due caution in interpreting the absolute quantification data. External calibration using quantifiable standards accounts for the instrumental parameters and physiochemical parameters, but does not account for the sample parameters, which include salt load and unknown impurities in the sample. Again, caution must be applied in interpreting these estimates of absolute quantities determined by external calibration.

For absolute quantification by isotope-dilution MS (Fig. 4), internal standards labeled with stable isotopes (i.e., isotopomers), such as deuterium (D), carbon (^{13}C), and nitrogen (^{15}N), are added to samples prior to LC–MS/MS analysis and ideally during the RNA hydrolysis step to account for losses at that stage. Since labeled and unlabeled nucleosides have identical physicochemical properties, the influence of instrumental (i), sample (ii), and physicochemical (iii) parameters are accounted for in the signal intensities of both the standard and analyte, including in particular contaminations that may suppress ionization. The use of internal standards thus allows highly accurate and precise comparisons among samples and experiments (Fig. 4).

The major limitation of absolute quantification using isotope-labeled internal standards is the availability of the standards. There are commercial sources for a few isotopically labeled ribonucleosides and several research groups have synthesized specific standards (Brandmayr et al., 2012; Hienzsch, Deiml, Reiter, & Carell, 2013). To overcome this limitation, Helm and coworkers developed a biosynthetic approach that is accessible to any laboratory equipped for LC–MS work. The approach involves growing *E. coli* (or the organism of interest) in minimal-medium containing [^{13}C]-glucose as the only carbon source; it is also possible to adapt the method to use [^{15}N]-labeled nutrients for additional isotopic labeling. The labeled RNA can then be isolated and hydrolyzed, with individual ribonucleosides purified by HPLC, or the mixture used to spike samples directly (Fig. 4). The initially unknown amounts of the various modifications in the mixture of isotope-labeled ribonucleosides can be determined and validated by a combination of external calibration, internal standard, and spike-in methods (Figs. 3 and 4). This accounts for variations of instrumental parameters, physicochemical parameters, and in particular of sample parameters. This was shown by comparing the RSD of the mass signal (RSD $\sim$24%) over 12 weeks with the RSD of the response factors (RSD $\sim$2%). Consequently, the method allows, in addition to comparison of the relative modification content of related samples, an absolute quantification, limited to those nucleosides available in weighable quantities

(Buchhaupt et al., 2014; Spenkuch et al., 2014). The method also extends the linear dynamic range for quantification by two orders-of-magnitude by equalizing ion suppression effects due to saturation (gray area in Fig. 4; Kellner, Ochel, et al., 2014).

2.7 Analysis of Stress-Induced Reprogramming of Modified Ribonucleosides in Total tRNA

The datasets of modified ribonucleosides in treated and control cells must next be subjected to systems-level interrogation to detect stress-specific patterns of change and links to up- and downstream regulatory mechanisms in translational aspects of the stress response. Emerging evidence points to tRNA modifications as regulatory elements that dynamically reprogram following stressful conditions to control selective translation of critical stress response proteins (Begley et al., 2007; Chan et al., 2010, 2012; Patil et al., 2012; Rezgui et al., 2013). In this section, the strategy for exposing cells and then assessing this dynamic repertoire of modified ribonucleosides is discussed as a series of steps, with the approach extended to interpreting stress-specific signature tRNA modification patterns that are mechanistically linked to codon-biased genes (Chan et al., 2012).

Step 1: Establish a biological endpoint for the exposure

The experimental design for quantitative studies of stress-induced reprogramming of tRNA modifications must include a quantifiable endpoint or anchor, such as cell death, apoptosis, or some other measurable phenotypic change, for determining the exposure dose and the exposure duration. This phenotypic anchor allows meaningful comparisons among mechanistically different stressors or stress-inducing conditions and among different time points following exposure. In such studies, cell sensitivity is defined in relatively detailed dose–response analyses to identify the doses producing, for example, 20%, 50%, and 80% response (e.g., Fig. 5). The coverage of low, medium, and high killing concentrations can facilitate the identification of any dose-specific tRNA modification changes.

Step 2: Postexposure processing to stop reactions

Using the established stress conditions, the quantitative platform can be applied to analyze changes in the levels of each modified ribonucleoside from cells exposed to a set of diverse toxicants. It is important to immediately process the exposed cells at the defined endpoint to stop all cellular processes, since the tRNA modification changes are still subject to change if biologically friendly conditions exist during the postexposure processing steps. To this end, the cells should either be snap-frozen in liquid nitrogen, possibly

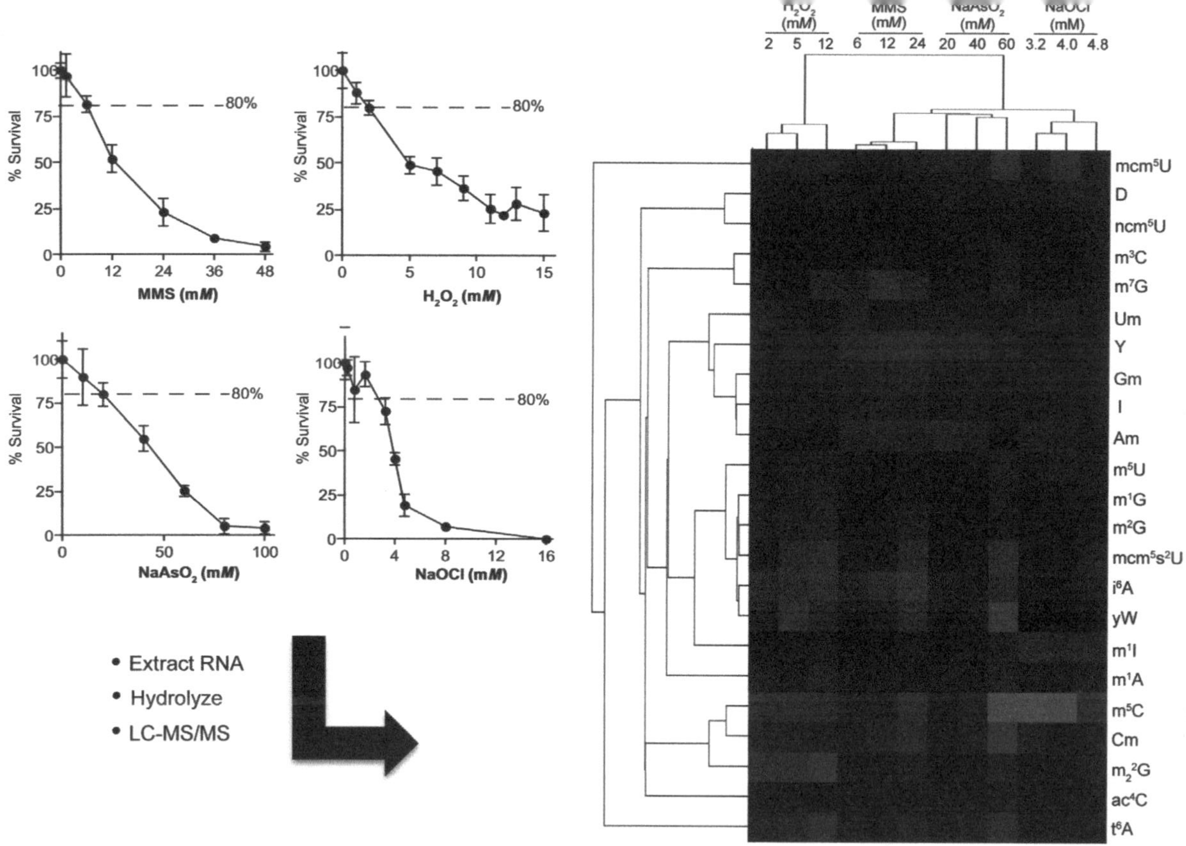

Figure 5 Stress-induced reprogramming of tRNA modifications analyzed by quantitative LC–MS/MS. To assess the reprogramming of tRNA modifications in cells exposed to different toxic stresses, cytotoxicity dose–response curves were determined for *S. cerevisiae* exposed H_2O_2, MMS, $NaAsO_2$, and NaOCl (left panel). Following exposure of cells to equitoxic doses of each toxicant (LC_{20}, LD_{50}, LD_{80}), tRNA was isolated and hydrolyzed, and each modified ribonucleoside was subjected to LC–MS/MS analysis. The data were used to calculate fold-change values for exposed cells compared to unexposed controls, with the fold-change data analyzed by hierarchical clustering analysis (right panel) to reveal stress-induced tRNA modification signatures. *Cytotoxicity plots and heatmap image were adapted from Chan et al. (2010).* (See the color plate.)

following a brief 4 °C centrifugation to create a cell pellet, or the cells should be subjected to denaturing conditions such as immediate addition of phenol-containing RNA extraction reagents.

Steps 3–6: RNA isolation, tRNA purification, tRNA hydrolysis, and LC–MS/MS analysis as described in Sections 2.1–2.6.

Step 7: Data normalization and calculation of fold-change values

Following LC–MS/MS analysis, the normalized signal intensity for each ribonucleoside in the various sample datasets is now used to calculate a fold-change value relative to the unexposed control sample. It is important to point out that, for large numbers of samples, the LC–MS/MS analysis should be performed for complete replicate exposure sets, including a control sample, to minimize day-to-day analytical variance. In addition to the presence of [15N]-dA, the fold-change calculation based on samples analyzed the same day removes any variance introduced by machine response changes.

Step 8: Multivariate statistical analysis to find patterns and mechanistic links

The fold-change values are now analyzed using multivariate statistic tools, such as hierarchical clustering and principal component analysis, either as mean values for each replicate dataset or as individual datasets. The intent of MVS analysis is to find patterns of covariance among the ribonucleoside fold-change data, which suggests biological relatedness of potential mechanistic significance. An example of hierarchical clustering of fold-change datasets is shown in Fig. 5 for budding yeast exposed to four mechanistically distinct toxicants. The MVS analysis distinguished both toxicant- and dose-specific patterns, with the patterns amounting to stress-specific tRNA modification signatures. The heat maps from hierarchical clustering and scores and loadings plots from PCA analysis reveal specific tRNA modifications as candidates for mechanistic linkage to subsequent biological response. Such stress-related patterns of tRNA modification reprogramming have been demonstrated to drive selective translation of codon-biased stress response genes (Chan et al., 2012). For example, m^5C is uniquely upregulated by H_2O_2 exposure (Fig. 5), with subsequent analysis revealing the presence of m^5C specifically increased at the wobble position of tRNALeu(CAA) (Chan et al., 2012). Wobble modifications are thought to be critical for enhancing wobble decoding and therefore affecting translational efficiency (Agris, 2004, 2008; Agris et al., 2007; Huang, Lu, & Bystrom, 2008; Kruger, Pedersen, Hagervall, & Sorensen, 1998; Murphy, Ramakrishnan, Malkiewicz, & Agris, 2004). Changes in the levels of wobble modifications in response to stress can therefore regulate the expression of certain response genes through enhanced decoding of specific codons. These genes, termed modification-tunable transcripts (MoTTs), have been demonstrated to contain statistically significant usage of specific degenerate codons (Begley et al., 2007; Dedon & Begley, 2014). In the instance of H_2O_2-induced increases in wobble m^5C in tRNALeu(CAA), there was an associated upregulation of proteins coded by genes enriched with its cognate codon, TTG, with loss of the enzyme responsible for inserting m^5C, Trm4, causing loss of this codon-biased translational response as well as hypersensitivity to the toxicity of H_2O_2 exposure (Chan et al., 2012). Similar observations have been made for Trm9-dependent wobble modifications mcm^5U and selective translation of UGU-enriched mRNAs (Begley et al., 2007). These observations demonstrate the strong mechanistic link between patterns of stress-altered modifications and selective translation of codon-enriched stress response genes. The major limitation of this approach of analyzing total tRNA populations is that the specific tRNA(s) containing stress-sensitive modified ribonucleosides cannot be identified, a problem resolved in the next section.

2.8 Analysis of Stress-Induced Changes in Modified Ribonucleosides at Specific Locations in Individual tRNA Species

To make the mechanistic link between a stress-sensitive modified ribonucleoside and a specific tRNA and its anticodon, it is critical to know the location of the modification within one or more specific tRNA species. Here, we describe two approaches to mapping tRNA modifications in a tRNA isoacceptor or isodecoder: affinity purification of individual tRNAs followed by MS-based sequencing analysis or MS-based analysis in RNase-digested total tRNA populations.

2.8.1 Affinity Purification of tRNAs

In order to study the function of individual modifications on a specific tRNA under stress conditions, it may be necessary to purify a target tRNA species from a complex mixture of RNAs. Here, we describe an affinity purification method using an oligo-based pull-down approach to isolate a single tRNA species from a crude tRNA mixture (Chan et al., 2012; Mandal et al., 2010; Suzuki & Suzuki, 2007). This approach exploits a solid-phase-bound oligodeoxynucleotide probe to hybridize to a unique sequence on the 3′ acceptor end of a specific.

Step 1: Attached biotinylated oligos to beads

Approximately 18 nmol of a 3′-end biotinylated 20-base oligodeoxynucleotide (IDT, Coralville, IA) that is complementary to the 3′ acceptor stem of a target tRNA (Fig. 6) is first bound to 0.2 ml of NeutrAvidin agarose beads (Thermo Scientific, Waltham, MA) for 10 min at ambient temperature.

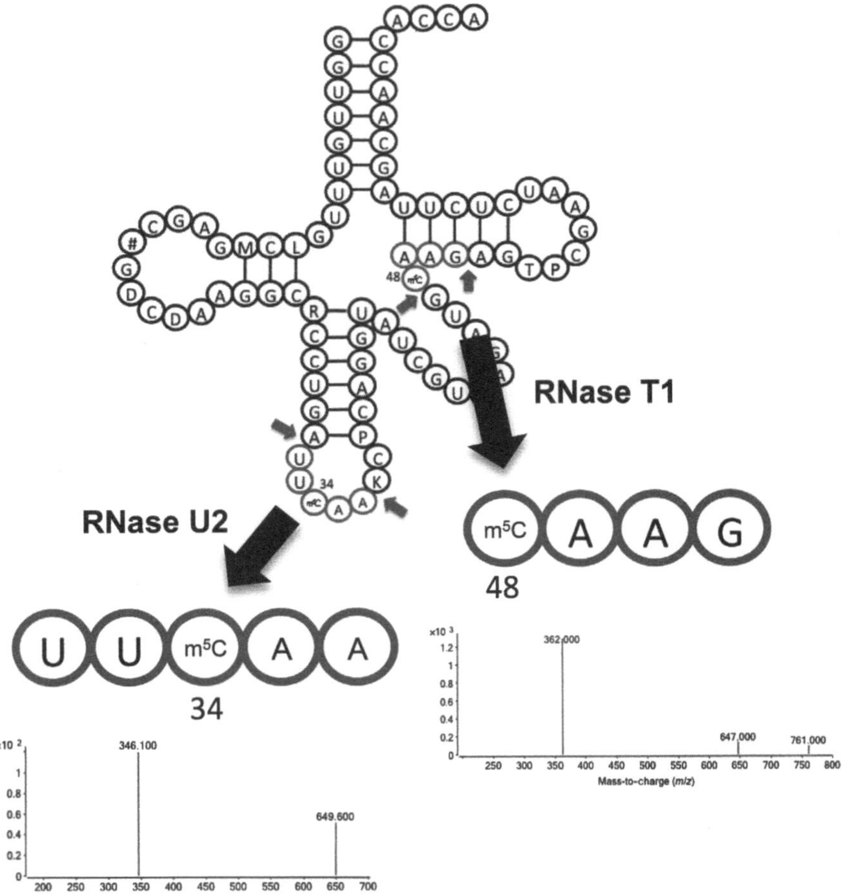

Figure 6 Quantitative mapping of modified ribonucleosides in a specific tRNA. To assess stress-induced changes in m^5C levels in $tRNA^{Leu(\underline{C}AA)}$, two RNase-generated oligonucleotides possessing m^5C-modified sites in this tRNA were identified using the Mongo Oligo Mass Calculator. RNase T1 yields a 4-mer oligonucleotide harboring either C or m^5C at position 48 ($\underline{C}$AAG), while RNase U2 produces a 5-mer oligonucleotide with C or m^5C at the wobble position 34 of (UU$\underline{C}$AA). The UUCAA oligonucleotide was sequenced and quantified by LC–MS/MS analysis with affinity-purified tRNA and the CAAG oligonucleotide was analyzed by LC–MS/MS analysis using total tRNA. (See the color plate.)

Step 2: Hybridize target RNA to bead-bound oligos

The oligodeoxynucleotide-modified beads are incubated with a crude tRNA sample in 6 × NHE (1.2 *M* NaCl, 30 m*M* HEPES-KOH, pH 7.5, and 15 m*M* EDTA) at 72 °C for 30 min, followed by cooling to ambient temperature over 80 min with occasional agitation.

Step 3: Washing to remove unbound RNA

The bead-bound complexes are then washed three times each with 3 × NHE, 1 × NHE, and 0.1 × NHE, respectively, until the UV absorbance (260 nm) of the wash falls below 0.01.

Step 4: Elution of target RNA

Finally, the target tRNA is eluted from the beads in three washings involving incubation in 0.1 × NHE at 65 °C for 5 min at each elution step. The eluent should consist of a relatively pure sample of a single tRNA species that can be used for quantitative analysis of all modified ribonucleosides present in the tRNA (Section 2.6), for relative quantification of stress-induced changes in modifications (Section 2.7), and for mapping modified ribonucleosides and stress-induced changes in the tRNA (see below). This method can also be scaled up or down depending on the quantity of starting material as well as the expected quantity of the final product.

2.8.2 tRNA Oligonucleotide Fingerprinting by RNase Digestion and LC-QTOF Analysis

A much simpler approach to quantitative mapping of modified ribonucleosides in specific tRNAs can be achieved by direct MS analysis of total tRNA. The most effective approach here is an oligofingerprinting method involving RNase digestion and MS analysis of the resulting oligonucleotides, as originally described by Limbach and coworkers (Castleberry & Limbach, 2010; Hossain & Limbach, 2007). This approach begins with defining the RNase digestion products arising in a mixture of tRNAs (Fig. 6), based on genome sequences of the full set of expressed tRNAs. Depending on the purity of the tRNA sample, either a single RNase or a cocktail of multiple RNases is used to generate oligonucleotides that are unique in terms of sequence. The oligofingerprinting method is applicable to both pure and population tRNA samples and does not require an initial affinity purification step. The only requirements are that the oligonucleotide of interest be efficiently ionized (i.e., not suppressed by other coeluting oligonucleotides), identifiable by a selectable m/z value in Q1, and amenable to CID fragmentation for both sequence confirmation and localization of the modified ribonucleoside (Fig. 6). It is best to have a synthetic oligonucleotide(s) as standards for the target oligonucleotide to allow HPLC elution and CID fragmentation patterns to be defined to facilitate analysis of the tRNA sample. It is often possible to synthesize the oligonucleotide with appropriately positioned modifications, depending upon the commercial availability of phosphoramidite versions of the modified ribonucleosides.

As shown in Fig. 6, this RNase-mapping approach is illustrated in our recent study assessing changes in the level of m^5C at two positions in *S. cerevisiae* $tRNA^{Leu(\underline{C}AA)}$ (Chan et al., 2012).

Step 1: Define RNase digestion products

We first used the Mongo Oligo Mass Calculator (v2.06; http://library.med.utah.edu/masspec/mongo.htm) to determine that RNase T1 would yield a 4-mer oligonucleotide harboring either C or m^5C at position 48 (CAAG), while RNase U2 to produce another unique 5-mer oligonucleotide with C or m^5C at the wobble position 34 of (UUCAA).

Step 2: RNase digestion

We began with ~2.5 μg of affinity-purified tRNALeu(CAA) (as described earlier) and digested it with 1 U RNase T1 (Ambion, Austin, TX) in 10 m*M* Tris buffer (pH 7.4) at 37 °C for 1 h. For RNase U2 digestion, we started with 0.5 mg of total tRNA and digested with 4 U of RNase U2 (Thermo Scientific, Waltham, MA) at 37 °C for 4 h. All oligoribonucleotides were dephosphorylated with alkaline phosphatase (10 U).

Step 3: LC–MS/MS analysis

In both cases, RNase-digested tRNA oligos were resolved by LC–MS/MS (Agilent 6410) using RP HPLC (C18 Hypersil GOLD aQ, 150 × 2.1 mm, 3 μm particle; Thermo Scientific) with an ESI source operated in negative ion mode. HPLC was performed with a gradient of acetonitrile in 8 m*M* ammonium acetate (0.2 ml/min, 45 °C): 0–2 min, 1%; 2–30 min, 1–15%; 30–31 min, 15–100%; 31–41 min, 100%. QQQ parameters: drying gas, 325 °C and 8 l/min; nebulizer, 30 p.s.i.; capillary voltage, 3800 V; dwell time, 200 ms. The Q1 and Q3 were set to unit resolution and the oligonucleotides containing m^5C were identified by comparison with standards and CID fragmentation patterns generated in a QTOF MS. A selected ion chromatogram for a particular charge state of each oligo (unexposed and exposed to H2O2) was obtained, and the summation of the mass spectra over a particular peak was used for relative quantification of changes in m^5C levels at positions 34 and 48 of tRNALeu(CAA) (Chan et al., 2012).

Step 4: Data analysis

These analyses revealed that H_2O_2 caused a twofold increase in m^5C at the wobble position and a slight decrease in m^5C at position 48, which was consistent with the observation of upregulation of proteins from genes enriched with the TTG codon recognized by m^5C-modified tRNALeu(CAA) (Chan et al., 2012).

3. DISCUSSION

The recent resurgence of interest in the dozens of cellular RNA modifications reflects in part a new perspective on understanding them as a system in the context of other signaling pathways and networks. This perspective has motivated several groups to explore the potential of tRNA modifications to operate as a system in translational control of the cell stress response (Bauer & Hermand, 2012; Begley et al., 2007; Chan et al., 2010, 2012; Patil et al., 2012; Rezgui et al., 2013). Here we present a platform for a highly quantitative and sensitive systems-level analysis of RNA modifications that is applicable to any form of RNA. In the case of tRNA, this approach entails five steps: (1) tRNA isolation and purification, (2) hydrolysis of tRNA into individual ribonucleosides, (3) liquid chromatographic resolution of ribonucleosides, (4) identification of modified ribonucleosides by high-mass accuracy spectrometric methods, (5) quantification of ribonucleosides by tandem quadrupole mass spectrometry, (6) multivariate statistical analysis of RNA modification patterns, and (7) mapping of stress-altered ribonucleosides to specific tRNAs and locations. All of these steps can be customized to accommodate the abundance and stability of the full set of modified ribonucleosides in different biological systems.

The key feature of this RNA modification analysis platform is quantitative rigor. This is essential for exploring the regulatory characteristics exhibited by tRNA modifications in cellular stress response and exploring the mechanistic links to translational control of the response. For example, even small stress-induced changes in a low-abundance modified ribonucleoside in total tRNA could indicate a mechanistically critical stress-specific modification in a single tRNA species. These small changes at the population level must be detectable, which requires sensitivity and precision in the analysis.

To ensure quantitative rigor, there are several features of this method that need to be considered. The first critical step is to stop all cellular processes immediately at the end of the exposure period by snap-freezing or adding denaturants. While the exact kinetics of RNA modification changes have not been thoroughly established, the potential for artifacts is high in light of the striking changes that occur in tRNA modification patterns in relatively brief stresses. A second important consideration is RNA isolation and purification. Since almost all the various RNA species present in an organism contain modified ribonucleosides, it is crucial to start the analysis with RNA of high quality and purity. The RNA isolation method should minimize degradation and maximize the yield of the target RNA species. The two-column RNA isolation method described here has been validated for yeast and bacterial cells in terms of the isolation of all types of RNA with yields consistent with expectation. The SEC and RP HPLC methods described here will resolve all of the canonical ncRNA species in a single sample, though contaminating RNA fragments from larger RNA species is unavoidable. Ultimately, affinity purification of individual tRNA species will provide the highest possible purity. Clearly, the importance of the RNA purity must be balanced by time and cost.

The next key step is the hydrolysis of RNA into individual ribonucleosides for LC–MS analysis. While most RNA modifications are stable during commonly used RNA isolation and hydrolysis conditions, the cases of ct^6A (Miyauchi et al., 2013) and oh^5C (Havelund et al., 2011) point to the potential for adventitious pH effects and reactions with buffer components. Further, the presence of appropriate additives such as deaminase inhibitors and antioxidants can significantly reduce the chance of forming biologically irrelevant products. These problems suggest that there may be modifications eluding discovery due to their instability during RNA isolation, purification, and hydrolysis.

It is also possible to miss even stable ribonucleosides if their physicochemical properties lead to poor resolution during the HPLC separation step prior to MS analysis. The choice of column and mobile phase can be empirically tuned to optimize resolution, detection, and quantification of even the most polar and nonpolar ribonucleosides, especially the low-abundance modified uridines important as wobble modifications. As with all other steps in this platform, it is important to consider potential confounders such as pH, adventitious reactions, and ionization suppression by mobile phase components, and coelution of isobaric ribonucleosides with similar CID fragmentation (e.g., methylated G). Overall, it is important to resolve individual ribonucleosides to reduce the number of coeluting analytes and hence increase the sensitivity of MS quantification.

The prerequisite for using this analytical platform for quantification of tRNA modifications is knowledge of as many ribonucleosides as possible in the biological system under study. The current databases do not provide organism-centric lists of modified ribonucleosides and ribonucleosides have not been characterized in most organisms, though nearly complete lists can be assembled for well-studied model organisms such as *E. coli*, *S. cerevisiae*, and humans (Czerwoniec et al., 2009). Experience points to ~70% conservation of the modification repertoire across different organisms, which reduces the work of defining the complete set. The approaches provided here, along with the growing literature on RNA modifications in different organisms, provide an efficient means to assemble ribonucleoside lists for any organism.

The LC–MS step presents its own challenges to obtaining biologically meaningful quantitative data for changes in the levels of modified ribonucleoside and to avoiding large variation among replicate datasets. For relative quantification, the key is to anchor (i.e., normalize) the MS signals of each ribonucleoside to the input tRNA concentration for every sample before using these normalized signals to compare between treated and control samples. These steps involve monitoring the total MS signal intensities of the canonical ribonucleosides, which reflect the amount of input tRNA. The use of the [15N]-dA internal standard also minimizes variation by providing a constant signal intensity to account for day-to-day variations in MS sensitivity. For absolute quantification, there are several issues presented by external calibration such as sample parameters that are beyond the scope of this method. Although the alternative method SIL-IS can overcome issues presented by external calibration, using SIL-IS for the analysis of the full ribonucleoside repertoire requires the corresponding synthetic isotopomer for every ribonucleoside, which is reasonably practical using biosynthetic standards isolated from cells grown in media with [^{13}C]-glucose. Working in concert with commercially available standards, these approaches overcome issues of different instrumental and sample parameters in absolute and relative quantification while allowing for comparison of relative modification content of related samples.

Ultimately, the population-based analyses of tRNA modification reprogramming lead to the identification of stress-sensitive modifications that need to be further characterized in terms of their location in specific tRNA molecules. The two approaches of affinity purification and population analysis of RNase digests provide a means to map ribonucleosides to specific tRNAs and to quantify stress-induced changes in the level of the modification at that site. This information is critical to making the mechanistic link between tRNA modification reprogramming and selective translation of codon-biased mRNAs. The proteomic and genomic tools needed to characterize the translational response are beyond the scope of this chapter, but are well described in the literature (Begley et al., 2007; Chan et al., 2012, 2015; Tumu, Patil, Towns, Dyavaiah, & Begley, 2012).

In summary, we have presented a detailed description of a platform for quantitatively rigorous analysis of the system of RNA modifications. The emphasis here is not on protocols, which are all detailed in the literature. Instead, we have focused on the critical determinants of analytical sensitivity, specificity, precision, and accuracy in an effort to ensure the most biologically meaningful data on mechanisms of translational control of cell stress response. The methods described here should find wide use in virtually any analysis involved RNA modifications.

ACKNOWLEDGMENTS

The authors gratefully acknowledge current and former members of the Dedon Laboratory for their published and unpublished contributions to this chapter. Financial support for the work presented in this chapter was provided by the Singapore-MIT Alliance for Research and Technology, the National Institute of Environmental Health Science (ES017010, ES022858, ES002109), the National Science Foundation (CHE-1308839, MCB-1412379), the Howard Hughes Medical Institute for fellowship support for C.G., the Deutsche Forschungsgemeinschaft for fellowship support for S.K., the SMA3 Programme for fellowship support for W.M.C. and Y.H.C., and the NIEHS Toxicology Training Grant (T32-ES007020) for fellowship support for E.G.P. This publication is made possible by the Singapore National Research Foundation under its Singapore-MIT Alliance for Research and Technology research enterprise. The views expressed herein are solely the responsibility of the authors and do not necessarily represent the official views of the foundation.

REFERENCES

Agris, P. F. (1991). Wobble position modified nucleosides evolved to select transfer RNA codon recognition: A modified-wobble hypothesis. *Biochimie, 73*(11), 1345–1349.

Agris, P. F. (2004). Decoding the genome: A modified view. *Nucleic Acids Research, 32*(1), 223–238.

Agris, P. F. (2008). Bringing order to translation: The contributions of transfer RNA anticodon-domain modifications. *EMBO Reports, 9*(7), 629–635.

Agris, P. F., Vendeix, F. A., & Graham, W. D. (2007). tRNA's wobble decoding of the genome: 40 years of modification. *Journal of Molecular Biology, 366*(1), 1–13.

Akhtar, S., Sarkar, S., Mishra, A., & Sarkar, D. (2011). A method to extract intact and pure RNA from mycobacteria. *Analytical Biochemistry, 417*(2), 286–288.

Alpert, A. J. (1990). Hydrophilic-interaction chromatography for the separation of peptides, nucleic acids and other polar compounds. *Journal of Chromatography, 499*, 177–196.

Arnvig, K., & Young, D. (2012). Non-coding RNA and its potential role in Mycobacterium tuberculosis pathogenesis. *RNA Biology, 9*(4), 427–436.

Aviv, H., & Leder, P. (1972). Purification of biologically active globin messenger RNA by chromatography on oligothymidylic acid-cellulose. *Proceedings of the National Academy of Sciences of the United States of America, 69*(6), 1408–1412.

Azarani, A., & Hecker, K. H. (2001). RNA analysis by ion-pair reversed-phase high performance liquid chromatography. *Nucleic Acids Research, 29*(2), E7.

Batey, R. T., & Kieft, J. S. (2007). Improved native affinity purification of RNA. *RNA, 13*(8), 1384–1389.

Bauer, F., & Hermand, D. (2012). A coordinated codon-dependent regulation of translation by Elongator. *Cell Cycle, 11*(24), 4524–4529. http://dx.doi.org/10.4161/cc.22689.

Begley, U., Dyavaiah, M., Patil, A., Rooney, J. P., DiRenzo, D., Young, C. M., et al. (2007). Trm9-catalyzed tRNA modifications link translation to the DNA damage response. *Molecular Cell, 28*(5), 860–870.

Biba, M., Welch, C. J., Foley, J. P., Mao, B., Vazquez, E., & Arvary, R. A. (2013). Evaluation of core-shell particle columns for ion-pair reversed-phase liquid chromatography analysis of oligonucleotides. *Journal of Pharmaceutical and Biomedical Analysis, 72*, 25–32.

Birkedal, U., Christensen-Dalsgaard, M., Krogh, N., Sabarinathan, R., Gorodkin, J., & Nielsen, H. (2015). Profiling of ribose methylations in RNA by high-throughput sequencing. *Angewandte Chemie International Edition in English, 54*(2), 451–455.

Bjork, G. R., Durand, J. M., Hagervall, T. G., Leipuviene, R., Lundgren, H. K., Nilsson, K., et al. (1999). Transfer RNA modification: Influence on translational frameshifting and metabolism. *FEBS Letters, 452*(1–2), 47–51.

Brandmayr, C., Wagner, M., Bruckl, T., Globisch, D., Pearson, D., Kneuttinger, A. C., et al. (2012). Isotope-based analysis of modified tRNA nucleosides correlates modification density with translational efficiency. *Angewandte Chemie International Edition in English, 51*(44), 11162–11165.

Brawerman, G., Mendecki, J., & Lee, S. Y. (1972). A procedure for the isolation of mammalian messenger ribonucleic acid. *Biochemistry, 11*(4), 637–641.

Bregeon, D., Colot, V., Radman, M., & Taddei, F. (2001). Translational misreading: A tRNA modification counteracts a +2 ribosomal frameshift. *Genes and Development, 15*(17), 2295–2306.

Bristow, A. W. (2006). Accurate mass measurement for the determination of elemental formula—A tutorial. *Mass Spectrometry Reviews, 25*(1), 99–111.

Buchhaupt, M., Sharma, S., Kellner, S., Oswald, S., Paetzold, M., Peifer, C., et al. (2014). Partial methylation at Am100 in 18S rRNA of baker's yeast reveals ribosome heterogeneity on the level of eukaryotic rRNA modification. *PLoS One, 9*(2), e89640.

Buck, M., Connick, M., & Ames, B. N. (1983). Complete analysis of tRNA-modified nucleosides by high-performance liquid chromatography: The 29 modified nucleosides of Salmonella typhimurium and Escherichia coli tRNA. *Analytical Biochemistry, 129*(1), 1–13.

Cantara, W. A., Crain, P. F., Rozenski, J., McCloskey, J. A., Harris, K. A., Zhang, X., et al. (2011). The RNA Modification Database, RNAMDB: 2011 update. *Nucleic Acids Research, 39*(Database issue), D195–D201.

Carlile, T. M., Rojas-Duran, M. F., Zinshteyn, B., Shin, H., Bartoli, K. M., & Gilbert, W. V. (2014). Pseudouridine profiling reveals regulated mRNA pseudouridylation in yeast and human cells. *Nature, 515*(7525), 143–146.

Castleberry, C. M., & Limbach, P. A. (2010). Relative quantitation of transfer RNAs using liquid chromatography mass spectrometry and signature digestion products. *Nucleic Acids Research, 38*(16), e162.

Chan, C. T., Chionh, Y. H., Ho, C. H., Lim, K. S., Babu, I. R., Ang, E., et al. (2011). Identification of N6, N6-dimethyladenosine in transfer RNA from Mycobacterium bovis Bacille Calmette-Guerin. *Molecules*, *16*(6), 5168–5181.

Chan, T. Y. C., Deng, W., Li, F., DeMott, M. S., Babu, I. R., Begley, T. J., et al. (2015). Highly predictive reprogramming of tRNA modifications is linked to selective expression of codon-biased genes. *Chemical Research in Toxicology*, *28*(5), 978–988.

Chan, C. T., Dyavaiah, M., DeMott, M. S., Taghizadeh, K., Dedon, P. C., & Begley, T. J. (2010). A quantitative systems approach reveals dynamic control of tRNA modifications during cellular stress. *PLoS Genetics*, *6*(12), e1001247. http://dx.doi.org/10.1371/journal.pgen.1001247.

Chan, C. T., Pang, Y. L., Deng, W., Babu, I. R., Dyavaiah, M., Begley, T. J., et al. (2012). Reprogramming of tRNA modifications controls the oxidative stress response by codon-biased translation of proteins. *Nature Communications*, *3*, 937.

Chen, Y. G., Kowtoniuk, W. E., Agarwal, I., Shen, Y., & Liu, D. R. (2009). LC/MS analysis of cellular RNA reveals NAD-linked RNA. *Nature Chemical Biology*, *5*(12), 879–881.

Chionh, Y. H., Ho, C.-H., Pruksakorn, D., Babu, R. I., Ng, C. S., Hia, F., et al. (2013). A multidimensional platform for the purification of non-coding RNA species. *Nucleic Acids Research*, *41*(17), e168.

Crain, P. F. (1990). Preparation and enzymatic hydrolysis of DNA and RNA for mass spectrometry. *Methods in Enzymology*, *193*, 782–790.

Crain, P. F., Rozenski, J., & McCloskey, J. A. (1999). The RNA modification database. *Nucleic Acids Research*, *27*, 196–197. Available from: http://library.med.utah.edu/RNAmods/.

Cunningham, L., Kittikamron, K., & Lu, Y. (1996). Preparative-scale purification of RNA using an efficient method which combines gel electrophoresis and column chromatography. *Nucleic Acids Research*, *24*(18), 3647–3648.

Czerwoniec, A., Dunin-Horkawicz, S., Purta, E., Kaminska, K. H., Kasprzak, J. M., Bujnicki, J. M., et al. (2009). MODOMICS: A database of RNA modification pathways. 2008 update. *Nucleic Acids Research*, *37*(Database issue), D118–D121.

Dalluge, J. J., Hashizume, T., Sopchik, A. E., McCloskey, J. A., & Davis, D. R. (1996). Conformational flexibility in RNA: The role of dihydrouridine. *Nucleic Acids Research*, *24*(6), 1073–1079.

Davis, D. R. (1995). Stabilization of RNA stacking by pseudouridine. *Nucleic Acids Research*, *23*(24), 5020–5026.

Dedon, P. C., & Begley, T. J. (2014). A system of RNA modifications and biased codon use controls cellular stress response at the level of translation. *Chemical Research in Toxicology*, *27*(3), 330–337.

Dong, M., Wang, C., Deen, W. M., & Dedon, P. C. (2003). Absence of 2'-deoxyoxanosine and presence of abasic sites in DNA exposed to nitric oxide at controlled physiological concentrations. *Chemical Research in Toxicology*, *16*(9), 1044–1055.

Dudley, E., Tuytten, R., Bond, A., Lemière, F., Brenton, A. G., Esmans, E. L., et al. (2005). Study of the mass spectrometric fragmentation of pseudouridine: Comparison of fragmentation data obtained by matrix-assisted laser desorption/ionisation post-source decay, electrospray ion trap multistage mass spectrometry, and by a method utilising electrospray quadrupole time-of-flight tandem mass spectrometry and in-source fragmentation. *Rapid Communications in Mass Spectrometry*, *19*(21), 3075–3085.

Dumelin, C. E., Chen, Y., Leconte, A. M., Chen, Y. G., & Liu, D. R. (2012). Discovery and biological characterization of geranylated RNA in bacteria. *Nature Chemical Biology*, *8*(11), 913–919.

Easton, L. E., Shibata, Y., & Lukavsky, P. J. (2010). Rapid, nondenaturing RNA purification using weak anion-exchange fast performance liquid chromatography. *RNA*, *16*(3), 647–653.

Emilsson, V., Naslund, A. K., & Kurland, C. G. (1992). Thiolation of transfer RNA in Escherichia coli varies with growth rate. *Nucleic Acids Research*, *20*(17), 4499–4505.
Farand, J., & Gosselin, F. (2009). De novo sequence determination of modified oligonucleotides. *Analytical Chemistry*, *81*(10), 3723–3730.
Fu, Y., Jia, G., Pang, X., Wang, R. N., Wang, X., Li, C. J., et al. (2013). FTO-mediated formation of N6-hydroxymethyladenosine and N6-formyladenosine in mammalian RNA. *Nature Communications*, *4*, 1798.
Fujii, S., Inagaki, K., Miyashita, S., Groombridge, A. S., Nagasawa, K., Chiba, K., et al. (2014). Separation and quantification of RNA molecules using size-exclusion chromatography hyphenated with inductively coupled plasma-mass spectrometry. *Electrophoresis*, *35*(9), 1315–1318.
Gehrke, C. W., & Kuo, K. C. (1989). Ribonucleoside analysis by reversed-phase high-performance liquid chromatography. *Journal of Chromatography*, *471*, 3–36.
Gehrke, C. W., Kuo, K. C., McCune, R. A., Gerhardt, K. O., & Agris, P. F. (1982). Quantitative enzymatic hydrolysis of tRNAs: Reversed-phase high-performance liquid chromatography of tRNA nucleosides. *Journal of Chromatography*, *230*(2), 297–308.
Gehrke, C. W., Kuo, K. C., & Zumwalt, R. W. (1980). Chromatography of nucleosides. *Journal of Chromatography*, *188*(1), 129–147.
Giessing, A. M., Jensen, S. S., Rasmussen, A., Hansen, L. H., Gondela, A., Long, K., et al. (2009). Identification of 8-methyladenosine as the modification catalyzed by the radical SAM methyltransferase Cfr that confers antibiotic resistance in bacteria. *RNA*, *15*(2), 327–336.
Grosjean, H., Droogmans, L., Roovers, M., & Keith, G. (2007). Detection of enzymatic activity of transfer RNA modification enzymes using radiolabeled tRNA substrates. *Methods in Enzymology*, *425*, 55–101.
Gu, C., Begley, T. J., & Dedon, P. C. (2014). tRNA modifications regulate translation during cellular stress. *FEBS Letters*, *588*(23), 4287–4296.
Guo, Y., & Gaiki, S. (2005). Retention behavior of small polar compounds on polar stationary phases in hydrophilic interaction chromatography. *Journal of Chromatography A*, *1074*(1–2), 71–80.
Havelund, J. F., Giessing, A. M., Hansen, T., Rasmussen, A., Scott, L. G., & Kirpekar, F. (2011). Identification of 5-hydroxycytidine at position 2501 concludes characterization of modified nucleotides in E. coli 23S rRNA. *Journal of Molecular Biology*, *411*(3), 529–536.
He, C. (2010). Grand challenge commentary: RNA epigenetics? *Nature Chemical Biology*, *6*(12), 863–865.
Hia, F., Chionh, Y. H., Pang, Y. L., DeMott, M. S., McBee, M. E., & Dedon, P. C. (2015). Mycobacterial RNA isolation optimized for non-coding RNA: High fidelity isolation of 5S rRNA from Mycobacterium bovis BCG reveals novel post-transcriptional processing and a complete spectrum of modified ribonucleosides. *Nucleic Acids Research*, *43*, e32 (pii: gku1317).
Hienzsch, A., Deiml, C., Reiter, V., & Carell, T. (2013). Total synthesis of the hypermodified RNA bases wybutosine and hydroxywybutosine and their quantification together with other modified RNA bases in plant materials. *Chemistry*, *19*(13), 4244–4248.
Hill, A. W., & Mortishire-Smith, R. J. (2005). Automated assignment of high-resolution collisionally activated dissociation mass spectra using a systematic bond disconnection approach. *Rapid Communications in Mass Spectrometry*, *19*(21), 3111–3118.
Hossain, M., & Limbach, P. A. (2007). Mass spectrometry-based detection of transfer RNAs by their signature endonuclease digestion products. *RNA*, *13*(2), 295–303.
Huang, B., Lu, J., & Bystrom, A. S. (2008). A genome-wide screen identifies genes required for formation of the wobble nucleoside 5-methoxycarbonylmethyl-2-thiouridine in Saccharomyces cerevisiae. *RNA*, *14*(10), 2183–2194.

Katsivela, E., & Hofle, M. G. (1995). Separation of transfer RNA and 5S ribosomal RNA using capillary electrophoresis. *Journal of Chromatography A*, *700*(1–2), 125–136.

Kellner, S., Burhenne, J., & Helm, M. (2010). Detection of RNA modifications. *RNA Biology*, 7(2), 237–247.

Kellner, S., Neumann, J., Rosenkranz, D., Lebedeva, S., Ketting, R. F., Zischler, H., et al. (2014). Profiling of RNA modifications by multiplexed stable isotope labelling. *Chemical Communications (Cambridge)*, *50*(26), 3516–3518.

Kellner, S., Ochel, A., Thuring, K., Spenkuch, F., Neumann, J., Sharma, S., et al. (2014). Absolute and relative quantification of RNA modifications via biosynthetic isotopomers. *Nucleic Acids Research*, *42*(18), e142.

Kieft, J. S., & Batey, R. T. (2004). A general method for rapid and nondenaturing purification of RNAs. *RNA*, *10*(6), 988–995.

Kim, I., McKenna, S. A., Viani Puglisi, E., & Puglisi, J. D. (2007). Rapid purification of RNAs using fast performance liquid chromatography (FPLC). *RNA*, *13*(2), 289–294.

Kim, Y. K., Yeo, J., Kim, B., Ha, M., & Kim, V. N. (2012). Short structured RNAs with low GC content are selectively lost during extraction from a small number of cells. *Molecular Cell*, *46*(6), 893–895.

Kind, T., & Fiehn, O. (2006). Metabolomic database annotations via query of elemental compositions: Mass accuracy is insufficient even at less than 1 ppm. *BMC Bioinformatics*, 7, 234.

Kinger, A. K., Verma, A., & Tyagi, J. S. (1993). A method for the isolation of pure intact RNA from mycobacteria. *Biotechniques*, *14*(5), 724–725.

Kirino, Y., & Mourelatos, Z. (2007). Mouse Piwi-interacting RNAs are 2'-O-methylated at their 3' termini. *Nature Structural and Molecular Biology*, *14*(4), 347–348.

Klootwijk, J., & Planta, R. J. (1973). Analysis of the methylation sites in yeast ribosomal RNA. *European Journal of Biochemistry*, *39*(2), 325–333.

Köhrer, C., & Rajbhandary, U. L. (2008). The many applications of acid urea polyacrylamide gel electrophoresis to studies of tRNAs and aminoacyl-tRNA synthetases. *Methods*, *44*(2), 129–138.

Koubek, J., Lin, K. F., Chen, Y. R., Cheng, R. P., & Huang, J. J. (2013). Strong anion-exchange fast performance liquid chromatography as a versatile tool for preparation and purification of RNA produced by in vitro transcription. *RNA*, *19*(10), 1449–1459 (rna.038117.113 [pii]).

Kowtoniuk, W. E., Shen, Y., Heemstra, J. M., Agarwal, I., & Liu, D. R. (2009). A chemical screen for biological small molecule-RNA conjugates reveals CoA-linked RNA. *Proceedings of the National Academy of Sciences of the United States of America*, *106*(19), 7768–7773.

Kruger, M. K., Pedersen, S., Hagervall, T. G., & Sorensen, M. A. (1998). The modification of the wobble base of tRNAGlu modulates the translation rate of glutamic acid codons in vivo. *Journal of Molecular Biology*, *284*(3), 621–631.

Kuschel, M., & Ausserer, W. (2002). Use of lab-on-a-chip technology for protein sizing and quantitation. *Journal of Biomolecular Techniques*, *13*(3), 172–178.

Lewicki, P. P., & Sinskey, A. J. (1970). Precision of RNA separation by polyacrylamide gel electrophoresis. *Analytical Biochemistry*, *33*(2), 273–278.

Li, Z., Dou, X., Ni, Y., Chen, Q., Cheng, S., & Yamaguchi, Y. (2012). Is pulsed electric field still effective for RNA separation in capillary electrophoresis? *Journal of Chromatography A*, *1229*, 274–279.

Li, Z., Dou, X., Ni, Y., Sumitomo, K., & Yamaguchi, Y. (2010). Acetic acid denaturing pulsed field capillary electrophoresis for RNA separation. *Electrophoresis*, *31*(21), 3531–3536.

Lim, H. K., Chen, J., Sensenhauser, C., Cook, K., & Subrahmanyam, V. (2007). Metabolite identification by data-dependent accurate mass spectrometric analysis at resolving power of 60,000 in external calibration mode using an LTQ/Orbitrap. *Rapid Communications in Mass Spectrometry*, *21*(12), 1821–1832.

Little, J. L., Cleven, C. D., & Brown, S. D. (2011). Identification of "known unknowns" utilizing accurate mass data and chemical abstracts service databases. *Journal of the American Society for Mass Spectrometry*, *22*(2), 348–359.

Little, J. L., Williams, A. J., Pshenichnov, A., & Tkachenko, V. (2012). Identification of "known unknowns" utilizing accurate mass data and ChemSpider. *Journal of the American Society for Mass Spectrometry*, *23*(1), 179–185.

Luo, Y., Eldho, N. V., Sintim, H. O., & Dayie, T. K. (2011). RNAs synthesized using photocleavable biotinylated nucleotides have dramatically improved catalytic efficiency. *Nucleic Acids Research*, *39*(19), 8559–8571.

Machnicka, M. A., Milanowska, K., Osman Oglou, O., Purta, E., Kurkowska, M., Olchowik, A., et al. (2013). MODOMICS: A database of RNA modification pathways—2013 update. *Nucleic Acids Research*, *41*(Database issue), D262–D267.

Mandal, D., Kohrer, C., Su, D., Babu, I. R., Chan, C. T., Liu, Y., et al. (2014). Identification and codon reading properties of 5-cyanomethyl uridine, a new modified nucleoside found in the anticodon wobble position of mutant haloarchaeal isoleucine tRNAs. *RNA*, *20*(2), 177–188.

Mandal, D., Kohrer, C., Su, D., Russell, S. P., Krivos, K., Castleberry, C. M., et al. (2010). Agmatidine, a modified cytidine in the anticodon of archaeal tRNA(Ile), base pairs with adenosine but not with guanosine. *Proceedings of the National Academy of Sciences of the United States of America*, *107*(7), 2872–2877.

McCarthy, S. M., Gilar, M., & Gebler, J. (2009). Reversed-phase ion-pair liquid chromatography analysis and purification of small interfering RNA. *Analytical Biochemistry*, *390*(2), 181–188.

McLafferty, F. W. (1993). *Interpretation of mass spectra*. Mill Valley, CA: University Science Books.

Miyauchi, K., Kimura, S., & Suzuki, T. (2013). A cyclic form of N^6-threonylcarbamoyladenosine as a widely distributed tRNA hypermodification. *Nature Chemical Biology*, *9*(2), 105–111.

Motorin, Y., & Helm, M. (2010). tRNA stabilization by modified nucleotides. *Biochemistry*, *49*(24), 4934–4944.

Murphy, F. V., 4th, Ramakrishnan, V., Malkiewicz, A., & Agris, P. F. (2004). The role of modifications in codon discrimination by tRNA(Lys)UUU. *Nature Structural and Molecular Biology*, *11*(12), 1186–1191.

Nameki, N. (1995). Identity elements of tRNA(Thr) towards Saccharomyces cerevisiae threonyl-tRNA synthetase. *Nucleic Acids Research*, *23*(15), 2831–2836.

Nameki, N., Asahara, H., Shimizu, M., Okada, N., & Himeno, H. (1995). Identity elements of Saccharomyces cerevisiae tRNA(His). *Nucleic Acids Research*, *23*(3), 389–394.

Ohara, T., Sakaguchi, Y., Suzuki, T., Ueda, H., Miyauchi, K., & Suzuki, T. (2007). The 3' termini of mouse Piwi-interacting RNAs are 2'-O-methylated. *Nature Structural and Molecular Biology*, *14*(4), 349–350.

Patil, A., Chan, C. T., Dyavaiah, M., Rooney, J. P., Dedon, P. C., & Begley, T. J. (2012). Translational infidelity-induced protein stress results from a deficiency in Trm9-catalyzed tRNA modifications. *RNA Biology*, *9*(7), 990–1001.

Perry, R. P., La Torre, J., Kelley, D. E., & Greenberg, J. R. (1972). On the lability of poly(A) sequences during extraction of messenger RNA from polyribosomes. *Biochimica et Biophysica Acta*, *262*(2), 220–226.

Pomerantz, S. C., & McCloskey, J. A. (1990). Analysis of RNA hydrolyzates by liquid chromatography-mass spectrometry. *Methods in Enzymology*, *193*, 796–824.

Putz, J., Florentz, C., Benseler, F., & Giege, R. (1994). A single methyl group prevents the mischarging of a tRNA. *Natural Structural Biology*, *1*(9), 580–582.

Rederstorff, M., & Hüttenhofer, A. (2010). Small non-coding RNAs in disease development and host-pathogen interactions. *Current Opinion in Molecular Therapeutics*, *12*(6), 684–694.

Reynolds, W. F., & Enriquez, R. G. (2002). Choosing the best pulse sequences, acquisition parameters, postacquisition processing strategies, and probes for natural product structure elucidation by NMR spectroscopy. *Journal of Natural Products*, *65*(2), 221–244.

Rezgui, V. A., Tyagi, K., Ranjan, N., Konevega, A. L., Mittelstaet, J., Rodnina, M. V., et al. (2013). tRNA tKUUU, tQUUG, and tEUUC wobble position modifications fine-tune protein translation by promoting ribosome A-site binding. *Proceedings of the National Academy of Sciences of the United States of America*, *110*(30), 12289–12294.

Rudinger, J., Florentz, C., & Giege, R. (1994). Histidylation by yeast HisRS of tRNA or tRNA-like structure relies on residues −1 and 73 but is dependent on the RNA context. *Nucleic Acids Research*, *22*(23), 5031–5037.

Russell, S. P., & Limbach, P. A. (2013). Evaluating the reproducibility of quantifying modified nucleosides from ribonucleic acids by LC-UV-MS. *Journal of Chromatography, B: Analytical Technologies in the Biomedical and Life Sciences*, *923–924*, 74–82.

Salazar, J. C., Ambrogelly, A., Crain, P. F., McCloskey, J. A., & Söll, D. (2004). A truncated aminoacyl-tRNA synthetase modifies RNA. *Proceedings of the National Academy of Sciences of the United States of America*, *101*(20), 7536–7541.

Schroeder, A., Mueller, O., Stocker, S., Salowsky, R., Leiber, M., Gassmann, M., et al. (2006). The RIN: An RNA integrity number for assigning integrity values to RNA measurements. *BMC Molecular Biology*, 7, 3.

Shanagar, J. (2005). Purification of a synthetic oligonucleotide by anion exchange chromatography: Method optimisation and scale-up. *Journal of Biochemical and Biophysical Methods*, *64*(3), 216–225.

Skeidsvoll, J., & Ueland, P. M. (1996). Analysis of RNA by capillary electrophoresis. *Electrophoresis*, *17*(9), 1512–1517.

Sparkman, D. O. (2007). *Mass spectrometry desk reference. Vol. 18* (2nd ed.). Pittsburgh, PA: Global View Publishing.

Spenkuch, F., Hinze, G., Kellner, S., Kreutz, C., Micura, R., Basche, T., et al. (2014). Dye label interference with RNA modification reveals 5-fluorouridine as non-covalent inhibitor. *Nucleic Acids Research*, *42*(20), 12735–12745.

Srisawat, C., & Engelke, D. R. (2001). Streptavidin aptamers: Affinity tags for the study of RNAs and ribonucleoproteins. *RNA*, 7(4), 632–641.

Srisawat, C., Goldstein, I. J., & Engelke, D. R. (2001). Sephadex-binding RNA ligands: Rapid affinity purification of RNA from complex RNA mixtures. *Nucleic Acids Research*, *29*(2), E4.

Su, D., Chan, C. T., Gu, C., Lim, K. S., Chionh, Y. H., McBee, M. E., et al. (2014). Quantitative analysis of ribonucleoside modifications in tRNA by HPLC-coupled mass spectrometry. *Nature Protocols*, *9*(4), 828–841.

Suzuki, T., Ikeuchi, Y., Noma, A., & Sakaguchi, Y. (2007). Mass spectrometric identification and characterization of RNA-modifying enzymes. *Methods in Enzymology*, *425*, 211–229.

Suzuki, T., & Suzuki, T. (2007). Chaplet column chromatography: Isolation of a large set of individual RNAs in a single step. *Methods in Enzymology*, *425*, 231–239.

Taniguchi, T., Miyauchi, K., Nakane, D., Miyata, M., Muto, A., Nishimura, S., et al. (2013). Decoding system for the AUA codon by tRNAIle with the UAU anticodon in Mycoplasma mobile. *Nucleic Acids Research*, *41*(4), 2621–2631.

Tissot, C. (2008). *Analysis of miRNA content in total RNA preparations using the Agilent 2100 bioanalyzer*. Palo Alto, CA: Agilent Technologies. http://www.chem.agilent.com/Library/applications/5989-7870EN.pdf.

Tumu, S., Patil, A., Towns, W., Dyavaiah, M., & Begley, T. J. (2012). The gene-specific codon counting database: A genome-based catalog of one-, two-, three-, four- and five-codon combinations present in Saccharomyces cerevisiae genes. *Database (Oxford), 2012*, bas002.

Tuytten, R., Lemiere, F., Van Dongen, W., Witters, E., Esmans, E. L., Newton, R. P., et al. (2008). Development of an on-line SPE-LC-ESI-MS method for urinary nucleosides: Hyphenation of aprotic boronic acid chromatography with hydrophilic interaction LC-ESI-MS. *Analytical Chemistry, 80*(4), 1263–1271.

Urbonavicius, J., Qian, Q., Durand, J. M., Hagervall, T. G., & Bjork, G. R. (2001). Improvement of reading frame maintenance is a common function for several tRNA modifications. *EMBO Journal, 20*(17), 4863–4873.

Varricchio, F., & Ernst, H. J. (1975). Separation of low molecular weight RNA by polyacrylamide gel electrophoresis. *Analytical Biochemistry, 68*(2), 485–492.

Waghmare, S. P., Pousinis, P., Hornby, D. P., & Dickman, M. J. (2009). Studying the mechanism of RNA separations using RNA chromatography and its application in the analysis of ribosomal RNA and RNA:RNA interactions. *Journal of Chromatography A, 1216*(9), 1377–1382.

Walker, S. C., Scott, F. H., Srisawat, C., & Engelke, D. R. (2008). RNA affinity tags for the rapid purification and investigation of RNAs and RNA-protein complexes. *Methods in Molecular Biology, 488*, 23–40.

Wang, X., & He, C. (2014). Dynamic RNA modifications in posttranscriptional regulation. *Molecular Cell, 56*(1), 5–12.

Ward, A. M., Gunaratne, J., & Garcia-Blanco, M. A. (2014). Identification of dengue RNA binding proteins using RNA chromatography and quantitative mass spectrometry. *Methods in Molecular Biology, 1138*, 253–270.

Warren, W. J., & Vella, G. (1995). Principles and methods for the analysis and purification of synthetic deoxyribonucleotides by high-performance liquid chromatography. *Molecular Biotechnology, 4*(2), 179–199.

Werner, A. (1993). Reversed-phase and ion-pair separations of nucleotides, nucleosides and nucleobases: Analysis of biological samples in health and disease. *Journal of Chromatography, 618*(1–2), 3–14.

Yarian, C., Townsend, H., Czestkowski, W., Sochacka, E., Malkiewicz, A. J., Guenther, R., et al. (2002). Accurate translation of the genetic code depends on tRNA modified nucleosides. *Journal of Biological Chemistry, 277*(19), 16391–16395.

Zhang, Q., Pu, R., Du, Y., Han, Y., Su, T., Wang, H., et al. (2012). Non-coding RNAs in hepatitis B or C-associated hepatocellular carcinoma: Potential diagnostic and prognostic markers and therapeutic targets. *Cancer Letters, 321*(1), 1–12.

Zhao, H. Q., Wang, X., Li, H. M., Yang, B., Yang, H. J., & Huang, L. (2013). Characterization of nucleosides and nucleobases in natural Cordyceps by HILIC-ESI/TOF/MS and HILIC-ESI/MS. *Molecules, 18*(8), 9755–9769.

Zheng, G., Dahl, J. A., Niu, Y., Fu, Y., Klungland, A., Yang, Y. G., et al. (2013). Sprouts of RNA epigenetics: The discovery of mammalian RNA demethylases. *RNA Biology, 10*(6), 915–918.

Zhou, G., Pang, H., Tang, Y., Yao, X., Ding, Y., Zhu, S., et al. (2014). Hydrophilic interaction ultra-performance liquid chromatography coupled with triple-quadrupole tandem mass spectrometry (HILIC-UPLC-TQ-MS/MS) in multiple-reaction monitoring (MRM) for the determination of nucleobases and nucleosides in ginkgo seeds. *Food Chemistry, 150*, 260–266.

CHAPTER FOUR

Recognition of Specified RNA Modifications by the Innate Immune System

Tatjana Eigenbrod*, Patrick Keller†, Steffen Kaiser†, Katharina Rimbach*, Alexander H. Dalpke*, Mark Helm†,[1]

*Department of Infectious Diseases, Medical Microbiology and Hygiene, University of Heidelberg, Heidelberg, Germany
†Institute of Pharmacy and Biochemistry, Johannes Gutenberg-University of Mainz, Mainz, Germany
[1]Corresponding author: e-mail address: mhelm@uni-mainz.de

Contents

Abstract

Microbial nucleic acids have been described as important activators of human innate immune responses by triggering so-called pattern recognition receptors (PRRs) that are expressed on innate immune cells, including plasmacytoid dendritic cells and monocytes. Although host and microbial nucleic acids share pronounced chemical and structural similarities, they significantly differ in their posttranscriptional modification profile, allowing the host to discriminate between self and nonself. In this regard, ribose 2′-*O*-methylation has been discovered as suppressor of RNA-induced PRR activation. Although 2′-*O*-methylation occurs with higher frequencies in eukaryotic than in prokaryotic RNA, the immunosuppressive properties of 2′-*O*-methylated nucleotides may be misused by certain bacteria as immune evasion mechanism. In the course of identifying inhibitory RNA modifications, our groups have synthesized and comparatively analyzed a series of differentially modified RNAs, so-called modivariants, for their immune stimulatory capacities. In this chapter, we will detail the protocols for the design and synthesis of RNA modivariants by molecular cut-and-paste techniques (referred to as molecular surgery) and describe testing of their immune stimulatory properties upon transfection into peripheral blood mononuclear cells.

Methods in Enzymology, Volume 560
ISSN 0076-6879
http://dx.doi.org/10.1016/bs.mie.2015.03.006

1. INTRODUCTION

The innate immune system is an ancient part of the mammalian immune system which is activated within minutes after pathogen encounter. Innate immune cells include granulocytes, macrophages, myeloid dendritic cells, and plasmacytoid dendritic cells (pDC), which are all able to respond to infectious organisms immediately and without any delay. Recognition of microbes is mediated by germ-line encoded, nonvariable receptors, termed pattern recognition receptors (PRRs). An important group of PRRs are Toll-like receptors (TLRs) that are either expressed on the cell surface or in the endosomal compartment. Microbial sensing of pathogens by PRRs is achieved by the recognition of the so-called "pathogen-associated molecular patterns" (PAMPs), which are highly conserved structure that can be found exclusively in microbes, including bacteria, viruses, parasites, and fungi. Typical PAMPs comprise molecules like flagellin, cell wall components of Gram-positive and Gram-negative bacteria, or nucleic acids of both viral and bacterial origin (reviewed in Kawai & Akira, 2010; Mogensen, 2009). Nucleic acids of microbial origin can be differentiated from eukaryotic host nucleic acids due to differences in structure, nucleoside modifications, and intracellular localization.

In this regard, bacterial RNA has gained increasing attention during the past years as potent immune stimulator. Intracellular delivery of bacterial RNA into innate immune cells results in cell-type-specific activation of endosomal TLRs and subsequent production of proinflammatory cytokines. In both murine and human pDCs, bacterial RNA triggers secretion of interferon-α (IFN-α) in a TLR7-dependent manner (Eberle, Sirin, Binder, & Dalpke, 2009; Gehrig et al., 2012; Jockel et al., 2012; Mancuso et al., 2009). By contrast, human monocytes sense bacterial RNA independently of TLR7 but probably make use of TLR8 (Cervantes et al., 2013; Hornung et al., 2002). The latter respond to bacterial RNA stimulation with production of NF-κB dependent cytokines including IL-6 and TNF. In murine macrophages and myeloid DCs, TLR13 has recently been identified as sensor for bacterial RNA that specifically detects as highly conserved sequence within bacterial 23S rRNA (Hidmark, von Saint Paul, & Dalpke, 2012; Li & Chen, 2012; Oldenburg et al., 2012). However, TLR13 is not expressed in humans.

Although host and microbial nucleic acids share pronounced chemical and structural similarities, they significantly differ in their posttranscriptional modification profile. Differences in both the kind and extent of RNA modifications allow the host to discriminate between self and nonself origin (Kariko, Buckstein, Ni, & Weissman, 2005; Koski et al., 2004). In general, eukaryotic RNA is more abundantly modified than prokaryotic RNA and more than 150 different modifications of either the nucleobase or the ribose have been identified. Among all RNA species, tRNAs show the highest degree of modified nucleotides, with modification rates of up to 25%. Experiments using *in vitro*-transcribed RNA with randomly incorporated nucleotide modifications indicated that m^6A, m^5C, m^5U, s^2U, and Ψ modifications suppressed TLR7 and TLR8 activation when these receptors where overexpressed in HEK293 cells (Kariko et al., 2005). A TLR7 antagonistic effect in both murine and human pDCs has been described for 2′-*O*-methylation of the ribose, a modification that occurs with higher abundancies in eukaryotic than in prokaryotic RNA. Of note, the immunosuppressive properties of 2′-*O*-methylated nucleotides may be misused by certain bacteria as a potential immune evasion mechanism (Robbins et al., 2007; Sioud, 2009). Indeed, a detailed analysis of the TLR7 stimulatory properties of different *E. coli* tRNA species identified a 2′-*O*-methylated guanosine (Gm) at position 18 as necessary and sufficient for suppressing TLR7 activation in human pDC (Gehrig et al., 2012; Jockel et al., 2012). Remarkably, this modification is most active in an optimized dinucleotide context (Kaiser, Rimbach, Eigenbrod, Dalpke, & Helm, 2014), and not only renders RNA nonstimulatory but also exerts a dominant-inhibitory effect on otherwise stimulatory RNA species by acting as TLR7 antagonist.

In the course of the identification of Gm as the TLR-inhibiting entity, our groups have synthesized and tested a series of differentially modified RNAs, the so-called modivariants, to gauge the recognition by components of the human immune system. This chapter details the protocols of the two crucial components of this approach. The first step consists in the design and synthesis of modivariants by molecular cut-and-paste techniques sometimes referred to as molecular surgery (Hengesbach, Kobitski, et al., 2008; Hengesbach, Meusburger, Lyko, & Helm, 2008; Kurschat, Muller, Wombacher, & Helm, 2005). Of note, this step requires access to microgram amounts of both native tRNA and the corresponding *in vitro* transcripts. The second step is the testing of modivariants after transfection into peripheral blood mononuclear cells (PBMCs) (Fig. 1).

Under physiological conditions, RNA can gain access to nucleic acid-sensing TLRs after phagocytic engulfment of microbial species and

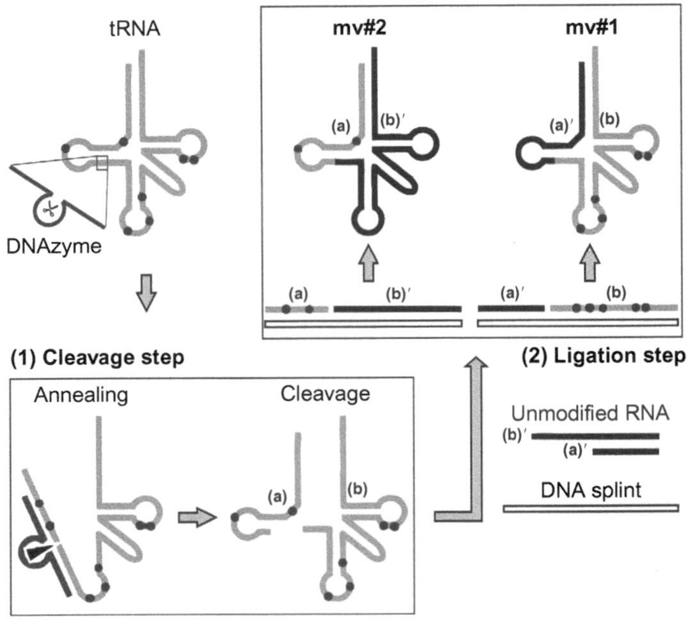

Figure 1 Schematic view of the molecular surgery approach based on DNAzymes (adjusted from Gehrig et al. (2012)). The DNAzyme (darker grey line) is capable of cutting between determined dinucleotide targets after its annealing to the RNA molecule. The cleavage reaction generates modified fragments (lighter grey lines with dots indicating modifications) which can be hybridized together with unmodified fragments (black lines) to a complementary DNA splint (white line). This procedure yields modivariants mv#1 and mv#2 after ligation and digestion of DNAzyme/DNA splint.

subsequent degradation within the phagolysosome. This process of intracellular delivery of bacterial RNA can be mimicked *in vitro* by stimulating cells with purified RNA that has been complexed with cationic transfection reagents. Packaging of bacterial RNA into such lipoplexes is essential for induction of host cell responses as free RNA does not efficiently trigger immune cell activation. A commonly used model system to investigate RNA-induced immune responses is the stimulation of freshly isolated human PBMCs. Apart from B- and T-lymphocytes, the PBMC fraction contains monocytes (approx. 3–12% of total blood cells) as well as pDCs (less than 1% of total blood cells). The use of freshly collected blood for PBMC isolation is recommended if pDC-dependent immune responses (e.g., IFN-α production) are to be measured. Stimulation of PBMCs with encapsulated bacterial RNA and measurement of secretion of IFN-α (within PBMCs exclusively produced in a TLR7-dependent manner by pDCs;

Eberle et al., 2009; Gehrig et al., 2012; Jockel et al., 2012) is an easy method to determine the "immunestimulatory" potential of bacterial nucleic acid preparations.

2. PREPARATION OF tRNA MODIVARIANTS

While in general, all unmodified tRNA molecules obtained, e.g., by *in vitro* transcription, have proven to stimulate TLR7, certain native tRNA species, such as, e.g., *E. coli* $tRNA^{Ty}r$, do not. Since the only difference in these two species consists in the presence of posttranscriptional modifications, the question arises, which single modification, or which subset of modifications are responsible for the lack of TLR7-mediated immunestimulation. This question can be addressed by tested modivariants, RNA molecules containing selected modifications, and these modivariants can be synthesized by recombining modified and unmodified fragments obtained by site-specific cleavage from the respective tRNAs. The molecular tools of choice for site-specific cleavage are short single-stranded catalytic DNA sequences, the so-called deoxyribozymes (DNAzymes). In general, DNAzymes build up a catalytic core in their tertiary structure that executes the cleavage reaction of the RNA substrates (Cruz, Withers, & Li, 2004). In addition to that, DNAzymes consist of two flanking sequences which act as substrate-specific components by binding their reverse complementary RNA target sequence via Watson–Crick pairing. Since these flanking sequences can be chosen freely, DNAzymes enable site-specific cleavage of RNA. Thus, some of these catalytic sequences, namely 10–23 and 8–17 DNAzymes, have been optimized and used in several applications (Buchhaupt, Peifer, & Entian, 2007; Hengesbach, Meusburger, et al., 2008). For both groups, the presence of divalent metal cations (Mg^{2+} or Mn^{2+}) is required for their cleavage activity. Furthermore, 10–23 DNAzymes are sensitive to exchange of nucleotides within the catalytic core, whereas 8–17 DNAzymes remain catalytically active even after several changes within their catalytic loop (Cruz et al., 2004; Zaborowska, Furste, Erdmann, & Kurreck, 2002; Zaborowska, Schubert, Kurreck, & Erdmann, 2005).

Cleavage of RNA substrates by DNAzymes occurs in the phosphate backbone with variable efficiency (Meusburger, Hengesbach, & Helm, 2011). The 10–23 DNAzymes for instance cleave A–U and G–U combinations well, whereas their cleavage activities for G–C and A–C are dramatically decreased (Cruz et al., 2004), some of them require inosine incorporation for their activity (Cairns, King, & Sun, 2003). The 8–17 DNAzymes in turn are able to cleave all nucleotide junctions with exception of the Y–U combination (Meusburger et al., 2011). However, it has to be mentioned that one disadvantage of DNAzymes is their limited reaction rate in comparison to, e.g., restriction enzymes. The cleavage rate strongly depends on the dinucleotide sequence around the cleavage site, and for each dinucleotide combination, optimized DNAzyme have been proposed (as depicted in Table 1; Meusburger et al., 2011).

Corresponding flanking sequences can be chosen freely, provided that they are reverse complementary to their target RNA sequence.

The following protocol describes the synthesis of modivariants by means of DNAzymes and DNA-splinted ligation.

First, native or *in vitro*-transcribed tRNAs are hybridized to corresponding DNAzymes which cleave the tRNA site specifically. The resulting RNA fragments contain a cyclic phosphate at the 3′-end of the cleavage site and a hydroxyl-function at the freed 5′-end (Baum & Silverman, 2008). By taking into account that these structures are not suitable for enzymatic ligations where a 5′-phosphate and a 3′-OH are required, a phosphorylation step prior to ligation is necessary. Therefore, after DNAzyme-mediated cleavage, the resulting oligonucleotides must be dephosphorylated at the 3′ and phosphorylated at the 5′, respectively, by

Table 1 Proposed Motifs for DNAzyme Design

Target Dinucleotides	Catalytic Sequence (5′–3′)	DNAzyme Type
AA	GTCAGCTGACTCGAA	8–17
AG	TGTCAGCGACACGAA	8–17
AU	AGGCTAGCTACAACGA	10–23
CA	GTCAGCTGACTCGAA	8–17
CG	TGTCAGCGACACGAA	8–17
GA	GTCAGCTGACTCGAA	8–17
GC	IGGCTAGCTACAACGA	10–23
GG	TGTCAGCGACACGAA	8–17
GU	AGGCTAGCTACAACGA	10–23
UA	GTCAGCTGACTCGAA	8–17
UG	TGTCAGCGACACGAA	8–17

incubation with T4 polynucleotide kinase, which is able to execute both reactions. The desired PAGE-purified RNA fragments (native and/or unmodified) are then hybridized to a DNA template encompassing the full length of the tRNA sequence in reverse complement, such that the fragment align in the proper orientation. The subsequent ligation process is initiated by addition of T4 DNA ligase. In this step, it is also possible to ligate commercially synthesized (unmodified) RNA oligonucleotides instead of cleaved RNA or more than two fragments. The template DNA and DNAzymes are then digested by DNase I. Last, full-length RNA products are purified by denaturing polyacrylamide gel electrophoresis.

Materials

- Millipore grad water (for dilution of all nucleic acids and preparation of all reaction buffers)
- Target RNA (*in vitro* transcript or purified native RNA)
- DNAzymes stock solution at 100 μ*M* (0.2 μmol synthesis scale, IBA GmbH, http://www.iba-go.com)
- 5 × DNAzyme cleavage buffer (50 m*M* $MgCl_2$, 250 m*M* Tris–HCl (pH 7.5))
- DNase I, 50 U/μl (RNase-free, Fermentas)
- T4 polynucleotide kinase, 10 U/μl (PNK, Fermentas)
- 5 × KL (kinase and ligase) buffer (250 m*M* Tris–Cl (pH 7.4), 50 m*M* $MgCl_2$, 25 m*M* DTT, 10 m*M* ATP; store up to several months at −20 °C)
- 100 μ*M* DNA template for ligation (splint; IBA GmbH, http://www.iba-go.com; for more information, see Kurschat et al., 2005)
- 100 μ*M* Synthetic/native RNA fragments
- T4 DNA ligase HC, 30 U/μl (Fermentas)
- Ethanol, absolute (VWR)
- 80% (v/v) Ethanol
- Acrylamide gel casting solutions (e.g., Rotiphorese Sequencing Gel system, Carl Roth GmbH):

  Sequencing gel concentrate solution: 25% 19:1 acrylamide/bisacrylamide/8.3 *M* urea

  Sequencing gel buffer concentrate: 8.3 *M* urea/10 × TBE

  Sequencing gel diluter solution: 8.3 *M* urea
- Repellent solution: 2% dichlorodimethylsilane (Sigma) in $CHCl_3$
- 10% (w/v) Ammonium persulfate (Sigma)
- *N,N,N,N*-tetramethylethan-1,2-diamine (TEMED; Sigma)
- 10 × TBE buffer (Invitrogen)

- Loading buffer (10% (v/v) 10 × TBE buffer 90% formamide (p.A. grade; AppliChem, http://www.applichem.com), 0.01% bromophenol blue, 0.01% xylene cyanol; store up to several months at 4 °C)
- 0.5 *M* Ammonium acetate (Sigma)

Instruments

- Thermoshaker (e.g., Eppendorf)
- 200 × 200 × 5-mm Gel casting plates with 1-mm spacers (Biometra, http://www.biometra.de/)
- Gel comb (Biometra, http://www.biometra.de/)
- Polyacrylamide gel electrophoresis apparatus with power source
- Instruments for visualization of gel bands:

  UV handlamp or UV-transilluminator

  25 TLC plastic sheets 20 × 20 cm cellulose (Merck)

  Typhoon Variable Mode Imager, equipped with suitable lasers (GE Healthcare)
- Nanosep 0.45 μ*M* membrane filters (Pall)
- UV spectrometer (e.g., Nanodrop 2000, Thermo Fisher)

DNAzyme design

1. Identify your target dinucleotide in your target RNA sequence and the corresponding catalytic sequence (see Table 1).
2. Add reverse complementary flanking sequences to the DNAzyme so that site-directed cleavage of your target RNA is ensured.
3. Synthesize or order the desired DNAzyme.

DNAzyme cleavage and digestion (see Notes 1–5 in Section 2)

4. Dilute 20 μg of either native or *in vitro* transcript tRNA in DNAzyme cleavage buffer and add a 10 × molar excess of DNAzyme and H_2O so that a final volume of 250 μl is obtained.
5. Subject mixture to 20 repetitions of the following cleavage cycle: Denaturation for 30 s at 80 °C, hybridization and folding by temperature decrease of 0.5 °C/s to 37 °C and subsequent cleavage for 5 min at 37 °C.
6. Add DNase I (4–6 U/μg DNA substrate) directly to the reaction mixture and incubate for 1 h at 37 °C to digest the utilized DNAzyme.

   After this step, the efficiency of DNAzyme digestion should be verified. This can be done at the same time as the isolation of RNA fragments (see isolation of RNA fragments by PAGE in Section 2).

Phosphorylation of RNA fragments

7. Incubate 27 μ*M* of each appropriate RNA fragment (diluted in Milli-Q filtered water) with 0.75 U/μl T4 polynucleotide kinase in KL buffer for 60 min at 37 °C.

The phosphorylation reaction can be performed in 150 µl or various volumes depending on the amount of phosphorylated oligonucleotide needed. Note that it is relevant to provide ATP in excess (included in KL buffer; see Materials in Section 2).

The phosphorylated RNA may be used directly for enzymatic ligation without a further purification step or may be stored at −20 °C until use.

DNA-splinted ligation of RNA oligonucleotides

8. Dilute phosphorylated RNA fragments with the DNA splint in equal stoichiometry (8 µ*M* final concentration) in KL buffer (final volume of 500 µl) and heat the RNA/DNA mix in a denaturation step for 4 min at 75 °C in a thermoshaker.

   Typically, ligations are performed in volumes up to 1 ml.
9. Hybridize RNA fragments to the DNA template by slow subsequent cooling down to room temperature over 15 min.
10. Begin the ligation by adding 1.5 U/µl T4 DNA ligase and incubate overnight at 16 °C in a thermoshaker.
11. Start digestion of DNA splint by addition of 0.5 U/µL (final concentration) DNase I to the ligation mixture and subsequent incubation for 30 min at 37 °C.
12. Stop the reaction by placing on ice or subsequent PAGE purification.

    The reaction mixture can be stored for several days at −20 °C.

Isolation of RNA fragments by PAGE

13. Prepare a 10% or a 15% denaturing polyacrylamide gel premix solution (see Table 2).

    For tRNA modivariants up to 100 nt, 10% solutions are suitable.
14. Purge the spacers, glass plates, and comb thoroughly with water and ethanol. Use repellent solution to treat both glass plates in order to improve delamination of the gel later. Assemble plates and spacers (bottom and side), place the assembly on a horizontal support.
15. To prepare a 200 × 200 × 1-mm gel, provide 100 ml denaturing PAGE premix solution, add 400 µl of 10% (w/v) ammonium persulfate and

Table 2 Examples for Denaturing PAGE Premixes[a]

	10% Gel (100 ml)	15% Gel (100 ml)
Sequencing gel concentrate solution	40 ml	60 ml
Sequencing gel buffer concentrate	10 ml	10 ml
Sequencing gel diluter solution	50 ml	30 ml

[a]Volumes for Rotiphorese Sequencing Gel system (Carl Roth GmbH).

50 µl TEMED, stir well, and immediately pour the gel solution between the horizontally placed glass plates. Prevent forming bubbles by adjusting the angle of the plates so that the solution flows slowly and carefully insert desired comb.

16. Let the poured gel polymerize for at least 30 min.
17. Gently remove the bottom spacer and the comb after acrylamide is polymerized, and immediately rinse the gel wells with 1 × TBE buffer using a syringe. Place the gel sandwich into the electrophoresis apparatus by clamping the plates to the upper buffer reservoir.
18. Fill the upper and the lower buffer reservoir with 1 × TBE buffer, making sure that the buffer replaces all air between the glass plates at the bottom of the gel. Again rinse the gel wells with buffer.
19. Prerun the gel for about 30 min at 10–15 W depending on gel percentage.

    This step assures equal buffer distribution and prewarms the gel in order to reduce band "smiling." The amount of heat generated depends on the power and gel size. Note that the absolute values of power, voltage, and current will vary during electrophoresis.
20. Add 1 volume of loading buffer (see Materials in Section 2) to the resuspended RNA samples from step 12.
21. Rinse the gel wells one more time and then load an adequate amount of each sample into its respective well.
22. Start the gel at 10–15 W and wait until the bromophenol blue is close to the bottom of the gel (this is the case after ~150 min).
23. After electrophoresis, disassemble the gel apparatus and open the gel sandwich with care. Remove the upper glass plate, turn the gel over, and remove the second glass plate.
24. Place the gel on a fluorescent TLC plate and visualize the bands by UV shadowing with a 254-nm UV lamp.

    Bands containing more than 10 µg RNA are subsequently dark in contrast to the green fluorescing TLC plate.

    For smaller amounts of RNA PAGE gels, SYBR-gold (Invitrogen) as staining solution may be used. These gels can be investigated by means of a Typhoon scanner.
25. Excise the correct product band with a clean razor blade, mince the gel into small pieces and transfer it into a clean 1.5 ml tube.

Elute full-length RNA from gel

26. In order to destroy the gel structure, freeze the gel pieces containing the full-length RNA product at −20 or −80 °C before starting the elution procedure. This step increases the elution yield of RNA.

27. After thawing add 450 µl of 0.5 *M* ammonium acetate to the sample and elute overnight at 20–25 °C by agitation at 600 rpm in a thermoshaker.
    For a more complete elution, a second elution step with 100–200 µl of 0.5 M ammonium acetate may be performed.
28. Filter the eluate by centrifugation through Nanosep membrane filters (see Materials in Section 2) to remove small gel fragments.
29. By adding 2.5 volumes of −80 °C absolute ethanol to the filtrate, vortexing and microcentrifuging (45 min at ~13,000 × *g*, −4 °C) the RNA can be precipitated. Remove ethanol supernatant.
30. Purge the RNA pellets carefully with cold 80% ethanol, dry them for 10–15 min in a vacuum concentrator (e.g., SpeedVac) or by exposure to air to remove residual ethanol, and resuspend in 10–100 µl Milli-Q filtered water.
    Avoid excessive drying under vacuum, as this may cause the RNA to stick to the tube walls.
31. Determine the concentration of the full-length RNA product by absorption at 254 nm (UV spectrometer, e.g., Nanodrop 2000) or via Typhoon scanner.

Notes

1. In general, cleavage conditions (such as temperature, duration, denaturation, and hybridization) should be adjusted to the target RNA. In addition, a stoichiometric excess of DNAzyme (2–10-fold) can help to cleave structured RNAs, e.g., tRNA, more efficiently.
2. Some DNAzymes reveal a better cleavage activity in the presence of Mn^{2+}.
3. Cleavage cycles with up to 20 repetitions can increase cleavage by generating more active DNAzyme–target RNA complexes (dissociation and reassociation step). Nevertheless extended cycling is able to forward degradation processes.
4. Normally native tRNAs are not cleaved very well due to their modification content. Therefore, the use of DNAzymes modified with 2′-OMe or LNA can be an alternative (Schubert et al., 2003; Vester et al., 2002).
5. For testing immune stimulatory properties, it is recommended to synthesize modivariant yields in the range of several micrograms (about 1–5).

3. PREPARATION AND STIMULATION OF HUMAN PBMC

In our experience, the use of freshly drawn blood is superior to the use of buffy coats which can be obtained during the process of blood product preparation. Specifically, viability of pDCs might be decreased if using older blood products. Specific regulations of the local authorities with respect to blood donations have to be considered and precautions have to be taken due to infectious risks when working with blood.

Materials

- RPMI 1640 with stable glutamine (e.g., Biochrome, order number FG1215)
- Opti-MEM® Reduced Serum medium (Life Technologies)
- Phosphate-buffered PBS, pH 7.4, sterile
- Fetal calf serum, heat-inactivated at 56 °C for 1 h (different providers). *In our experience, it is recommended to test different batches of FCS for background stimulation due to endogenous content of endotoxin. Endotoxin concentrations should be indicated by the providers and vary from batch to batch. It is recommended to use low-endotoxin FCS from one batch for comparative analysis.*
- Ficoll human (e.g., Pancoll human, density 1.077 g/l, Pan Biotech)
- DOTAP (*N*-[1-(2,3-Dioleoyloxy)]-*N*,*N*,*N*-trimethylammonium propane methylsulfate, Carl Roth GmbH) *DOTAP is used as a delivery agent to shuttle bacterial RNA into innate immune cells. Others transfection reagents might be suitable as well but cellular cytotoxicity has to be considered. Within the field, use of DOTAP has turned out to be the favored method which is described here.*
- Heparin (e.g., Ratiopharm)
- 0.9% NaCl, sterile

4. PREPARATION OF HUMAN PBMCs

Isolation of human PBMC is performed by Ficoll density gradient separation. All of the following steps have to be performed with sterile/autoclaved materials in a laminar flow cabinet Class II.

1. Fill a sterile 50-ml syringe with 2 ml of sterile 0.9% NaCl supplemented with heparin (final concentration of heparin = 125 I.E./ml).
2. Collect 50 ml blood from a healthy volunteer by peripheral venous puncture (local regulations regarding blood donation including vote from an ethical committee have be considered).
3. Transfer blood in a sterile vessel (e.g., 175 cm^2 cell culture flask) and add 50 ml of PBS (1:1 dilution).
4. Pipet 15 ml of Ficoll medium in each of four 50 ml falcon tubes.
5. Carefully layer diluted whole blood over the Ficoll medium. This is done by gently pipetting the blood along the wall of the falcon tube

onto the Ficoll with the falcon tube held at a slight angle. It is critical to obtain a clear separation between blood and Ficoll medium.

6. Carefully place the falcons in a precooled (4 °C) centrifuge with swinging bucket rotor without disturbing the layering. Program centrifuge with maximum acceleration but switch off the break function (to allow slow swing out at the end of the program). Centrifuge at 700 × *g* for 20 min at 4 °C.
7. Carefully remove the falcon tubes from the centrifuge.
8. The PBMC layer will be visible as whitish layer between the Ficoll medium and the serum (top layer). Remove the PBMC layer that is located at the Ficoll–serum interphase and transfer it into a new 50 ml falcon. This can be done using a 10-ml sterile plastic pipet and gentle aspiration of the visible layer under visual control. Aspiration of Ficoll should be avoided.
9. Add cold PBS to a final volume of 50 ml per falcon tube.
10. Centrifuge at 365 × *g* for 10 min at 4 °C.
11. Discard supernatants of each of the falcon tubes by decantation. Resuspend the cell pellets in some of the remaining PBS. Combine the cell pellets of two falcon tube and add again 50 ml PBS.
12. Centrifuge at 215 × *g* for 15 min at 4 °C.
13. Discard supernatants as above. Resuspend cells in remaining PBS and combine the cell pellets of two falcon tubes. Add again PBS to a final volume of 50 ml.
14. Centrifuge at 365 × *g* for 10 min at 4°C
15. Discard supernatant and resuspend cells in 10 ml cold complete cell culture medium. Complete culture medium is RPMI 1640 containing stable glutamine supplemented with 10% heat-inactivated FCS.
16. Count viable cells with a hemocytometer by mixing a small sample of the cells with trypan blue solution. Dead cells will appear blue. *Typical yield: 50–80 Mio PBMCs per 50 ml blood.*
17. Adjust PBMCs to a concentration of 2×10^6/ml in complete cell culture medium.
18. PBMC should be used immediately for stimulation or can be stored in a falcon at 4 or 37 °C for up to 1 or 2 h.

5. STIMULATION OF HUMAN PBMC

The quantities below are calculated for the transfection of three wells of a flat-bottom 96-well plate (e.g., Corning® Costar). All of the following steps have to be performed with sterile/autoclaved materials in a laminar flow cabinet Class II. Use filter tips only! For efficient intracellular delivery of bacterial RNA into immune cells, complexation of RNA with liposomal transfection reagents is mandatory.

1. Place 50 μl of Opti-MEM® into a 1.5-ml Eppendorf tube.
2. Add 500 ng of bacterial RNA.
3. Add 1.5 μl of DOTAP. The optimal ratio of DOTAP to RNA may need to be validated in preliminary experiments and is essential for successful transfection.
4. Mix gently by pipetting mixture up-and-down (two to three times).
5. Incubate the mixture for 10–15 min at RT.
6. Add cell culture medium (RPMI 1640 supplemented with 10% heat-inactivated FCS) to the mixture to obtain a total volume of 333 μl.
7. Transfer 100 μl of the transfection mixture in one well of a flat-bottom 96-well plate. The prepared mixture is sufficient for three wells.
8. Add 200 μl of freshly isolated PBMCs ($c = 2 \times 10^6$ cells/ml, resuspended in RPMI 1640 supplemented with 10% heat-inactivated FCS) in each well on top of the transfection mixture. This will result in a final RNA concentration of 500 ng/ml and a total volume of 300 μl per well. A total of 400,000 PBMCs will be stimulated per well.
9. Gently rock the plate by hand for even distribution.
10. Incubate cells overnight (at least 12 h) at 37 °C and 5% CO_2 in an incubator with humidified atmosphere.
11. The next day centrifuge the 96-well plate at 300 × *g* for 5 min at 4 °C.
12. Transfer 200 μl of cell culture supernatant to a round-bottom 96-well plate to store it for future analysis. Tissue culture grade 96-well plates are not required for this purpose.
13. Supernatants can either be used directly for cytokine analysis or can be stored at −20 °C for 3–6 months.
14. Cytokine measurement can be performed for classical pDC readouts, i.e., IFN-α, or more monocyte-specific cytokines like TNF, IL-12p40, or IL-6. Commercially available ELISA kits to conveniently determine cytokine concentrations can, for example, be obtained from Affymetrix eBioscience (human IFN-α, order number BMS216MST) or from Becton Dickinson for detection of human IL-6, TNF, and IL-12p40 (BD OptEIA™ Kits). All steps are performed according to the manufacturer's instructions. If half area ELISA plates (e.g., Greiner Bio-One) are used, volumes for antibodies and samples can be divided by two. *Typical range of cytokines upon overnight RNA*

stimulation: *IFN-α*: *300–2000 pg/ml*; *TNF*: *2000–10,000 pg/ml*; *IL-6*: *3000–15,000 pg/ml*; *IL-12p40*: *500–5000 pg/ml*.

Due to substantial donor variation, it might be useful to add a defined stimulus as positive control and for normalization purposes. Use of R848 that triggers both, TLR7 and TLR8, thus stimulating pDCs as well as monocytes is recommended. Moreover, a negative control using medium only as well as a control with DOTAP only are necessary. Typically, stimulations are done in duplicates or triplicates, ELISA measurement of cytokines also in duplicates.

ACKNOWLEDGMENT

This work was supported by the German Research Foundation (DFG) grant DA592/5 (to A.D.) and HE 3397/9 (to M.H.).

REFERENCES

Baum, D. A., & Silverman, S. K. (2008). Deoxyribozymes: Useful DNA catalysts *in vitro* and *in vivo*. *Cellular and Molecular Life Sciences*, *65*(14), 2156–2174. http://dx.doi.org/10.1007/s00018-008-8029-y.

Buchhaupt, M., Peifer, C., & Entian, K. D. (2007). Analysis of 2'-O-methylated nucleosides and pseudouridines in ribosomal RNAs using DNAzymes. *Analytical Biochemistry*, *361*(1), 102–108. http://dx.doi.org/10.1016/j.ab.2006.11.001.

Cairns, M. J., King, A., & Sun, L. Q. (2003). Optimisation of the 10–23 DNAzyme-substrate pairing interactions enhanced RNA cleavage activity at purine-cytosine target sites. *Nucleic Acids Research*, *31*(11), 2883–2889.

Cervantes, J. L., La Vake, C. J., Weinerman, B., Luu, S., O'Connell, C., Verardi, P. H., et al. (2013). Human TLR8 is activated upon recognition of Borrelia burgdorferi RNA in the phagosome of human monocytes. *Journal of Leukocyte Biology*, *94*(6), 1231–1241. http://dx.doi.org/10.1189/jlb.0413206.

Cruz, R. P., Withers, J. B., & Li, Y. (2004). Dinucleotide junction cleavage versatility of 8–17 deoxyribozyme. *Chemistry & Biology*, *11*(1), 57–67. http://dx.doi.org/10.1016/j.chembiol.2003.12.012.

Eberle, F., Sirin, M., Binder, M., & Dalpke, A. H. (2009). Bacterial RNA is recognized by different sets of immunoreceptors. *European Journal of Immunology*, *39*(9), 2537–2547. http://dx.doi.org/10.1002/eji.200838978.

Gehrig, S., Eberle, M. E., Botschen, F., Rimbach, K., Eberle, F., Eigenbrod, T., et al. (2012). Identification of modifications in microbial, native tRNA that suppress immunostimulatory activity. *The Journal of Experimental Medicine*, *209*(2), 225–233. http://dx.doi.org/10.1084/jem.20111044.

Hengesbach, M., Kobitski, A., Voigts-Hoffmann, F., Frauer, C., Nienhaus, G. U., & Helm, M. (2008). RNA intramolecular dynamics by single-molecule FRET. *Current Protocols in Nucleic Acid Chemistry*. http://dx.doi.org/10.1002/0471142700.nc1112s34 (Chapter 11, Unit 11.12).

Hengesbach, M., Meusburger, M., Lyko, F., & Helm, M. (2008). Use of DNAzymes for site-specific analysis of ribonucleotide modifications. *RNA*, *14*(1), 180–187. http://dx.doi.org/10.1261/rna.742708.

Hidmark, A., von Saint Paul, A., & Dalpke, A. H. (2012). Cutting edge: TLR13 is a receptor for bacterial RNA. *Journal of Immunology, 189*(6), 2717–2721. http://dx.doi.org/10.4049/jimmunol.1200898.

Hornung, V., Rothenfusser, S., Britsch, S., Krug, A., Jahrsdorfer, B., Giese, T., et al. (2002). Quantitative expression of toll-like receptor 1–10 mRNA in cellular subsets of human peripheral blood mononuclear cells and sensitivity to CpG oligodeoxynucleotides. *Journal of Immunology, 168*(9), 4531–4537.

Jockel, S., Nees, G., Sommer, R., Zhao, Y., Cherkasov, D., Hori, H., et al. (2012). The 2'-O-methylation status of a single guanosine controls transfer RNA-mediated Toll-like receptor 7 activation or inhibition. *The Journal of Experimental Medicine, 209*(2), 235–241. http://dx.doi.org/10.1084/jem.20111075.

Kaiser, S., Rimbach, K., Eigenbrod, T., Dalpke, A. H., & Helm, M. (2014). A modified dinucleotide motif specifies tRNA recognition by TLR7. *RNA, 20*(9), 1351–1355. http://dx.doi.org/10.1261/rna.044024.113.

Kariko, K., Buckstein, M., Ni, H., & Weissman, D. (2005). Suppression of RNA recognition by Toll-like receptors: The impact of nucleoside modification and the evolutionary origin of RNA. *Immunity, 23*(2), 165–175. http://dx.doi.org/10.1016/j.immuni.2005.06.008.

Kawai, T., & Akira, S. (2010). The role of pattern-recognition receptors in innate immunity: Update on Toll-like receptors. *Nature Immunology, 11*(5), 373–384. http://dx.doi.org/10.1038/ni.1863.

Koski, G. K., Kariko, K., Xu, S., Weissman, D., Cohen, P. A., & Czerniecki, B. J. (2004). Cutting edge: Innate immune system discriminates between RNA containing bacterial versus eukaryotic structural features that prime for high-level IL-12 secretion by dendritic cells. *Journal of Immunology, 17*(7), 3989–3993.

Kurschat, W. C., Muller, J., Wombacher, R., & Helm, M. (2005). Optimizing splinted ligation of highly structured small RNAs. *RNA, 11*(12), 1909–1914. http://dx.doi.org/10.1261/rna.2170705.

Li, X. D., & Chen, Z. J. (2012). Sequence specific detection of bacterial 23S ribosomal RNA by TLR13. *Elife, 1*, e00102. http://dx.doi.org/10.7554/eLife.00102.

Mancuso, G., Gambuzza, M., Midiri, A., Biondo, C., Papasergi, S., Akira, S., et al. (2009). Bacterial recognition by TLR7 in the lysosomes of conventional dendritic cells. *Nature Immunology, 10*(6), 587–594. http://dx.doi.org/10.1038/ni.1733.

Meusburger, M., Hengesbach, M., & Helm, M. (2011). A post-labeling approach for the characterization and quantification of RNA modifications based on site-directed cleavage by DNAzymes. *Methods in Molecular Biology, 718*, 259–270. http://dx.doi.org/10.1007/978-1-61779-018-8_16.

Mogensen, T. H. (2009). Pathogen recognition and inflammatory signaling in innate immune defenses. *Clinical Microbiology Reviews, 22*(2), 240–273. http://dx.doi.org/10.1128/CMR.00046-08 (Table of Contents).

Oldenburg, M., Kruger, A., Ferstl, R., Kaufmann, A., Nees, G., Sigmund, A., et al. (2012). TLR13 recognizes bacterial 23S rRNA devoid of erythromycin resistance-forming modification. *Science, 337*(6098), 1111–1115. http://dx.doi.org/10.1126/science.1220363.

Robbins, M., Judge, A., Liang, L., McClintock, K., Yaworski, E., & MacLachlan, I. (2007). 2'-O-methyl-modified RNAs act as TLR7 antagonists. *Molecular Therapy, 15*(9), 1663–1669. http://dx.doi.org/10.1038/sj.mt.6300240.

Schubert, S., Gul, D. C., Grunert, H. P., Zeichhardt, H., Erdmann, V. A., & Kurreck, J. (2003). RNA cleaving '10-23' DNAzymes with enhanced stability and activity. *Nucleic Acids Research, 31*(20), 5982–5992.

Sioud, M. (2009). Deciphering the code of innate immunity recognition of siRNAs [Review]. *Methods in Molecular Biology, 487*, 41–59. http://dx.doi.org/10.1007/978-1-60327-547-7_2.

Vester, B., Lundberg, L. B., Sorensen, M. D., Babu, B. R., Douthwaite, S., & Wengel, J. (2002). LNAzymes: Incorporation of LNA-type monomers into DNAzymes markedly increases RNA cleavage. *Journal of the American Chemical Society*, *124*(46), 13682–13683.

Zaborowska, Z., Furste, J. P., Erdmann, V. A., & Kurreck, J. (2002). Sequence requirements in the catalytic core of the "10-23" DNA enzyme. *The Journal of Biological Chemistry*, *277*(43), 40617–40622. http://dx.doi.org/10.1074/jbc.M207094200.

Zaborowska, Z., Schubert, S., Kurreck, J., & Erdmann, V. A. (2005). Deletion analysis in the catalytic region of the 10–23 DNA enzyme. *FEBS Letters*, *579*(2), 554–558. http://dx.doi.org/10.1016/j.febslet.2004.12.008.

CHAPTER FIVE

Kinetic Analysis of tRNA Methyltransferases

Ya-Ming Hou[1], Isao Masuda
Department of Biochemistry and Molecular Biology, Thomas Jefferson University, Philadelphia, Pennsylvania USA
[1]Corresponding author: e-mail address: ya-ming.hou@jefferson.edu

Contents

Abstract

Transfer RNA (tRNA) molecules contain many chemical modifications that are introduced after transcription. A major form of these modifications is methyl transfer to bases and backbone groups, using *S*-adenosyl methionine (AdoMet) as the methyl donor. Each methylation confers a specific advantage to tRNA in structure or in function. A remarkable methylation is to the G37 base on the 3′-side of the anticodon to generate m^1G37-tRNA, which suppresses frameshift errors during protein synthesis and is therefore essential for cell growth in all three domains of life. This methylation is catalyzed by TrmD in bacteria and by Trm5 in eukaryotes and archaea. Although TrmD and Trm5 catalyze the same methylation reaction, kinetic analysis reveals that these two enzymes are unrelated to each other and are distinct in their reaction mechanism. This chapter summarizes the kinetic assays that are used to reveal the distinction between TrmD and Trm5. Three types of assays are described, the steady-state, the pre-steady-state, and the single-turnover assays, which collectively provide the basis for mechanistic investigation of AdoMet-dependent methyl transfer reactions.

Methods in Enzymology, Volume 560
ISSN 0076-6879
http://dx.doi.org/10.1016/bs.mie.2015.04.012

1. INTRODUCTION

Naturally occurring tRNA molecules contain many posttranscriptionally modified bases and backbones, where the standard nucleotides (A, C, G, and U) are modified with the addition of various chemical moieties. Each posttranscriptional modification is synthesized in a specific enzymatic pathway on the primary tRNA transcripts. Modifications that occur in the tRNA elbow region mainly contribute to the tertiary folding of the nucleic acid and those in the anticodon region mainly contribute to the protein synthesis activity on the ribosome. These and other posttranscriptional modifications can also play a role in the cellular response to stress (Yi & Pan, 2011). Although other types of cellular RNA (e.g., mRNA, rRNA, microRNA, piwi-interacting RNA, and small interfering RNA) also contain posttranscriptional modifications (Yi & Pan, 2011), tRNA is distinguished with the most diversity and abundance. A major form of posttranscriptional modifications in tRNA is methylation. For example, of the 76 nucleotides in the standard sequence framework of *E. coli* tRNA, up to nine different types of methylation can occur (Fig. 1A; Table 1), whereas other types of modifications (pseudouridylation, thiolation, dihydrouridylation, etc.) are much less frequent. Given the dominance of methylation, the question of how the addition of a single methyl group to a nucleotide can modulate tRNA structure and activity becomes increasingly important.

To address the role of each methylation to tRNA, it is necessary to understand the reaction mechanism of the methyltransferase and to relate it to biology. One consideration is the methyl donor of the methyltransferase. The greatest majority of methyltransferases use AdoMet as the methyl donor. The preference for AdoMet over other methyl donors, such as folate, reflects the favorable energetics resulting from a nucleophilic attack on the positively charged methyl group of the sulfonium center (Fig. 1B). The energy release upon methyl transfer from AdoMet is more than twice of the energy release upon hydrolysis of ATP to ADP and Pi (Cantoni, 1975). The target atoms performing the nucleophilic attack on the methyl group of AdoMet are diverse, including nitrogen, oxygen, carbon, and sulfur. AdoMet itself is synthesized by condensation of methionine with ATP by methionine adenosyltransferase (or SAM synthetase) (Markham, Hafner, Tabor, & Tabor, 1980), and the product of methyl transfer is *S*-adenosyl homocysteine (AdoHcy). The ratio of AdoMet versus AdoHcy is defined

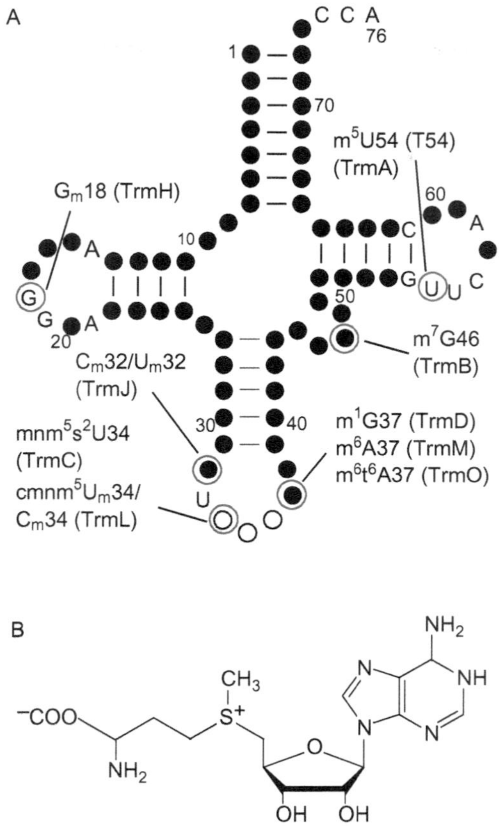

Figure 1 AdoMet-dependent methyl transfer to *E. coli* tRNA. (A) The cloverleaf structure of tRNA is made up of nucleotides as circles. Conserved nucleotides are shown in letters, nonconserved nucleotides are shown as closed circles, and the anticodon nucleotides are shown as open circles. The position of each AdoMet-dependent methylation (catalyzed by the *E. coli* enzyme in parentheses) is indicated. The nucleotide numbering is based on the standard sequence framework of tRNA (Sprinzl, Horn, Brown, Ioudovitch, & Steinberg, 1998). (B) The chemical structure of AdoMet. (See the color plate.)

as the methylation potential and it is subject to change according to the energy state of a cell. In mammals, the reduction of the methylation potential lengthens the circadian rhythm (Fustin et al., 2013), indicating a broad impact on cell physiology. This observation then raises the question of whether the circadian lengthening is due to the loss of methylation at specific sites in a specific subset of tRNA. It is possible that, upon the reduction of the methylation potential, certain types of tRNA methylation remain while others decrease or disappear, leading to global changes in the tRNA

Table 1 AdoMet-Dependent tRNA Methyltransferases in *E. coli*

Enzyme (synonyms)	Modification	Chemical structure	Class
TrmA (rT)	m^{5}U54 (T54)	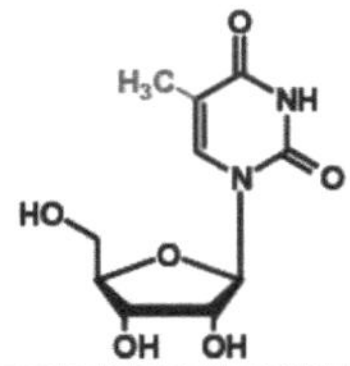	I
TrmB (TrmI, YggH)	m^{7}G46	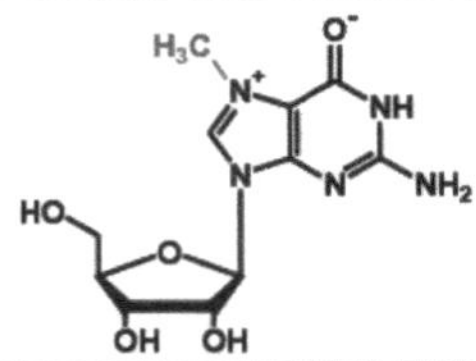	I
TrmC (MnmC, YfcK)	mnm^{5}s^{2}U34	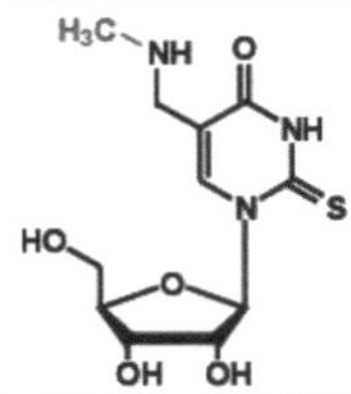	I
TrmD	m^{1}G37	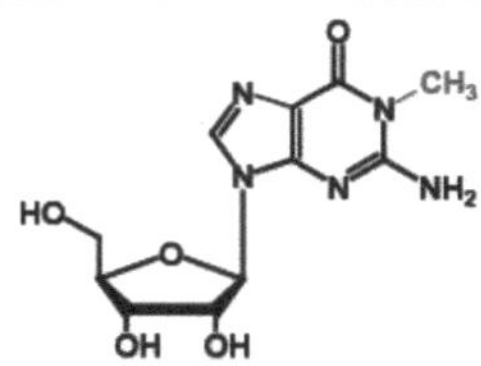	IV
TrmH (SpoU)	G$_m$18	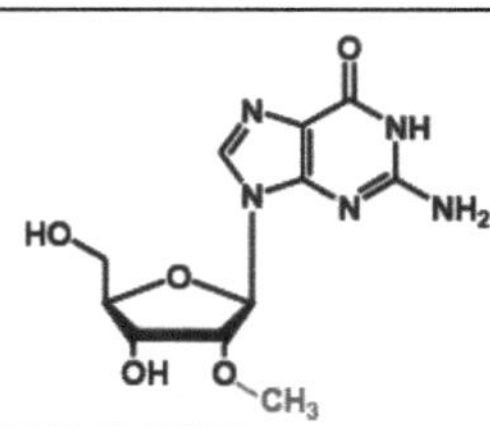	IV
TrmJ (YfhQ)	C$_m$32/U$_m$32	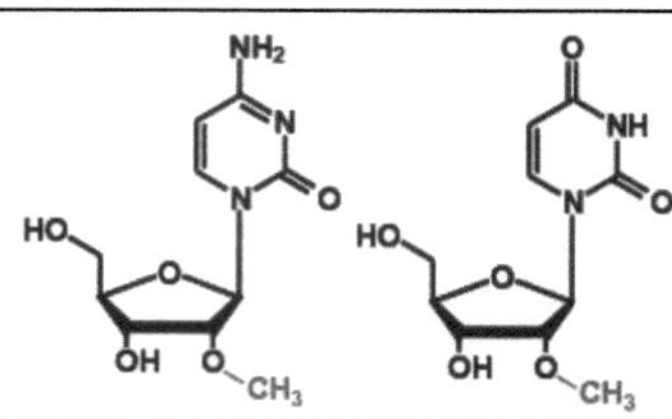	IV

Table 1 AdoMet-Dependent tRNA Methyltransferases in *E. coli*—cont'd

Enzyme (synonyms)	**Modification**	**Chemical structure**	**Class**
TrmL (YibK)	$C_m34/cmnm^5U_m34$	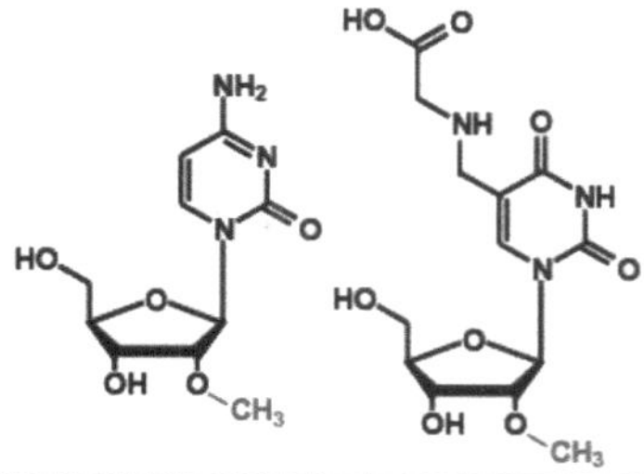	IV
TrmM (YfiC)	m^6A37	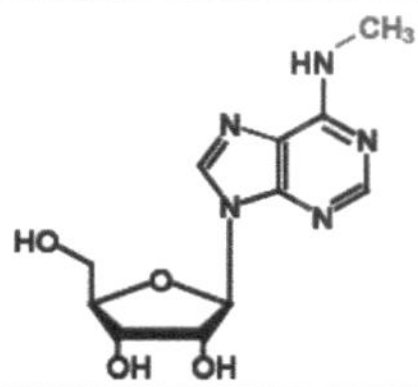	I
TrmO (TsaA, YaeB)	m^6t^6A37	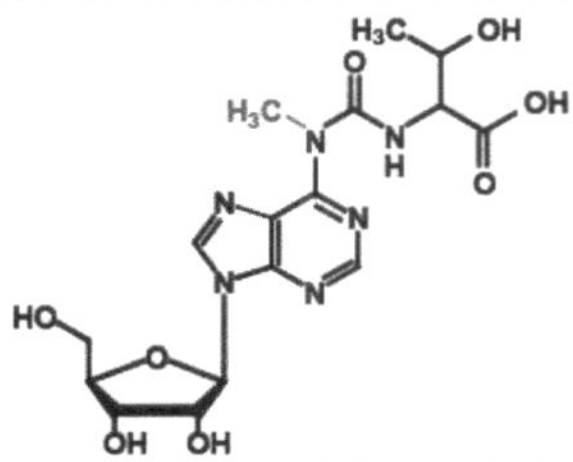	VIII

References for each enzyme are as follows: *Ec*TrmA (Ny & Bjork, 1980), *Ec*TrmB (De Bie et al., 2003), *Ec*TrmC (Taya & Nishimura, 1973), *Ec*TrmD (Bystrom & Bjork, 1982b), *Ec*TrmH (Persson, Jager, & Gustafsson, 1997), *Ec*TrmJ (Purta et al., 2006), *Ec*TrmL (Benitez-Paez, Villarroya, Douthwaite, Gabaldon, & Armengod, 2010), *Ec*TrmM (Golovina et al., 2009), and *Ec*TrmO (Kimura et al., 2014). Note that *Ec*TrmO has a unique single-sheeted β-barrel structure that does not belong to any of the five known classes of methyltransferases and is designated as class VIII (Kimura et al., 2014). Only the first reference is given for each enzyme.

structure and activity. To address the dynamics of such changes in cells, a prerequisite is an understanding of the sensitivity of each tRNA methyltransferase to alterations of the AdoMet level in kinetic analysis.

A second consideration is the structural conformation of AdoMet when bound to a tRNA methyltransferase. To date, at least five classes (class I–V) of structurally distinct AdoMet-dependent methyltransferases have been identified (Schubert, Blumenthal, & Cheng, 2003). The distinction among the five classes is the topological structural fold that binds AdoMet and as such the conformation of AdoMet in each fold. This wide diversity among AdoMet-dependent methyltransferases is paralleled only by the diversity among ATP-dependent protein kinases and phosphoryl transferases

(Schubert et al., 2003). The discovery of different AdoMet conformations among methyltransferases suggests the existence of distinct reaction mechanisms. For example, the class I fold is the most common of the five classes and it includes the greatest majority of tRNA methyltransferases, including TrmA for synthesis of m^5U54 and TrmB for synthesis of m^7G46 in *E. coli* (Table 1) (Hou & Perona, 2010). The class I fold binds AdoMet in the open space of a dinucleotide-fold (also known as the Rossmann fold) (Schubert et al., 2003). A signature of the class I fold is that AdoMet adopts a straight conformation, where the adenosine and methionine moieties are extended in opposite direction from each other (Schubert et al., 2003). Among tRNA methyltransferases, besides the class I fold, the other major type is the class IV fold, which is characterized by binding AdoMet to the bottom of a deep cleft in a topologically knotted protein fold. The class IV fold is rather rare among protein structures and it is made up of three passages of the protein backbone in and out of a loop in a structure known as the trefoil-knot fold (Nureki et al., 2002, 2004). A signature of the class IV fold is that AdoMet adopts a sharp bent conformation, where the two component moieties are spatially facing each other almost at a right angle (Schubert et al., 2003). *E. coli* tRNA methyltransferases that have the class IV fold include TrmD for synthesis of m^1G37, TrmH for synthesis of G_m18 ($G_m = 2'$-O-methyl), TrmJ for synthesis of C_m32/U_m32, and TrmL for synthesis of C_m34/U_m34 (Table 1).

TrmD presents an important case for why kinetic analysis of methyl transfer is critical for understanding the reaction mechanism. TrmD is broadly conserved in the bacterial domain (Bystrom & Bjork, 1982a, 1982b), while its counterpart in the eukaryotic and archaeal domain is Trm5 (Christian, Evilia, Williams, & Hou, 2004; Christian, Gamper, & Hou, 2013). Both TrmD and Trm5 are essential for cell growth (Baba et al., 2006; Bjork et al., 2001), because their reaction product m^1G37 occurring on the 3′-side of the tRNA anticodon is necessary to suppress +1-frameshift errors on the ribosome. Unlike mis-sense errors, +1-frameshift errors are deleterious, resulting in premature termination of protein synthesis and leading to cell death. However, while TrmD and Trm5 catalyze the same reaction, they are fundamentally distinct (Ahn et al., 2003; Christian et al., 2004; Elkins et al., 2003; Goto-Ito, Ito, Kuratani, Bessho, & Yokoyama, 2009; Goto-Ito et al., 2008). The distinction is manifested in their structure: while TrmD is an obligated dimer that uses the class IV fold for AdoMet binding, Trm5 is an active monomer that uses the class I fold (Christian et al., 2004). More importantly, the distinction is manifested in their kinetics: while the slow step of the TrmD reaction is the chemistry of methyl transfer, that of the Trm5 reaction is after methyl transfer and is associated with release of the m^{1}G37-tRNA product (Christian, Lahoud, Liu, & Hou, 2010). Indeed, it is with the understanding of their kinetic distinction that we now gain insight into how their structural distinction affects methyl transfer. For example, while TrmD recognizes AdoMet using a rigid lock-and-key mode, Trm5 uses an induced-fit mode (Lahoud et al., 2011). While TrmD binds tRNA by recognizing only the anticodon stem-loop, Trm5 recognizes the entire tRNA L-shape with an emphasis on the elbow region (Christian & Hou, 2007; Goto-Ito et al., 2009). While TrmD discriminates the target G37 base by carefully reading all three functional groups on the Watson–Crick face, Trm5 is more relaxed and reads only two of the three functional groups (Sakaguchi et al., 2012). Finally, while both enzymes involve a general base to abstract the N^1 proton of G37 to activate the nucleophile, TrmD differs from Trm5 by engaging a catalytic Mg^{2+} ion for the proton abstraction (Christian, Lahoud, Liu, Hoffmann, et al., 2010; Sakaguchi, Lahoud, Christian, Gamper, & Hou, 2014). Together, these mechanistic distinctions show that the structural difference between TrmD and Trm5 has resulted in the kinetic distinction of their methyl transfer. Importantly, this correlation between structure and mechanism would not have been realized without the development of kinetic assays to monitor the methyl transfer reaction of each enzyme.

Here, we describe three kinetic assays that were used to distinguish TrmD from Trm5. Combined, these assays provide information for each enzyme on the rate constant of catalytic turnover (k_{cat}), the substrate concentration (K_m) that permits methyl transfer proceeding at the half-maximum turnover rate, the rate constant of methyl transfer in a single turnover (k_{chem}), the substrate binding affinity (K_d) for the methyltransferase in a single turnover, and the rate-limiting step in multiple rounds of turnover. A summary of these kinetic parameters for *E. coli* TrmD (*Ec*TrmD) and the archaeal *Methanococcus jannaschii* Trm5 (*Mj*Trm5) is shown in Table 2. The central component of these assays is the radioactive [^{3}H-methyl]-AdoMet as the methyl donor (abbreviated as ^{3}H-AdoMet hereafter), which permits the incorporation of [^{3}H-methyl] to synthesize the product m^{1}G37-tRNA. The radioactivity in the product tRNA is an integral part of the nucleic acid and therefore is acid precipitable, whereas that in the free AdoMet substrate is not acid precipitable and can be washed away. Measuring the radioactivity of the product tRNA, usually collected on filter pads, then quantifies the amount of product synthesis. This quantitative information is valuable for insight into the kinetic and catalytic mechanism. An advantage of this method is that because the radiolabel is placed on AdoMet, the substrate G37-tRNA can be prepared without label in one transcription reaction.

Table 2 Kinetic Parameters of *Ec*TrmD and *Mj*Trm5

Parameter	*Ec*TrmD	*Mj*Trm5	Assays	References
K_m (tRNA), μM	3.1 ± 0.1	0.70 ± 0.03	Steady-state	Elkins et al. (2003); Masuda, Sakaguchi, Liu, Gamper, and Hou (2013); Christian, Gamper, and Hou (2013)
k_{cat}, s^{-1}	0.09 ± 0.01	0.017 ± 0.002	Steady-state	Christian and Hou (2007); Christian, Gamper, and Hou (2013)
		0.020 ± 0.007	Pre-steady-state	Christian, Lahoud, Liu, and Hou (2010)
K_d (tRNA), μM	0.44 ± 0.04	1.4 ± 0.1	Single turnover	Christian, Lahoud, Liu, and Hou (2010)
K_d (AdoMet), μM		0.8 ± 0.1	Single turnover	Christian, Lahoud, Liu, and Hou (2010)
K_d (m^1G37-tRNA), μM		0.2 ± 0.1	Pre-steady-state	Christian, Lahoud, Liu, and Hou (2010)
K_d (AdoHcy), μM		0.020 ± 0.005	Pre-steady-state	Christian, Lahoud, Liu, and Hou (2010)
k_{chem}, s^{-1}	0.09 ± 0.01	0.12 ± 0.03	Pre-steady-state	Christian, Lahoud, Liu, and Hou (2010)
		0.12 ± 0.01	Single turnover	Christian, Lahoud, Liu, and Hou (2010)

In contrast, there exists an alternative and also quantitative assay (Swinehart, Henderson, & Jackman, 2013), which places the label on the substrate G37-tRNA while using unlabeled AdoMet as the methyl donor. In this alternative method, the G37-tRNA substrate is prepared with a site-specifically placed ^{32}P at the 5′-end of G37, so that the label is associated only with G37 in the substrate or m^1G37 in the product (Jackman, Montange, Malik, & Phizicky, 2003). After methyl transfer, the substrate and product tRNAs are digested to single 5′-monophosphate nucleotides, and G37 and m^1G37 are separated by one-dimensional thin-layer chromatography and quantified individually on a phosphor imager screen. A major challenge of this alternative assay, however, is that the site-specific labeling of

5′-^{32}P-G37 in tRNA is a multi-step process, involving synthesis of a tRNA fragment starting with G37, 5′-^{32}P-labeling of this fragment, and joining this labeled fragment with another tRNA fragment to reconstitute the full-length molecule. While this multi-step labeling procedure was successful for the G37-tRNA substrate, it may be difficult for other tRNA substrates, possibly involving ligation of three or more fragments (Sakaguchi et al., 2012).

2. METHODOLOGY

1. Steady-state assays
2. Pre-steady-state assays
3. Single-turnover assays

3. STEADY-STATE ASSAYS

For decades, kinetic analysis of methyl transfer to tRNA substrates has been dominated by steady-state assays. In these assays, the enzyme is in catalytic amounts, the tRNA substrate is in excess, the AdoMet concentration is saturating, and the reaction proceeds over multiple rounds of turnover. Fitting the data of the initial rate of methyl transfer (V_0) as a function of the tRNA substrate concentration to the Michaelis–Menten equation yields the catalytic turnover k_{cat} and the Michaelis constant for the tRNA substrate K_m (tRNA). Conversely, analysis of the initial rate (V_0) as a function of the AdoMet concentration, while tRNA is saturating, yields the catalytic turnover k_{cat} and the Michaelis constant K_m (AdoMet). In these two measurements, the k_{cat} value should be similar, which is a validation for the reliability of the assay. Using such steady-state assays, the kinetic parameters of *Ec*TrmD were determined and used to evaluate the corresponding parameters of mutant enzymes harboring amino acid substitutions (Elkins et al., 2003). This analysis revealed the broad landscape of the enzyme, showing the residues important for synthesis of m^{1}G37-tRNA. Because steady-state assays require only minimal amounts of the enzyme, they are easy to prepare and are usually the first assays to perform for a methyl transfer reaction.

3.1 Preparation of a Target tRNA Transcript

The substrate tRNA is prepared as a transcript from run-off transcription of a DNA template encoding a native or mutant sequence. The DNA template sequence can be cloned into a plasmid and made available for run-off transcription by digestion with a restriction enzyme to expose the 3′-end. Alternatively, the DNA template can be constructed from overlapping oligonucleotides (Hou, 2012; Zhang, Liu, Slater, & Hou, 2008). The template sequence is usually placed behind the promoter of the highly processive T7 RNA polymerase and the transcription by this enzyme gives the best yield for the range of tRNA lengths (70–90 nucleotides). An over-producer strain of T7 RNA polymerase is available (Tabor, Huber, & Richardson, 1987) and the enzyme can be purified in-house and titrated to the level where 10 μL of the enzyme gives visible precipitation of pyrophosphate-Mg^{2+} conjugates in less than 1 h at 37 °C. The pyrophosphate is released from NTP upon incorporation of NMP during active transcription. The observation of such precipitation is usually a good indication of strong transcription. If no precipitation is observed in more than 1 h, then add more T7 RNA polymerase to the transcription reaction.

An important note of T7 RNA polymerase is that it initiates transcription only with G (guanosine, GMP, or GTP). Therefore, substrate sequences that do not start with a G are prepared with a hammerhead ribozyme sequence between the promoter and the first nucleotide of the tRNA gene (Fechter, Rudinger, Giege, & Theobald-Dietrich, 1998; Liu, Snaders, Pascal, & Hou, 2012; Pham et al., 2014). Transcription of this hybrid construct gives rise to a fusion, where the synthesized ribozyme in the upstream portion will self-cleave to liberate the downstream substrate tRNA with a 5′-OH end. Another consideration is the propensity of T7 RNA polymerase to generate heterogeneous 3′-ends, which can affect some posttranscriptional modification reactions (Roovers et al., 2006). This problem can be reduced, but not completely eliminated, by two methods. One is to use the "foot" mutant of T7 RNA polymerase, which lacks the C-terminal F882-A883 residues of the wild-type enzyme and as a result exhibits reduced processivity (Mookhtiar, Peluso, Muller, Dunn, & Coleman, 1991). The other is to introduce two consecutive 2′-*O*-methyl backbone modifications to the 5′-terminus of the noncoding strand of the DNA template to reduce the processivity of T7 RNA polymerase (Kao, Zheng, & Rudisser, 1999). Notably, introduction of the 2′-*O*-methyl backbone modifications are not easily achieved with plasmid DNA templates, but are readily accommodated by chemical synthesis of oligonucleotides for constructing DNA templates. The synthesized tRNA transcripts are separated from DNA templates by denaturing 12% polyacrylamide/7 *M* urea (12% PAGE/7 *M* urea) gels, identified by UV shadowing, extracted from gel materials into the TE buffer

(10 mM Tris–HCl, pH 8.0, 1 mM EDTA), collected by 3 volumes of ethanol precipitation, washed by 70% ethanol, and resuspended and stored in TE.

3.2 Methyl Transfer in Steady-State Conditions

The representative enzymes are *Ec*TrmD and *Mj*Trm5, each of which is well characterized and has a high-resolution crystal structure in complex with AdoMet or the product AdoHcy (Elkins et al., 2003; Goto-Ito et al., 2008, 2009). Additionally, *Ec*TrmD is homologous to *Haemophilus influenza* TrmD, which has crystal structures in the apo form and in the form with AdoMet or AdoHcy (Ahn et al., 2003), while *Mj*Trm5 is homologous to human Trm5 (*Homo sapiens* Trm5, *Hs*Trm5) with similar kinetic parameters (Christian et al., 2013). The chosen assay temperature is 37 °C for *Ec*TrmD and *Hs*Trm5 and 55 °C for *Mj*Trm5. The G37-tRNA substrate for each enzyme is the transcript of *Ec*tRNA^{Leu}, *Hs*tRNA^{Cys}, and *Mj*tRNA^{Cys}, respectively. The reaction buffer for *Ec*TrmD is 0.1 M Tris–HCl (pH 8.0), 24 mM NH_4Cl_2, 6 mM $MgCl_2$, 4 mM dithiothreitol, 0.1 mM EDTA, and 0.024 mg/mL bovine serum albumin. The same reaction buffer applies to *Mj*Trm5 and *Hs*Trm5, except that NH_4Cl_2 is increased to 100 mM. To determine the kinetic parameters for the tRNA substrate, the substrate concentration varies while AdoMet is saturating; conversely, to determine the kinetic parameters for AdoMet, the concentration of the methyl donor varies, while tRNA is saturating. In either case, the substrate must exist in excess of the enzyme (preferably more than 50-fold), so that the enzyme (in nM) can catalyze multiple turnovers while the reaction consumes no more than 5% of the substrate.

1. Prepare the working stock of ^{3}H-AdoMet by mixing as follows:

   Purchase ^{3}H-AdoMet commercial solution (Perkin Elmer, NET155H, 60 Ci/mmol, 6.6 μM, 0.55 μCi/μL).

   Purchase unlabeled AdoMet (1 mM): dissolve 0.57 mg of AdoMet in 1.0 mL water with 1 μL concentrated H_2SO_4 (final concentration of the acid = 12 mM). Store the solution at −20 °C.

   Mix 200 μL ^{3}H-AdoMet commercial solution with 90 μL unlabeled AdoMet (1 mM) to give a final concentration of ~300 μM AdoMet with a specific activity of 2650 dpm/pmol.
2. Calculate the specific activity of the working stock ^{3}H-AdoMet from step 1.

   The working stock consists of 200 μL of the commercial ^{3}H-AdoMet (6.6 μM, 0.55 μCi/μL) and 90 μL of unlabeled AdoMet (1 mM).

Total concentration of the commercial ^{3}H-AdoMet in the mixture: 4.55 μ*M*.

Total concentration of unlabeled AdoMet in the mixture: 310.34 μ*M*.

Combined concentration of AdoMet: 314.89 μ*M*.

Total μCi in the mixture: 0.55 μCi/μL × 200 μL = 110 μCi.

Specific activity in dpm/pmol: [110 μCi × (2.2 × 10^6 dpm/μCi)]/[314.89 pmol/μL × 290 μL] = 2650 dpm/pmol.

3. As an initial determination of the kinetic parameters for the G37-tRNA substrate for *Ec*TrmD or *Mj*Trm5, set up a series of reactions, where the concentration of G37-tRNA increases by twofold from 0.1 to 25 μ*M*. This range of concentration usually covers the K_m (tRNA) (3–5 μ*M* for *Ec*TrmD and 0.5 μ*M* for *Mj*Trm5) (Christian et al., 2013; Elkins et al., 2003; Masuda, Sakaguchi, Liu, Gamper, & Hou, 2013). Include in the series a reaction without tRNA as a control. Once an initial K_m (tRNA) is determined, one should repeat the K_m (tRNA) analysis by ranging the G37-tRNA concentration between 1/5 and 5 × of the initial K_m (tRNA) value.
4. Each reaction in the series above should have the following components:

Reaction components	Volume to add (μL)	Final concentration (μ*M*)
tRNA/TE	10.0	0.1–25
Heat-cool (HC buffer)	2.0	
5 × buffer	5.0	1.0 ×
^{3}H-AdoMet (310 μ*M*, 2650 dpm/pmol)	3.0	37.2
*Ec*TrmD or *Mj*Trm5 (25 n*M*)	5.0	0.005
Total	25.0	

Heat the series of reactions containing only tRNA/TE at 80 °C for 3 min, spin briefly, add the HC buffer (10 m*M* Tris–HCl, pH 8.0, 10 m*M* $MgCl_2$), and anneal the tRNA at 37 °C for 15 min.

5. Add the 5 × buffer, the working stock ^{3}H-AdoMet, and initiate the reaction by adding *Ec*TrmD or *Mj*Trm5. At the specified time points (e.g., 2, 4, 6, 8, and 10 min), remove 5 μL from the reaction, spot it

onto a 1 cm^2 Whatman 3MM filter pad, and place the filter pad into a beaker containing 5% trichloroacetic acid (TCA).

6. After all pads are in TCA, shake the solution for 10 min at 4 °C to wash off unincorporated ^{3}H-AdoMet, while allowing the synthesized m^{1}G37-tRNA to precipitate in TCA. Decant and repeat the 5% TCA wash.
7. Wash all filter pads with 95% ethanol by shaking for 10 min at 4 °C in the beaker. Repeat the ethanol wash one more time.
8. Wash all filter pads with ether. Agitate gently by hand and let the ether solution sit at room temperature for 5 min under a fume hood. Decant ether and dry the filter pads under the fume hood for 15 min.
9. Transfer each filter pad to a scintillation solution in a vial and measure the amount of radioactivity using a liquid scintillation counter.
10. Calculate the amount of m^{1}G37-tRNA synthesis based on the specific activity of the working stock of ^{3}H-AdoMet.
11. Correct the ^{3}H counting by measuring the quenching factor using the following procedure: Take a 5 μL aliquot at the final time point of a reaction and pass it through a quick spin column to remove unincorporated ^{3}H-AdoMet, which stays with the column. Directly transfer the eluate 5 μL (which contains counts only associated with the methylated tRNA) into the liquid scintillation fluid and measure the counts. The ratio of the direct measurement of this count over the count on the TCA precipitated filter pad at the same time point reveals the quenching factor, which should be used to correct the fraction of methylation. For the protocol described here, the quenching factor is usually 4.
12. For each enzyme concentration, fit the data of the time course to the linear equation:

$$y = ax + b \tag{1}$$

where y is the pmol of synthesis of m^{1}G37-tRNA, x is the time, a is the initial rate (V_0) of synthesis in pmol/s, and b is the intercept on the y axis, which arises from background counts of the reaction due to nonspecific binding of ^{3}H-AdoMet to filter pads. The value of b should be closely similar to the value obtained from the control reaction containing no tRNA.

13. Plot the initial rate V_0 as a function of the G37-tRNA concentration by fitting the data to the Michaelis–Menten equation:

$$V_0 = \frac{V\text{max}\,[\mathrm{S}]}{(K_\mathrm{m} + [\mathrm{S}])} \text{ or } V_0 = \frac{k_\mathrm{cat}\,[\mathrm{E}_0]\,[\mathrm{S}]}{(K_\mathrm{m} + [\mathrm{S}])} \tag{2}$$

where V_0 is the initial rate of synthesis of m[1]G37-tRNA (in pmol/s), S is the substrate G37-tRNA concentration (in μM), V_{max} is the maximum rate (in pmol/s), k_{cat} is the catalytic turnover (in s^{-1}), E_0 is the enzyme quantity in the reaction (in pmol), and K_m is the Michaelis constant for the G37-tRNA substrate (in μM).

3.3 Experimental Considerations

To determine the kinetic parameters of AdoMet, one should design a series of reactions similar to steps 3 and 4 in Section 3.2 but vary the AdoMet concentration in 0.1–25 μM, while the tRNA substrate is saturating (~10 × of K_m (tRNA)). The suggested range of AdoMet concentration covers the K_m (AdoMet) for *Ec*TrmD (5–8 μM) and *Mj*Trm5 (0.5 μM) (Christian et al., 2013; Lahoud et al., 2011). By having a working stock of ^{3}H-AdoMet (step 2), one can use this stock to make appropriate dilution to the desired concentration. However, an important difference here from the determination of K_m (tRNA) is that the radioactivity of AdoMet varies with each dilution. To address this difference, one needs to have a no-tRNA control reaction for each AdoMet concentration to provide the background counts for correction of the nonspecific radioactivity bound to filter pads at each concentration. The counts after correction of the no-tRNA control for each AdoMet concentration, followed by correction for the filter-quenching factor, are then converted to pmol of product synthesis. This information is then used to determine the K_m (AdoMet) as described in steps 11–13.

The steady-state parameters are each a composite term of multiple turnovers. Therefore, k_{cat} does not mean the rate constant of methyl transfer and K_m does not mean the binding affinity of the substrate to the enzyme. The kinetic assays described below, pre-steady-state and single-turnover assays, are designed to complement steady-state assays.

4. PRE-STEADY-STATE ASSAYS

The recent development of pre-steady-state assays has shed new light on the distinction between TrmD and Trm5. In pre-steady-state assays, the G37-tRNA substrate is maintained in 10-fold molar excess of the enzyme and both the substrate and the enzyme are at the μM level to permit one turnover of methyl transfer on the enzyme, followed by multiple rounds of steady-state turnover. It was in fact in pre-steady-state assays that the kinetics of TrmD was distinguished from Trm5. Specifically, while the TrmD reaction occurred linearly over time, the Trm5 reaction occurred

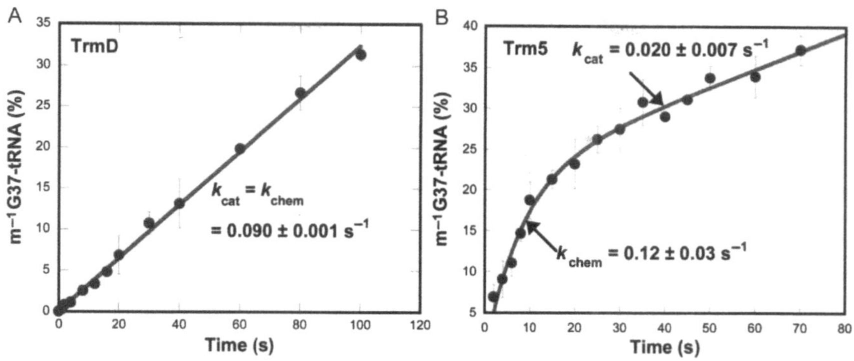

Figure 2 Pre-steady-state kinetics of m[1]G37-tRNA synthesis. (A) Monitoring the time course of synthesis upon mixing *Ec*TrmD (1 μ*M*) with *Ec*tRNA^{Leu} (10 μ*M*) and AdoMet (30 μ*M*) at 37 °C. The time-dependent synthesis is calculated as the amount of synthesis per active site of the enzyme (%) and is fit to Eq. (1) to determine the slope (a = slope and b = 0). (B) Monitoring the time course of synthesis upon mixing *Mj*Trm5 (1 μ*M*), *Mj*tRNA^{Cys} (10 μ*M*), and AdoMet (25 μ*M*). The time-dependent synthesis is calculated as the amount of synthesis per active site of the enzyme (%) and is fit to Eq. (3) to determine k_{chem} and k_{cat}. *This figure is adapted from Figure 1 in Christian, Lahoud, Liu, and Hou (2010). Reprinted with permission from Elsevier.*

in a rapid burst phase followed by a slower and linear phase (Fig. 2) (Christian, Lahoud, Liu, & Hou, 2010). Fitting the TrmD data to a linear equation revealed a slope of $0.09 \pm 0.01\ s^{-1}$, similar to the value of k_{cat} in steady state. Fitting the Trm5 data to a burst equation revealed a rate constant of $0.12 \pm 0.03\ s^{-1}$ for the first turnover and a rate constant of $0.020 \pm 0.007\ s^{-1}$ for steady-state turnover k_{cat}. The rate constant for the first turnover is associated with the chemistry of methyl transfer (k_{chem}) under conditions of rapid equilibrium binding and it can report either the chemical step or the pre-chemistry active-site rearrangement step of the enzyme. Because TrmD maintains the same rate constant in the first turnover as in steady-state turnover, its catalytic cycle is limited by k_{chem}. In contrast, because Trm5 exhibits a faster k_{chem} in the first turnover relative to the steady-state k_{cat}, its catalytic cycle is limited by k_{cat}, which is associated with the release of the m[1]G37-tRNA product. Indeed, by treating m[1]G37-tRNA as an inhibitor of Trm5, pre-steady-state assays showed that increasing concentration of the inhibitor progressively decreased the burst amplitude, a parameter that indicates the active fraction of the enzyme (Christian, Lahoud, Liu, & Hou, 2010). Similarly, increasing concentration of AdoHcy also decreased the burst amplitude (Christian, Lahoud, Liu, & Hou, 2010). The analysis of the K_d of each inhibitor for Trm5 then revealed that the enzyme has a tighter binding affinity to AdoHcy relative to AdoMet and a tighter binding affinity to m^1G37-tRNA relative to G37-tRNA. These results support the notion that the tight binding of Trm5 to either product contributes to the slow product release in catalytic turnover.

4.1 Methyl Transfer in Pre-Steady-State Conditions

Using *Ec*TrmD and *Mj*Trm5 as the examples, the reaction buffer for each is as described in Section 3.2. Because methyl transfer of the first turnover for *Ec*TrmD ($k_{obs}=0.09\pm0.01\ s^{-1}$) or *Mj*Trm5 ($k_{obs}=0.12\pm0.03\ s^{-1}$) occurs on the time scale of $0.1\ s^{-1}$ or faster (Christian, Lahoud, Liu, & Hou, 2010), meaning that it takes less than 10–11 s to complete the first turnover, the pre-steady-state measurement should be carried out on a rapid mixing and quench instrument. Our lab uses the KinTek RQF-3 model (KinTek Corp, Texas, www.kintek-corp.com), which operates with a computer panel to control rapid mixing of the contents of two syringes and time-dependent quenching of the reaction. For methyl transfer that proceeds on a slower timescale, the mixing and quenching can be performed without the instrument. We have found that data obtained from the same protocol performed with or without the instrument are similar, with the standard deviation less than 20%.

1. Take an aliquot of the unmodified transcript of G37-tRNA (200 pmol) and adjust the volume with TE to 15 μL. Heat the tRNA solution for 3 min at 80 °C, which is above the estimated melting temperature of the tRNA transcript. Quickly spin the solution and add 5 μL of the HC buffer. Anneal the G37-tRNA at 37 °C for 15 min. The resulting stock concentration of G37-tRNA is 100 μ*M*.
2. Prepare the G37-tRNA solution for syringe #1 of the RQF-3 instrument:

Syringe #1	Volume	2× Concentration	1× Concentration
G37-tRNA transcript (100 μ*M*)	60.0 μL	20.0 μ*M*	10.0 μ*M*
ddH_2O	151.0 μL		
5× buffer	60.0 μL		1×
Working stock of ^{3}H-AdoMet	29.0μL	30.0 μ*M*	15.0 μ*M*
Total	300.0 μL		

Mix G37-tRNA and ddH_2O, add the 5 × buffer, add the working stock of ^{3}H-AdoMet, and place the solution on ice until loading onto syringe #1. Note that the working stock of ^{3}H-AdoMet is prepared as described in steps 1 and 2 of Section 3.2. When performing pre-steady-state assays, both the tRNA and enzyme solutions are made in the 2 × concentration, which upon mixing will result in the 1 × concentration of each component.

3. Prepare the enzyme solution for syringe #2 of the RQF-3 instrument:

Syringe #2	**Volume**	**2 × Concentration**	**1 × Concentration**
*Ec*TrmD or *Mj*Trm5 (400 μ*M*)	1.5 μL	2.0 μ*M*	1.0 μ*M*
ddH_2O	238.5 μL		
5 × buffer	60.0 μL		1 ×
Total	300.0 μL		

Mix water and the 5 × buffer and add the enzyme. Place the solution on ice until ready to load onto syringe #2.

4. On the RQF-3 instrument, fill the large syringe on each side with 1 × buffer and the middle syringe with 5% TCA. Fill syringe #1 with 300 μL of the tRNA solution to one sample loop and syringe #2 with 300 μL of the enzyme solution to the second sample loop. Enter time points on the control panel. Upon hitting the start button, the 1 × buffer pushes 15 μL from syringe #1 and 15 μL from syringe #2 into the reaction loop. At each of the specified time lapse (e.g., 0, 1, 3, 5, 7, 9, 12, 15, 20, 25, 30, 40, 50, 60, 80, 100, and 120 s), the reaction is quenched with 54 μL of 5% TCA.
5. Collect the quenched solution of each time point in an Eppendorf tube. Spot 20 μL of the quenched solution onto a 1 cm^2 Whatman 3MM filter pad. Wash all filter pads with 5% TCA twice, followed by 95% ethanol twice, and followed by ether once. Dry filter pads in a hood and measure radioactivity of each filter pad in a scintillation counter. Correct the counts for the quenching factor, and calculate the synthesis of m^{1}G37-tRNA based on the corrected counts. Calculate the fractional conversion from G37 to m^{1}G37-tRNA and plot the data versus time. These procedures are as described in steps 6–12 of Section 3.2.

6. Fit the data to the burst equation as shown below:

$$y = y_0 + A \times \left(1 - e^{-k_1 \times t}\right) + k_2 \times E_0 \times t \quad (3)$$

where y_0 is the y intercept, A is the amplitude of the initial exponential phase (representing the active fraction of the enzyme), k_1 is the observed rate constant of the initial exponential regression, k_2 is the apparent rate constant of the steady-state phase, and t is the time in seconds (Dupasquier, Kim, Halkidis, Gamper, & Hou, 2008; Zhang, Perona, Ryu, Francklyn, & Hou, 2006).

4.2 Experimental Considerations

The appearance of burst kinetics implies slow product release after methyl transfer. To test this hypothesis, one can use the burst phase to determine the enzyme affinity for the product. For example, to determine the enzyme affinity for the m^1G37-tRNA product, we purified the product from the methyl transfer reaction by using RNase H to selectively cleave the substrate G37-tRNA (Hou, 2012; Hou, Li, & Gamper, 2006). The purified m^1G37-tRNA was used to form a series of Trm5-product complexes, which was tested for the forward methyl transfer reaction in pre-steady-state assays. We showed that while increasing concentration of the product had little effect on the rate of the burst or steady-state phase, it progressively decreased the amplitude of the burst phase (Christian, Lahoud, Liu, & Hou, 2010), indicating that the dissociation of the product from Trm5 was slow and it limited the availability of the enzyme for the forward reaction. A fit of the data to a quadruple equation (Eq. 4) revealed a K_d (m^1G37-tRNA) of 0.2 ± 0.1 μM.

$$y = \frac{(E_0 + S + K_d) - \sqrt{(E_0 + S + K_d)^2 - 4 \times E_0 \times S}}{2} \quad (4)$$

where y represents the methylated tRNA product after one turnover; E_0 is the active enzyme concentration; S is the tRNA concentration; K_d is the dissociation constant for tRNA (Dupasquier et al., 2008; Zhang et al., 2006). The K_d (m^1G37-tRNA) is lower than the K_d (G37-tRNA) of the enzyme–substrate complex that was determined from single-turnover kinetic assays (Christian, Lahoud, Liu, & Hou, 2010).

The burst amplitude can be further exploited to determine the parameters for AdoMet, such as K_d (AdoMet) and K_d (AdoHcy). For example, we determined K_d (AdoMet) by preforming a series of Trm5-AdoMet complexes. Analysis of the series of complexes for methyl transfer in pre-steady-state assays showed progressive increases of the burst amplitude with increasing concentration of the methyl donor. Fitting the amplitude data as a function of AdoMet concentration to a hyperbolic equation (Eq. 5) revealed a K_d (AdoMet) of $0.44 \pm 0.09\ \mu M$. Conversely, we determined K_d (AdoHcy) by preforming a series of Trm5-AdoHcy complexes and showed that increasing concentration of AdoHcy progressively decreased the burst amplitude in pre-steady-state assays. Fitting the amplitude data as a function of AdoHcy concentration to a hyperbolic equation (Eq. 6) revealed a K_d (AdoHcy) of 20 ± 5 nM, which is nearly 20-fold lower than the K_d (AdoMet). Interestingly, between the two products of the methyl transfer reaction, we noticed that the K_d (AdoHcy) is 20-fold lower than the K_d (m^1G37-tRNA), indicating that it is AdoHcy that has the highest affinity for Trm5 and that it is the release of AdoHcy that controls the overall catalytic turnover for Trm5.

$$y = A \times S/(S + K_d) \tag{5}$$

where A is maximum amplitude in burst kinetics, S is the AdoMet concentration (Dupasquier et al., 2008; Zhang et al., 2006).

$$y = y_0 - A \times I/(I + K_d) \tag{6}$$

where y_0 is the initial burst amplitude in the absence of AdoHcy; A is the scaling constant for amplitude change upon addition of AdoHcy; I is the concentration of AdoHcy (Dupasquier et al., 2008; Zhang et al., 2006).

5. SINGLE-TURNOVER ASSAYS

In single-turnover assays, the enzyme is in excess of the tRNA substrate and the AdoMet concentration is saturating, so that the reaction proceeds only once. This single-turnover condition offers two advantages. First, because the observed rate constant (k_{obs} or k_{app}) reports the kinetics of just one methyl transfer, the plot of k_{obs} versus concentration leads to the determination of kinetic parameters intrinsically associated with the methyl transfer. Second, because the enzyme and AdoMet are both saturating relative to the tRNA substrate, the enzyme–substrate affinity is measured under the condition of rapid equilibrium binding, so that the determined K_d reflects the true thermodynamic binding affinity.

5.1 Methyl Transfer in Single-Turnover Conditions

1. Prepare the tRNA solution for syringe #1 of the RQF-3 instrument:

Syringe #1	Volume	2 × Concentration	1 × Concentration
G37-tRNA transcript (100 μ*M*)	1.5 μL	0.5 μ*M*	0.25 μ*M*
ddH$_2$O	209.5 μL		
5 × buffer	60.0 μL		1 ×
Working stock of ^{3}H-AdoMet	29.0 μL	30.0 μ*M*	15.0 μ*M*
Total	300.0 μL		

   Mix G37-tRNA and ddH$_2$O, add the tRNA, and denature the mixture by heating at 85 °C for 3 min. Briefly spin down the heated solution and then add the 5 × buffer to allow annealing of the tRNA at 37 °C for 15 min. Add the working stock of ^{3}H-AdoMet and place the solution on ice until loading onto syringe #1. Note that the working stock of ^{3}H-AdoMet is prepared as described in steps 1 and 2 of Section 3.2. When performing single-turnover assays, both the tRNA and enzyme solutions are made in the 2 × concentration, which upon mixing will result in the 1 × concentration of each component.

2. In single-turnover assays, the reaction rate is driven by the enzyme concentration, not by the tRNA concentration (Dupasquier et al., 2008; Zhang et al., 2006; Liu et al., 2007; Liu, Gamper, Liu, Cooperman, & Hou, 2011). Therefore, a series of reactions are designed with increasing concentration of the enzyme, so that the plot of k_{obs} versus concentration gives the K_d of the enzyme for the tRNA substrate (K_d (tRNA)). The six reactions designed below are to provide an initial evaluation of the K_d (tRNA). Once an initial K_d (tRNA) is obtained, the experiments should be repeated to vary the enzyme concentration in the range of 1/5 and 5 × of the K_d.
3. Using *Ec*TrmD as an example, prepare a series of the enzyme solutions (ranging from 2 to 32 μ*M* as the 2 ×) for syringe #2 of the RQF-3 instrument. For each concentration, prepare a 300 μL solution. A total of six concentrations are prepared.

	1	2	3	4	5	6
2× *Ec*TrmD	2 μ*M*	4 μ*M*	8 μ*M*	16 μ*M*	24 μ*M*	32 μ*M*
1× *Ec*TrmD	1 μ*M*	2 μ*M*	4 μ*M*	8 μ*M*	12 μ*M*	16 μ*M*
*Ec*TrmD stock (400 μ*M*)	1.5 μL	3.0 μL	6 μL	12 μL	18 μL	24 μL
5× buffer	60 μL	60 μL	60 μL	60 μL	60 μL	60 μL
ddH_2O	238.5 μL	237 μL	234 μL	228 μL	222 μL	216 μL
Total	300 μL	300 μL	300 μL	300 μL	300 μL	300 μL

Mix ddH_2O and the 5× buffer, add the enzyme solution. Place on ice until loading onto syringe #2.

4. On the RQF-3 instrument, fill the 1× buffer solution, 5% TCA solution, the tRNA solution, and one enzyme solution to the appropriate syringes as described in step 4 of Section 4.1. After the specified time lapse, the reaction is quenched with 54 μL of 5% TCA. Put the quenched solution in an Eppendorf tube and collect samples up to 17 time points (e.g., 0, 1, 3, 5, 7, 9, 12, 15, 20, 25, 30, 40, 50, 60, 80, 100, and 120 s).
5. Spot 20 μL of each aliquot onto a 1 cm^2 Whatman 3MM filter pad. Wash all filter pads with 5% TCA twice, with 95% ethanol twice, and with ether once. Dry filter pads in a hood and measure radioactivity of each filter pad in a scintillation counter. Correct the counts for the quenching factor, calculate the synthesis of m^1G37-tRNA based on the specific activity of ^{3}H-AdoMet, convert the data into fractional conversion from G37 to m^1G37-tRNA, and plot the data versus time. These procedures are as described in steps 6–12 of Section 3.2.
6. Data points for each time course are fit to the single exponential equation:

$$y = y_0 + A \times \left(1 - e^{-k_{app} \times t}\right) \tag{7}$$

where y_0 is the y intercept, A is the scaling constant, k_{app} is the apparent rate constant, and t is the time in seconds to determine k_{obs} (or k_{app}) (Dupasquier et al., 2008; Zhang et al., 2006). The data of k_{obs} versus enzyme concentration for single-turnover analysis of m^1G37-tRNA synthesis are fit to the hyperbolic equation:

$$y = k_{chem} \times E_0/(E_0 + K_d) \tag{8}$$

where k_{chem} is the rate constant for the steps associated with the methyl transfer chemistry, K_d is the enzyme affinity for the tRNA substrate (K_d (tRNA)), and E_0 is the enzyme concentration (Dupasquier et al., 2008; Zhang et al., 2006). Note that in single-turnover assays using chemical quench, k_{chem} can be the chemistry of methyl transfer or the pre-chemistry rearrangement of the enzyme active site. The latter possibility can be addressed by dynamics experiments using fluorescent tRNA (Liu, Betteridge, & Hou, 2009) in stopped-flow methodologies.

5.2 Experimental Considerations

In single-turnover assays, control experiments should be performed in which enzyme, tRNA, and AdoMet are mixed in different orders and in different syringes. For example, syringe #1 may contain just the tRNA solution, while syringe #2 may contain AdoMet and the enzyme solution. If all of the mixing control experiments return the same k_{obs} value, then this is evidence that the rate of methyl transfer is independent of the mixing order, and that the reaction is proceeding under a rapid equilibrium binding condition.

For TrmD, which lacks burst kinetics, the determination of K_d (AdoMet) cannot take advantage of the burst amplitude as in the case of Trm5. Instead, the value of K_d (AdoMet) for TrmD can be determined using single-turnover assays, in which the enzyme concentration is in excess of the AdoMet substrate, while the tRNA concentration is saturating. For example, prepare the tRNA solution for syringe #1 of RQF-3 at the 2 × concentration of 20 μM of G37-tRNA and 1 μM of the working stock of ^{3}H-AdoMet in 1 × buffer. The specific activity of ^{3}H-AdoMet is 3400 dpm/pmol, higher than the specific activity (2650 dpm/pmol) for determination of K_d (tRNA), to compensate for the lower amount of AdoMet used in the reaction. Prepare a range of TrmD concentration for syringe #2 as in step 3 of Section 5.1. After rapid mixing of the tRNA solution with an enzyme solution, monitor the synthesis of m^{1}G37-tRNA over time, and determine the k_{obs} from the single exponential equation (Eq. 7) for the specific enzyme concentration. Fitting the data of k_{obs} as a function of enzyme concentration to the hyperbola equation (Eq. 8) will then determine the K_d (AdoMet) for the enzyme. This assay has also been successfully applied to *Hs*Trm5 without relying on the burst kinetics of the enzyme (Christian et al., 2013).

6. CONCLUSIONS

All tRNA molecules contain multiple posttranscriptional modifications by AdoMet-dependent methyl transfer reactions. These methylation reactions are subject to regulation, due to the variability of AdoMet levels depending on the energy state of a cell. However, because different methyltransferases bind AdoMet differently, their reactions are likely to be subject to different regulation. To understand the regulation of each methyltransferase and how the regulation contributes to the biology of a tRNA, kinetic analysis of the methyl transfer reaction is necessary. This chapter describes three types of kinetic assays that collectively form the foundation to address the regulation. The steady-state assays are important for establishing a global view of the methyl transfer reaction and the data provide the basis for comparing data of other assays. The pre-steady-state assays are performed under conditions closely similar to those in a cell and as such the data are relevant to biology, particularly with respect to the rate-limiting step in catalytic turnovers. The single-turnover assays are performed upon rapid equilibrium binding of a methyltransferase to its substrates, so the data are most appropriate for correlating with the structural information of the enzyme. Although the three assays are described for the TrmD and Trm5 reactions, they are easily adaptable to other AdoMet-dependent methyltransferases.

ACKNOWLEDGMENTS

This work was supported by NIH grants R01GM108972 and R01GM81601 to Y. M. H. The authors thank Megumi Shigematsu for help with Fig. 1.

REFERENCES

Ahn, H. J., Kim, H. W., Yoon, H. J., Lee, B. I., Suh, S. W., & Yang, J. K. (2003). Crystal structure of tRNA(m1G37)methyltransferase: Insights into tRNA recognition. *The EMBO Journal*, *22*, 2593–2603.

Baba, T., Ara, T., Hasegawa, M., Takai, Y., Okumura, Y., Baba, M., et al. (2006). Construction of Escherichia coli K-12 in-frame, single-gene knockout mutants: The Keio collection. *Molecular Systems Biology*, *2*(2006), 0008.

Benitez-Paez, A., Villarroya, M., Douthwaite, S., Gabaldon, T., & Armengod, M. E. (2010). YibK is the 2'-O-methyltransferase TrmL that modifies the wobble nucleotide in Escherichia coli tRNA(Leu) isoacceptors. *RNA*, *16*, 2131–2143.

Bjork, G. R., Jacobsson, K., Nilsson, K., Johansson, M. J., Bystrom, A. S., & Persson, O. P. (2001). A primordial tRNA modification required for the evolution of life? *The EMBO Journal*, *20*, 231–239.

Bystrom, A. S., & Bjork, G. R. (1982a). Chromosomal location and cloning of the gene (trmD) responsible for the synthesis of tRNA (m1G) methyltransferase in Escherichia coli K-12. *Molecular & General Genetics*, *188*, 440–446.

Bystrom, A. S., & Bjork, G. R. (1982b). The structural gene (trmD) for the tRNA(m1G) methyltransferase is part of a four polypeptide operon in Escherichia coli K-12. *Molecular & General Genetics*, *188*, 447–454.

Cantoni, G. L. (1975). Biological methylation: Selected aspects. *Annual Review of Biochemistry*, *44*, 435–451.

Christian, T., Evilia, C., Williams, S., & Hou, Y. M. (2004). Distinct origins of tRNA(m1G37) methyltransferase. *Journal of Molecular Biology*, *339*, 707–719.

Christian, T., Gamper, H., & Hou, Y. M. (2013). Conservation of structure and mechanism by Trm5 enzymes. *RNA*, *19*, 1192–1199.

Christian, T., & Hou, Y. M. (2007). Distinct determinants of tRNA recognition by the TrmD and Trm5 methyl transferases. *Journal of Molecular Biology*, *373*, 623–632.

Christian, T., Lahoud, G., Liu, C., Hoffmann, K., Perona, J. J., & Hou, Y. M. (2010). Mechanism of N-methylation by the tRNA m1G37 methyltransferase Trm5. *RNA*, *16*, 2484–2492.

Christian, T., Lahoud, G., Liu, C., & Hou, Y. M. (2010). Control of catalytic cycle by a pair of analogous tRNA modification enzymes. *Journal of Molecular Biology*, *400*, 204–217.

De Bie, L. G., Roovers, M., Oudjama, Y., Wattiez, R., Tricot, C., Stalon, V., et al. (2003). The yggH gene of Escherichia coli encodes a tRNA (m7G46) methyltransferase. *Journal of Bacteriology*, *185*, 3238–3243.

Dupasquier, M., Kim, S., Halkidis, K., Gamper, H., & Hou, Y. M. (2008). tRNA integrity is a prerequisite for rapid CCA addition: Implication for quality control. *Journal of Molecular Biology*, *379*, 579–588.

Elkins, P. A., Watts, J. M., Zalacain, M., van Thiel, A., Vitazka, P. R., Redlak, M., et al. (2003). Insights into catalysis by a knotted TrmD tRNA methyltransferase. *Journal of Molecular Biology*, *333*, 931–949.

Fechter, P., Rudinger, J., Giege, R., & Theobald-Dietrich, A. (1998). Ribozyme processed tRNA transcripts with unfriendly internal promoter for T7 RNA polymerase: Production and activity. *FEBS Letters*, *436*, 99–103.

Fustin, J. M., Doi, M., Yamaguchi, Y., Hida, H., Nishimura, S., Yoshida, M., et al. (2013). RNA-methylation-dependent RNA processing controls the speed of the circadian clock. *Cell*, *155*, 793–806.

Golovina, A. Y., Sergiev, P. V., Golovin, A. V., Serebryakova, M. V., Demina, I., Govorun, V. M., et al. (2009). The yfiC gene of E. coli encodes an adenine-N6 methyltransferase that specifically modifies A37 of tRNA1Val(cmo5UAC). *RNA*, *15*, 1134–1141.

Goto-Ito, S., Ito, T., Ishii, R., Muto, Y., Bessho, Y., & Yokoyama, S. (2008). Crystal structure of archaeal tRNA(m(1)G37)methyltransferase aTrm5. *Proteins*, *72*, 1274–1289.

Goto-Ito, S., Ito, T., Kuratani, M., Bessho, Y., & Yokoyama, S. (2009). Tertiary structure checkpoint at anticodon loop modification in tRNA functional maturation. *Nature Structural & Molecular Biology*, *16*, 1109–1115.

Hou, Y. M. (2012). High-Purity enzymatic synthesis of site-specifically modified tRNA. In G. L. Conn (Ed.), *Preparation of RNA for biochemical and biophysical analysis: Methods and protocols*. Totowa, New Jersey: Humana Press.

Hou, Y. M., Li, Z., & Gamper, H. (2006). Isolation of a site-specifically modified RNA from an unmodified transcript. *Nucleic Acids Research*, *34*, e21.

Hou, Y. M., & Perona, J. J. (2010). Stereochemical mechanisms of tRNA methyltransferases. *FEBS Letters*, *584*, 278–286.

Jackman, J. E., Montange, R. K., Malik, H. S., & Phizicky, E. M. (2003). Identification of the yeast gene encoding the tRNA m1G methyltransferase responsible for modification at position 9. *RNA*, *9*, 574–585.

Kao, C., Zheng, M., & Rudisser, S. (1999). A simple and efficient method to reduce nontemplated nucleotide addition at the 3 terminus of RNAs transcribed by T7 RNA polymerase. *RNA*, *5*, 1268–1272.

Kimura, S., Miyauchi, K., Ikeuchi, Y., Thiaville, P. C., Crecy-Lagard, V., & Suzuki, T. (2014). Discovery of the beta-barrel-type RNA methyltransferase responsible for N6-methylation of N6-threonylcarbamoyladenosine in tRNAs. *Nucleic Acids Research*, *42*, 9350–9365.

Lahoud, G., Goto-Ito, S., Yoshida, K., Ito, T., Yokoyama, S., & Hou, Y. M. (2011). Differentiating analogous tRNA methyltransferases by fragments of the methyl donor. *RNA*, *17*, 1236–1246.

Liu, C., Betteridge, T., & Hou, Y. M. (2009). Fluorophore labeling to monitor tRNA dynamics. *Methods in Enzymology*, *469*, 69–93.

Liu, C., Gamper, H., Liu, H., Cooperman, B. S., & Hou, Y. M. (2011). Potential for interdependent development of tRNA determinants for aminoacylation and ribosome decoding. *Nature Communications*, *2*, 329.

Liu, C., Gamper, H., Shtivelband, S., Hauenstein, S., Perona, J. J., & Hou, Y. M. (2007). Kinetic quality control of anticodon recognition by a eukaryotic aminoacyl-tRNA synthetase. *Journal of Molecular Biology*, *367*, 1063–1078.

Liu, C., Sanders, J. M., Pascal, J. M., & Hou, Y. M. (2012). Adaptation to tRNA acceptor stem structure by flexible adjustment in the catalytic domain of class I tRNA synthetases. *RNA*, *18*, 213–221.

Markham, G. D., Hafner, E. W., Tabor, C. W., & Tabor, H. (1980). S-Adenosylmethionine synthetase from Escherichia coli. *The Journal of Biological Chemistry*, *255*, 9082–9092.

Masuda, I., Sakaguchi, R., Liu, C., Gamper, H., & Hou, Y. M. (2013). The temperature sensitivity of a mutation in the essential tRNA modification enzyme tRNA methyltransferase D (TrmD). *The Journal of Biological Chemistry*, *288*, 28987–28996.

Mookhtiar, K. A., Peluso, P. S., Muller, D. K., Dunn, J. J., & Coleman, J. E. (1991). Processivity of T7 RNA polymerase requires the C-terminal Phe882-Ala883-COO- or "foot" *Biochemistry*, *30*, 6305–6313.

Nureki, O., Shirouzu, M., Hashimoto, K., Ishitani, R., Terada, T., Tamakoshi, M., et al. (2002). An enzyme with a deep trefoil knot for the active-site architecture. *Acta Crystallographica Section D: Biological Crystallography*, *58*, 1129–1137.

Nureki, O., Watanabe, K., Fukai, S., Ishii, R., Endo, Y., Hori, H., et al. (2004). Deep knot structure for construction of active site and cofactor binding site of tRNA modification enzyme. *Structure*, *12*, 593–602.

Ny, T., & Bjork, G. R. (1980). Cloning and restriction mapping of the trmA gene coding for transfer ribonucleic acid (5-methyluridine)-methyltransferase in Escherichia coli K-12. *Journal of Bacteriology*, *142*, 371–379.

Persson, B. C., Jager, G., & Gustafsson, C. (1997). The spoU gene of Escherichia coli, the fourth gene of the spoT operon, is essential for tRNA (Gm18) 2'-O-methyltransferase activity. *Nucleic Acids Research*, *25*, 4093–4097.

Pham, J. S., Sakaguchi, R., Yeoh, L. M., De Silva, N. S., McFadden, G. I., Hou, Y. M., et al. (2014). A dual-targeted aminoacyl-tRNA synthetase in Plasmodium falciparum charges cytosolic and apicoplast tRNACys. *The Biochemical Journal*, *458*, 513–523.

Purta, E., van Vliet, F., Tkaczuk, K. L., Dunin-Horkawicz, S., Mori, H., Droogmans, L., et al. (2006). The yfhQ gene of Escherichia coli encodes a tRNA:Cm32/Um32 methyltransferase. *BMC Molecular Biology*, *7*, 23.

Roovers, M., Hale, C., Tricot, C., Terns, M. P., Terns, R. M., Grosjean, H., et al. (2006). Formation of the conserved pseudouridine at position 55 in archaeal tRNA. *Nucleic Acids Research*, *34*, 4293–4301.

Sakaguchi, R., Giessing, A., Dai, Q., Lahoud, G., Liutkeviciute, Z., Klimasauskas, S., et al. (2012). Recognition of guanosine by dissimilar tRNA methyltransferases. *RNA*, *18*, 1687–1701.

Sakaguchi, R., Lahoud, G., Christian, T., Gamper, H., & Hou, Y. M. (2014). A divalent metal ion-dependent N(1)-methyl transfer to G37-tRNA. *Chemistry & Biology*, *21*, 1351–1360.

Schubert, H. L., Blumenthal, R. M., & Cheng, X. (2003). Many paths to methyltransfer: A chronicle of convergence. *Trends in Biochemical Sciences*, *28*, 329–335.

Sprinzl, M., Horn, C., Brown, M., Ioudovitch, A., & Steinberg, S. (1998). Compilation of tRNA sequences and sequences of tRNA genes. *Nucleic Acids Research*, *26*, 148–153.

Swinehart, W. E., Henderson, J. C., & Jackman, J. E. (2013). Unexpected expansion of tRNA substrate recognition by the yeast m1G9 methyltransferase Trm10. *RNA*, *19*, 1137–1146.

Tabor, S., Huber, H. E., & Richardson, C. C. (1987). Escherichia coli thioredoxin confers processivity on the DNA polymerase activity of the gene 5 protein of bacteriophage T7. *The Journal of Biological Chemistry*, *262*, 16212–16223.

Taya, Y., & Nishimura, S. (1973). Biosynthesis of 5-methylaminomethyl-2-thiouridylate. I. Isolation of a new tRNA-methylase specific for 5-methylaminomethyl-2-thiouridylate. *Biochemical and Biophysical Research Communications*, *51*, 1062–1068.

Yi, C., & Pan, T. (2011). Cellular dynamics of RNA modification. *Accounts of Chemical Research*, *44*, 1380–1388.

Zhang, C. M., Liu, C., Slater, S., & Hou, Y. M. (2008). Aminoacylation of tRNA with phosphoserine for synthesis of cysteinyl-tRNA(Cys). *Nature Structural & Molecular Biology*, *15*, 507–514.

Zhang, C. M., Perona, J. J., Ryu, K., Francklyn, C., & Hou, Y. M. (2006). Distinct kinetic mechanisms of the two classes of aminoacyl-tRNA synthetases. *Journal of Molecular Biology*, *361*, 300–311.

CHAPTER SIX

Preparation of Human Nuclear RNA m^{6}A Methyltransferases and Demethylases and Biochemical Characterization of Their Catalytic Activity

Jianzhao Liu[*,†,‡], **Yanan Yue**[*,†,‡], **Chuan He**[*,†,‡,¶,1]
[*]Department of Chemistry, The University of Chicago, Chicago, Illinois, USA
[†]Institute for Biophysical Dynamics, The University of Chicago, Chicago, Illinois, USA
[‡]Howard Hughes Medical Institute, The University of Chicago, Chicago, Illinois, USA
[¶]Department of Biochemistry and Molecular Biology, The University of Chicago, Chicago, Illinois, USA
[1]Corresponding author: e-mail address: chuanhe@uchicago.edu

Contents

Methods in Enzymology, Volume 560
ISSN 0076-6879
http://dx.doi.org/10.1016/bs.mie.2015.03.013

Abstract

N^6-Methyladenosine (m^6A) represents the most prevalent internal modification in messenger and long noncoding RNAs. There has been a surge of interest toward understanding the biological significance of m^6A modification. In this chapter, we describe the methods for biochemically studying the recently uncovered m^6A methyltransferases (METTL3 and METTL14) and demethylases (FTO and ALKBH5). How to express these proteins, perform their biochemistry reactions against various RNA probes, and characterize the methylation and demethylation activity will be discussed.

1. INTRODUCTION

Discovered four decades ago, N^6-methyladenosine (m^6A) is the most abundant internal modification in polyadenylated messenger RNA (mRNA) and long noncoding RNA (lncRNA) in eukaryotes (Desrosiers, Friderici, & Rottman, 1974; Fu, Dominissini, Rechavi, & He, 2014). It is conserved in eukaryotes that range from yeast (Agarwala, Blitzblau, Hochwagen, & Fink, 2012; Bodi, Button, Grierson, & Fray, 2010; Clancy, Shambaugh, Timpte, & Bokar, 2002), plants (Zhong et al., 2008), flies (Hongay & Orr-Weaver, 2011) to mammals (Narayan & Rottman, 1988; Schibler, Kelley, & Perry, 1977; Wei, Gershowitz, & Moss, 1976), as well as in virus with a nuclear phase (Beemon & Keith, 1977; Krug, Morgan, & Shatkin, 1976). The modification site of m^6A was demonstrated to be confined in a consensus motif of Pu[G>A] m^6AC[U>A>C] (Pu=purine). Its abundance was estimated to be 0.1–0.4% of that of adenines (that is, ~3–5 m^6A sites per mRNA) in mammals (Harper, Miceli, Roberts, & Manley, 1990; Horowitz, Horowitz, Nilsen, Munns, & Rottman, 1984; Narayan & Rottman, 1988) and ~0.25% in meiotic *Saccharomyces cerevisiae* (Bodi et al., 2010). In higher eukaryotes, the m^6A modification plays critical roles because knockdown or knockout of the m^6A methyltransferase has been shown to result in development arrest (Hongay & Orr-Weaver, 2011; Zhong et al., 2008), cell apoptosis (Bokar, 2005), and inhibition of stem cell differentiation (Batista et al., 2014; Geula et al., 2015). The discovery of an m^6A demethylase FTO (fat mass and obesity-associated protein) in 2011 (Jia et al., 2011) sparked extensive research interest in the roles of RNA modifications in gene expression regulation. The field has been expanding rapidly with the first transcriptome-wide maps of m^6A-modified mRNA obtained (Dominissini et al., 2012; Meyer et al., 2012), and the identification of N^6-adenosine methyltransferases (writers) (Liu et al., 2014), m^6A

demethylase (erasers) (Zheng et al., 2013), and m^6A-binding proteins (readers) (Dominissini et al., 2012; Wang et al., 2014). These emerging new results began to define cellular pathways involving the m^6A modification and indicate possible functions in mRNA splicing, mRNA transport, mRNA stability, translation, immune tolerance, and stress response. In this chapter, we summarize methods for the preparation of human nuclear RNA m^6A methyltransferases and demethylases as well as methods for biochemical characterization of their catalytic activity.

2. EXPRESSION OF HUMAN NUCLEAR RNA m^6A METHYLTRANSFERASE CORE COMPLEX METTL3–METTL14–WTAP IN INSECT CELL EXPRESSION SYSTEM

2.1 Subcloning of METTL3, METTL14, and WTAP

The recombinant proteins METTL3 (GenBank Accession No. NP_062826.2), METTL14 (GenBank Accession No. NP_066012.1), and WTAP (GenBank Accession No. NP_001257460.1) could not be expressed in a bacteria system. We successfully expressed these proteins in insect cells with different N-terminal tags (Flag, His_6, and GST) by using a Bac-to-Bac baculovirus expression system (Invitrogen). Their corresponding cDNAs (Open Biosystems) were cloned into a pFastBac™ dual expression vector (Invitrogen) with choices of suitable restriction enzymes. Meanwhile, two genes can be simultaneously cloned into one expression vector in order to maintain the stoichiometry of the encoded proteins if they form a heterodimeric complex. This strategy works out very well for the expression of the METTL3–METTL14 complex (Liu et al., 2014).

2.2 Generate Recombinant Bacmids

1. Thaw the MAX Efficiency® DH10Bac™ competent cells on ice.
2. Dispense 50 μL of the cells into 15-mL round-bottom polypropylene tubes.
3. Add 1 μL (around 500 ng) of specific gene-containing pFastBac™ plasmid into the cells. Mix well by tapping the side of the tube.
4. Incubate on ice for 30 min.
5. Apply heat shock by partially immersing the tube in a 42 °C water bath for 45 s.
6. Chill on ice for 2 min.
7. Add SOC medium up to 1 mL.

8. Shake at 37 °C with medium agitation at 225 rpm for 4 h.
9. Plate 100 μL of a 1/100 diluted culture onto Bac-to-Bac plates (1 L SOB agar for plates contains: 10 g Agar, 20 g tryptone, 5 g yeast extract, 2 mL 5 *M* NaCl, 2.5 mL 1 *M* KCl, 10 mL 1 *M* $MgCl_2$, 10 mL 1 *M* $MgSO_4$, and remaining water for filling up to 1 L for autoclave. The resultant casting Bac-to-Bac plate contains 50 μg/mL Kanamycin, 7 μg/mL Gentamicin, 10 μg/mL Tetracycline, 100 μg/mL Blue-gal, and 40 μg/mL IPTG).
10. Incubate the plates for 24–48 h in darkness at 37 °C. Colonies are very small and blue colonies may not be discernible before 24 h incubation.
11. Inoculate a single, isolated white colony from the Bac-to-Bac plate into 2 mL LB medium containing 50 μg/mL Kanamycin, 7 μg/mL Gentamicin, and 10 μg/mL Tetracycline. Grow at 37 °C for 24 h shaking at 250 rpm.
12. Transfer 1.5 mL culture into a mcf tube.
13. Spin down and remove the supernatant.
14. Resuspend cells in 0.3 mL of cold P1.
15. Lyse cells by adding 0.3 mL of RT P2. Incubate at RT for 5 min.
16. Neutralize by adding 0.3 mL of P3. Incubate on ice for 5–10 min.
17. Centrifuge at 14,000 rpm for 10 min.
18. Prepare 1.5 mL tubes with 0.8 mL isopropanol.
19. Transfer the supernatant to 0.8 mL isopropanol.
20. Centrifuge at 14,000 rpm for 15 min.
21. Add 0.5 mL of 70% EtOH to the pellet. Invert the tube several times.
22. Centrifuge at 14,000 rpm for 5 min.
23. Remove as much of the supernatant as possible. Air-dry the bacmid pellet at RT for 5–10 min.
24. Redissolve the pellet in 40 μL TE (100 m*M* Tris, pH 8.0; 10 m*M* EDTA pH 8.0).

2.3 Produce Recombinant Baculovirus P0, P1, and P2 in SF9 Cell Line

1. Seed SF9 insect cells by splitting confluent 15-cm plate with 1:12 ratio into 6-well tissue culture plate.
2. Let sit for at least 1 h.
3. Make the following mixes A and B in mcf tubes.
   a. A: 12 μL Bacmid + 100 μL Grace's medium without supplement.
   b. B: 8 μL Cellfectin II + 100 μL Grace's medium without supplement.
4. Combine A and B (mix C).

5. Incubate at RT for 15–30 min.
6. Add 780 μL Grace's medium without supplement to mix C.
7. Wash cells in the 6-well plate with 2 mL Grace's medium without supplement.
8. Overlay mix C and incubate at 27 °C for 5 h.
9. Change medium to Grace's medium with supplement.
10. Harvest virus from ~2 mL supernatant at 72 h posttransfection and label it P0 virus.
11. Make one 15-cm plate of SF9 cell culture ready when the P0 virus is obtained.
12. Drain the medium from the prepared 15-cm plate of SF9 cell with 1 mL left and add 1 mL P0 virus into the plate. Gently shake every 20 min 3 ×, and then add 20 mL supplemented medium to the plate and let it grow for 72 h.
13. Harvest the P1 virus from the supernatant by centrifugation at 1000 rpm for 10 min. Store at 4 °C.
14. Culture SF9 cells in as many plates as planned. Perform the same procedure as above (P0–P1) in order to amplify the baculoviral stock by adding 1 mL P1 virus to each plate for infection. Harvest the P2 virus from the supernatant by centrifugation after 72 h.

2.4 Expression and Purification of Proteins

1. Infect SF9 cells with P2 baculovirus stock in order to express recombinant proteins. To each 15-cm culture plate, 1 mL P2 virus and 20 mL of fresh medium were added followed by incubation at 27 °C for 48 h.
2. Collect cells from 50 plates by using culture medium to flush cells away from the dish bottom and centrifuging at 5000 rpm for 10 min at 4 °C. Wash cells (~10–13 g pellet) with the cell lysis buffer containing 500 m*M* NaCl, 10 m*M* Tris pH 7.4, 5% glycerol, and 1 m*M* DTT. Then, centrifuge and resuspend cells in 40 mL of cell lysis buffer. In this step, cell pellets can be frozen in liquid nitrogen and stored at −80 °C for at least 3 months. All purification steps should be performed on ice or at 4 °C. Here, we show the purification procedure for Flag-tagged proteins.
3. Add 1 × protease cocktail inhibitor (Nacalai) to the cell suspension.
4. Break cells by sonication for 15 min with 30% amplitude, 15 s on, and 15 s off.
5. Centrifuge the cell lysate at 37,000 rpm for 40 min at 4 °C.

6. Add 5 mL of anti-Flag M2 affinity gel (Sigma) to the glass Econo-Column and prewash it with at least 50 mL of cell lysis buffer using gravity.
7. Load the supernatant of the cell lysate to the equilibrated anti-Flag M2 affinity beads. Resuspend and incubate at 4 °C for 2 h using a rotating mixer.
8. Let the cell lysate flow through using gravity.
9. Wash the beads with 5 × 50 mL of cell lysis buffer containing 1 m*M* DTT, 1 m*M* PMSF, 1 μg/mL leupeptin, 1 μg/mL pepstatin, and 2 μg/mL aprotinin using gravity.
10. Elute protein from the beads using 3 × Flag peptide (Sigma). Add 6 mL 0.2 mg/mL 3 × Flag peptide in cell lysis buffer to the beads and incubate for 30 min with a rotating mixer. Collect the eluate by gravity.
11. Repeat step 10 using another 4 mL 0.2 mg/mL 3 × Flag peptide for the second elution.
12. Concentrate the eluate to 2 mL and then apply it to the 120 mL Superdex-200 column with gel filtration running buffer containing 150 m*M* NaCl, 10 m*M* Tris pH 7.4, and 1 m*M* DTT.
13. Collect the peak fraction and concentrate the protein with an Amicon Ultra-4 filter (Millipore) to around 5–10 mg/mL. Run SDS-PAGE in order to characterize the purity of the protein. See Fig. 1 for gel filtration traces and SDS-PAGE images.
14. Add glycerol to the purified protein to 30% (vol/vol) and aliquot. Freeze the aliquots in liquid nitrogen and store at −80 °C.

3. EXPRESSION OF HUMAN NUCLEAR RNA m^6A DEMETHYLASES FTO AND ALKBH5

3.1 Subcloning and Expression of FTO and ALKBH5 in Bacteria

The human FTO gene (GenBank Accession No. NP_001073901.1) was subcloned into pET28a vector (Novagen) with choices of suitable restriction enzymes to generate a His_6-tagged fusion protein. The human ALKBH5 (GenBank Accession No. NP_060228.3) with deletion of the amino-terminal 66 amino acids was subcloned into a pMCSG19 vector with a His_6-tag by ligation-independent cloning strategy (Donnelly et al., 2006). Both FTO and ALKBH5 were successfully expressed in the BL21 (DE3) *Escherichia coli* strain and purified in a soluble form (Jia et al., 2011; Zheng et al., 2013).

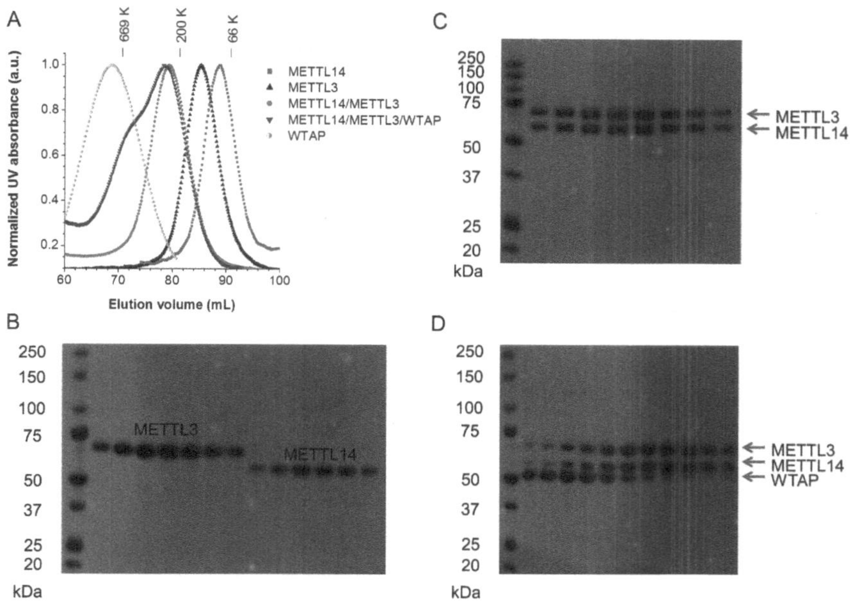

Figure 1 (A) Gel filtration traces of individual Flag-tagged METTL3, METTL14, and WTAP; coexpressed Flag-METTL14 and His_6-METTL3; and mixed Flag-METTL14, Flag-METTL3, and Flag-WTAP with equal molar amounts. All proteins were expressed in insect cells and purified by Flag antibody immunoprecipitation. Markers: 669 kDa (thyroglobulin, bovine), 200 kDa (β-amylase from sweet potato), and 66 kDa (BSA). (B–D) The SDS-PAGE images illustrate purified proteins from gel filtration fractions shown in panel A. The panels B–D correspond to METTL3 and METTL14, METTL3–METTL14 complex, and mixed METTL3/METTL14/WTAP, respectively. *Figure (A) is adapted from figure 1B in Liu et al. (2014), with permission from Nature Publishing Group.*

4. BIOCHEMICAL CHARACTERIZATION OF THE CATALYTIC ACTIVITY OF THE m^{6}A METHYLTRANSFERASES AND DEMETHYLASES

4.1 Methylation Reactions of RNA Probes

A typical 50 μL reaction mixture contains the following components: 0.15 nmol RNA probe, 0.15 nmol each recombinant protein (single METTL3, METTL14, WTAP, or their combinations with a molar ratio of 0.15 nmol/0.15 nmol for two components, 0.8 m*M* d_3-SAM, 80 m*M* KCl, 1.5 m*M* $MgCl_2$, 0.2 U/μL RNasin, 10 m*M* DTT, 4% glycerol, and 15 m*M* HEPES at pH 7.9). The reaction occurs at 16 °C overnight. Prior to the reaction, RNA probes were annealed using a program of: (i) 90 °C for 3 min and (ii) −2 °C/cycle for 40 cycles within 30 min. The reaction was incubated at 16 °C for 12 h or overnight. The resultant RNA was recovered by phenol/chloroform (low pH) extraction followed by ethanol precipitation.

4.2 Demethylation Reactions of RNA/DNA Probes

Here, an example of FTO demethylation reaction is shown. The demethylation assay was performed in 100 μL of the reaction mixture containing 1 nmol RNA/DNA probe with m^6A, FTO, 283 μ*M* of $(NH_4)_2Fe(SO_4)_2 \cdot 6H_2O$, 300 μ*M* of α-KG, 2 m*M* of L-ascorbic acid, 50 μg/mL of BSA, and 50 m*M* of HEPES buffer (pH 7.0). The reaction was incubated for ~3 h at room temperature or overnight at 16 °C and quenched by addition of 5 m*M* EDTA followed by heating for 5 min at 95 °C. The resultant RNA/DNA was recovered by phenol/chloroform (low pH) extraction followed by ethanol precipitation. Alternatively, the whole resultant reaction mixture could be directly subjected to digestion into free nucleosides for further quantification of the reaction yield.

4.3 LC–MS/MS Characterization of m^6A Methylation Activity

The LC–MS/MS method is very useful in order to characterize *in vitro* m^6A methylation reaction yield and the m^6A content in mRNA. Here, we include an example of the quantification of m^6A methyltransferase activity. Typically, 200–300 ng of recovered RNA from biochemistry reaction is digested by nuclease P1 (1–2 U, Wako USA) in 30 μL of buffer containing 25 m*M* NaCl and 2.5 m*M* $ZnCl_2$ at 42 °C for 2 h, followed by additions of NH_4HCO_3 (1 *M*, 3 μL) and alkaline phosphatase (0.5–1 U, Sigma) and incubation at 37 °C for 2 h (Fig. 2). The sample is then filtered (0.22 μm pore size, 4 mm diameter, Millipore), and 5 μL of the solution is injected into LC–MS/MS. The nucleosides are separated by reverse phase ultra-performance liquid chromatography on a C18 column with online mass spectrometry detection using Agilent 6460 QQQ triple-quadrupole LC mass spectrometer in positive electrospray ionization mode. The nucleosides are quantified by using the nucleoside to base ion mass transitions of 285–153 (d_3-m^6A) and 284–152 (G). The amount of reacted probe can be calculated through the newly formed d_3-m^6A, while G serves as an internal control with which the total amount of RNA probe can be calculated.

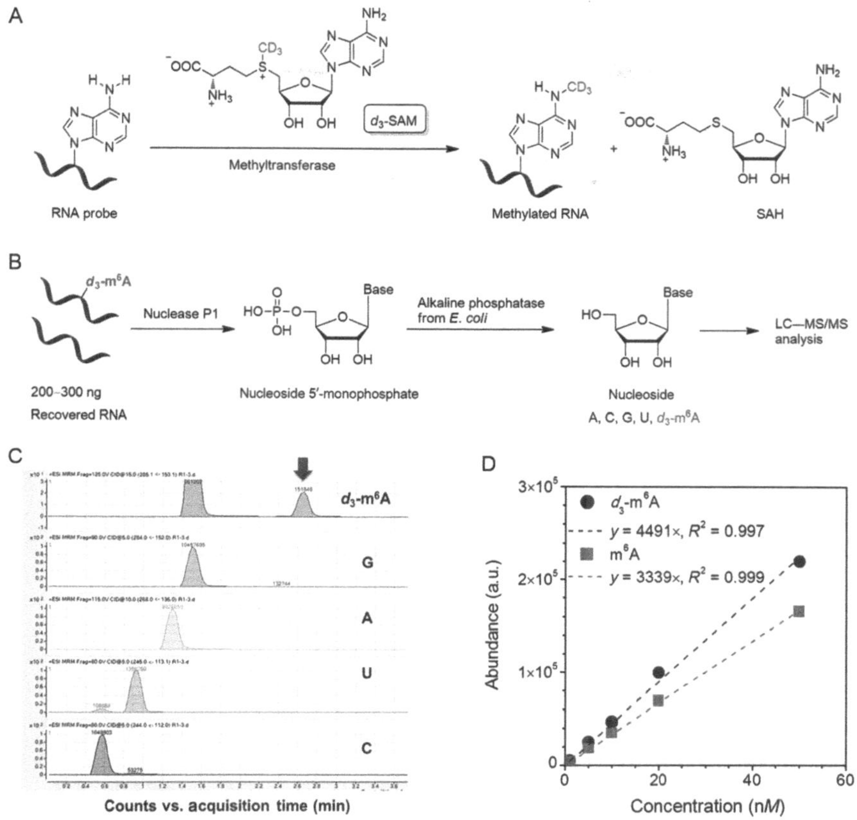

Figure 2 Typical procedure for quantification of m^{6}A methylation reaction yield by LC–MS/MS. (A) Schematic illustration of m^{6}A biochemistry reaction. (B) A flowchart of digestion of RNA for quantitative LC–MS/MS analysis. (C) On a LC–MS/MS system, the amount of each nucleoside was quantified by its integration area in the corresponding chromatogram. (D) The m^{6}A and d_3-m^{6}A standard calibration curves were obtained from five standard samples with different amounts of pure m^{6}A and d_3-m^{6}A, respectively. For *in vitro* biochemistry reaction, the d_3-m^{6}A calibration curve is applied to calculate the methylation reaction yield, while an m^{6}A calibration curve could be used to determine the m^{6}A content from cellular mRNA or other RNAs. *Figure (D) is adapted from supplementary figure 2B in Liu et al. (2014), with permission from Nature Publishing Group.* (See the color plate.)

4.4 HPLC Characterization of m^{6}A Demethylation Activity

The recovered RNA/DNA from the demethylation reactions were digested by nuclease P1 and alkaline phosphatase. The reaction condition is the same as the one indicated for the characterization of m^{6}A methylation activity. Digested free nucleosides were analyzed on a HPLC system equipped with an Agilent Eclipse XDB-C18 analysis column (150 × 4.6 mm) eluted with buffer A (25 m*M* NaH_2PO_4) and buffer B (acetonitrile) with a flow rate of 1 mL/min at room temperature. The detection wavelength was set at 266 nm. The demethylation yield can be estimated from the integration area of the m^6A peak. For more accurate characterization of demethylation yield, QQQ-LC–MS/MS can be used as described above.

4.5 Methyltransferase Activity and Selectivity Toward Different RNA Probes

Biochemical tests of the methyltransferase activity were performed for the individual recombinant protein or their combinations toward various synthetic RNA probes with or without the consensus sequence GGACU and the stem-loop secondary structure. The d_3-SAM (*S*-(5′-adenosyl)-L-methionine-d_3) with a deuterium-substituted methyl group was utilized as the cofactor to avoid potential m^6A contamination by RNA already present in the recombinant protein during the purification process and for accurate mass spectrometry quantification.

The RNA substrate was incubated with the recombinant protein at pH 7.9 and 16 °C overnight, followed by complete digestion to a single nucleoside by nuclease P1 and alkaline phosphatase, and subsequently analyzed by LC–MS/MS (Fig. 2). We used the molar ratio of the product d_3-m^6A to each probe to quantify the methylation efficiency. Results have been reported previously (Liu et al., 2014). We revealed that WTAP showed no methyltransferase activity with all probes tested, which is consistent with its lack of an active methyltransferase domain. Both METTL3 and METTL14 exhibited methyltransferase activity with METTL14 showing much higher *in vitro* activity (close to 10-fold with several probes) than METTL3. A noticeable synergistic effect was observed in that the combinations of METTL3/METTL14 dramatically enhanced the methyltransferase activity. This observation strongly implies that METTL3 and METTL14 interact with each other to enhance methyltransferase activity.

Recent m^6A profiling results revealed that most m^6A modifications in mammalian mRNAs occur in a consensus sequence of (Pu[G>A]m^6AC[U>A>C]). We examined potential sequence preference of METTL3 and METTL14 and found that they preferentially methylate substrates containing the consensus sequence, but exhibited no obvious preference to the stem-loop structure for m^6A deposition. These results support the notion that these methyltransferases have certain sequence specificity but show less structural preference to RNA substrates.

4.6 Demethylase Activity and Selectivity Toward Different RNA/DNA Probes

Several m^6A-containing single-stranded RNAs (ssRNAs) and DNAs (ssDNAs) were synthesized for the characterization of the FTO demethylation activity (Fig. 3). These probes were incubated with 0.2 mol equivalent of FTO for 3 h at room temperature. Afterward, the probes were digested into free nucleosides, which were further separated on a C18 column by HPLC. The HPLC trace profile showed that m^6A in ssRNA or ssDNA was completely converted to adenosine upon FTO treatment. The m^6A-containing double-stranded RNA (dsRNA) and DNA (dsDNA) were also tested and only negligible activity was observed under the same reaction conditions. When the FTO amount was increased to 1 mol equivalent at

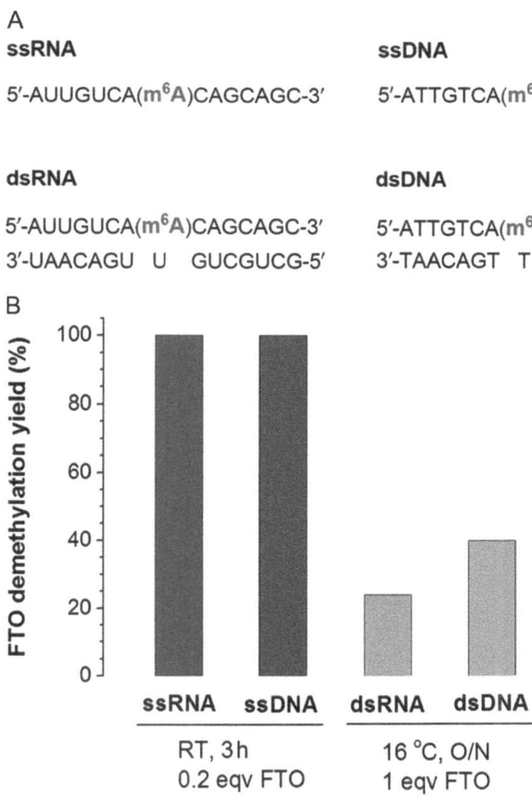

Figure 3 Examples showing demethylation activity of FTO against various RNA/DNA probes. (A) Sequences of RNA/DNA probes. (B) Comparison of the demethylation activity and selectivity among the probes. The demethylation yields were calculated by HPLC method (Jia et al., 2011).

16 °C overnight, demethylation yields of 24% and 40% for dsRNA and dsDNA were obtained, respectively (Fig. 3).

Similar to FTO, ALKBH5 also catalyzes the demethylation of m^6A in ssRNA. As shown in Fig. 4, the probe ssRNA-1 could be completely demethylated by an equal equivalent of ALKBH5 at 16 °C after overnight incubation. When the reaction was performed at room temperature for 2.5 h, the demethylation yield decreased to around 40%. Similar activity for ssDNA was observed. The strong preference of ALKBH5 for single-stranded substrates was shown by an almost unnoticeable demethylation activity toward m^6A in dsRNA (Fig. 4B). Most m^6A modifications in mammalian mRNA exist in a consensus sequence of (Pu[G/A]m^6AC[U/A/C]). Three ssRNAs (ssRNA1–3) with the currently known consensus sequences of (GGm^6ACU)/(CAm^6ACA) and a random sequence of (CUm^6AUU)

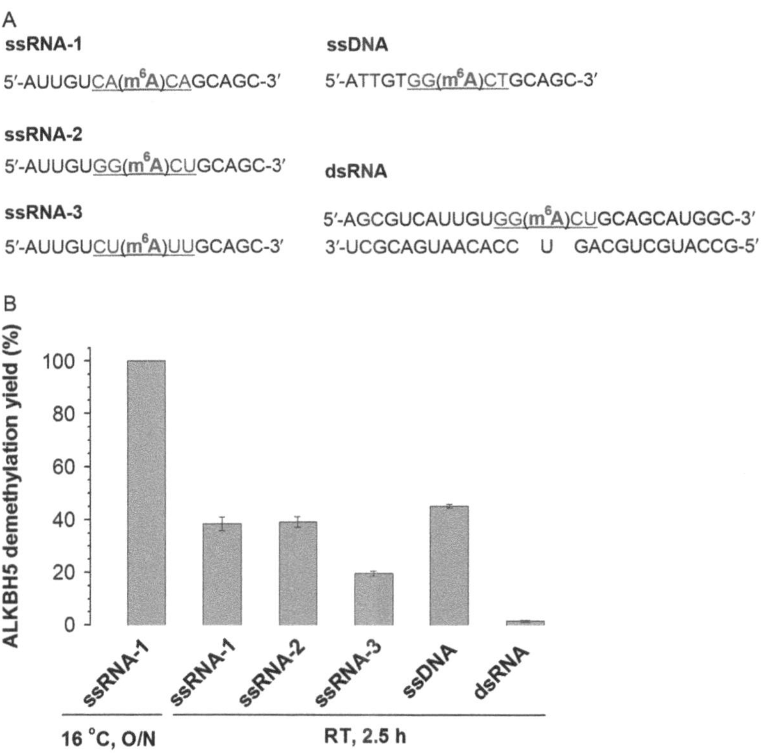

Figure 4 Examples showing demethylation activity of ALKBH5 against various RNA/DNA probes. (A) Sequences of RNA/DNA probes. (B) Comparison of the demethylation activity and selectivity among the probes. Error bars indicate ±SEM ($n=3$). *Figure (B) is adapted from supplementary figure 1F and 1G in Zheng et al. (2013), with permission from Elsevier.*

were synthesized and tested for demethylation activity (Fig. 4). While ALKBH5 showed similar activity toward the two consensus sequences (~40%), a reduced demethylation yield of ~20% was observed toward the nonconsensus sequence under the same conditions, indicating that ALKBH5 may possess a sequence preference in demethylation.

5. CONCLUSIONS

In summary, we have described the preparation of human nuclear RNA m^6A methyltransferases (METTL3 and METTL14) and demethylases (FTO and ALKBH5) as well as procedures to biochemically characterize their catalytic activity. The introduced characterization method could serve as a general strategy for future investigations of other RNA or DNA modification enzymes.

ACKNOWLEDGMENTS

This work was supported by National Institutes of Health (GM071440 to C.H.). S.F. Reichard contributed editing.

REFERENCES

Agarwala, S. D., Blitzblau, H. G., Hochwagen, A., & Fink, G. R. (2012). RNA methylation by the MIS complex regulates a cell fate decision in yeast. *PLoS Genetics*, *8*, e1002732.

Batista, P. J., Molinie, B., Wang, J., Qu, K., Zhang, J., Li, L., et al. (2014). m^6A RNA modification controls cell fate transition in mammalian embryonic stem cells. *Cell Stem Cell*, *15*, 707–719.

Beemon, K., & Keith, J. (1977). Localization of N6-methyladenosine in rous-sarcoma virus genome. *Journal of Molecular Biology*, *113*, 165–179.

Bodi, Z., Button, J. D., Grierson, D., & Fray, R. G. (2010). Yeast targets for mRNA methylation. *Nucleic Acids Research*, *38*, 5327–5335.

Bokar, J. A. (2005). The biosynthesis and functional roles of methylated nucleosides in eukaryotic mRNA. In H. Grosjean (Ed.), *Fine-tuning of RNA functions by modification and editing: Vol. 12* (pp. 141–177). Berlin: Springer-Verlag.

Clancy, M. J., Shambaugh, M. E., Timpte, C. S., & Bokar, J. A. (2002). Induction of sporulation in Saccharomyces cerevisiae leads to the formation of N6-methyladenosine in mRNA: A potential mechanism for the activity of the IME4 gene. *Nucleic Acids Research*, *30*, 4509–4518.

Desrosiers, R., Friderici, K., & Rottman, F. (1974). Identification of methylated nucleosides in messenger RNA from Novikoff hepatoma cells. *Proceedings of the National Academy of Sciences of the United States of America*, *71*, 3971–3975.

Dominissini, D., Moshitch-Moshkovitz, S., Schwartz, S., Salmon-Divon, M., Ungar, L., Osenberg, S., et al. (2012). Topology of the human and mouse m^6A RNA methylomes revealed by m^6A-seq. *Nature*, *485*, 201–206.

Donnelly, M. I., Zhou, M., Millard, C. S., Clancy, S., Stols, L., Eschenfeldt, W. H., et al. (2006). An expression vector tailored for large-scale, high-throughput purification of recombinant proteins. *Protein Expression and Purification*, *47*, 446–454.

Fu, Y., Dominissini, D., Rechavi, G., & He, C. (2014). Gene expression regulation mediated through reversible m6A RNA methylation. *Nature Reviews Genetics*, *15*, 293–306.

Geula, S., Moshitch-Moshkovitz, S., Dominissini, D., Mansour, A. A., Kol, N., Salmon-Divon, M., et al. (2015). m^6A mRNA methylation facilitates resolution of naïve pluripotency toward differentiation. *Science*, *347*, 1002–1006. http://dx.doi.org/10.1126/science.1261417.

Harper, J. E., Miceli, S. M., Roberts, R. J., & Manley, J. L. (1990). Sequence specificity of the human messenger-RNA N6-adenosine methylase in vitro. *Nucleic Acids Research*, *18*, 5735–5741.

Hongay, C. F., & Orr-Weaver, T. L. (2011). Drosophila Inducer of MEiosis 4 (IME4) is required for Notch signaling during oogenesis. *Proceedings of the National Academy of Sciences of the United States of America*, *108*, 14855–14860.

Horowitz, S., Horowitz, A., Nilsen, T. W., Munns, T. W., & Rottman, F. M. (1984). Mapping of N6-methyladenosine residues in bovine prolactin messenger-RNA. *Proceedings of the National Academy of Sciences of the United States of America*, *81*, 5667–5671.

Jia, G., Fu, Y., Zhao, X., Dai, Q., Zheng, G., Yang, Y., et al. (2011). N6-methyladenosine in nuclear RNA is a major substrate of the obesity-associated FTO. *Nature Chemical Biology*, *7*, 885–887.

Krug, R. M., Morgan, M. A., & Shatkin, A. J. (1976). Influenza viral mRNA contains internal N6-methyladenosine and 5′-terminal 7-methylguanosine in cap structures. *Journal of Virology*, *20*, 45–53.

Liu, J., Yue, Y., Han, D., Wang, X., Fu, Y., Zhang, L., et al. (2014). A METTL3-METTL14 complex mediates mammalian nuclear RNA N6-adenosine methylation. *Nature Chemical Biology*, *10*, 93–95.

Meyer, K. D., Saletore, Y., Zumbo, P., Elemento, O., Mason, C. E., & Jaffrey, S. R. (2012). Comprehensive analysis of mRNA methylation reveals enrichment in 3′ UTRs and near stop codons. *Cell*, *149*, 1635–1646.

Narayan, P., & Rottman, F. M. (1988). An in vitro system for accurate methylation of internal adenosine residues in messenger-RNA. *Science*, *242*, 1159–1162.

Schibler, U., Kelley, D. E., & Perry, R. P. (1977). Comparison of methylated sequences in messenger-RNA and heterogeneous nuclear-RNA from mouse L-cells. *Journal of Molecular Biology*, *115*, 695–714.

Wang, X., Lu, Z., Gomez, A., Hon, G. C., Yue, Y., Han, D., et al. (2014). N6-methyladenosine-dependent regulation of messenger RNA stability. *Nature*, *505*, 117–120.

Wei, C. M., Gershowitz, A., & Moss, B. (1976). 5′-Terminal and internal methylated nucleotide-sequences in HeLa-cell messenger-RNA. *Biochemistry*, *15*, 397–401.

Zheng, G., Dahl, J. A., Niu, Y., Fedorcsak, P., Huang, C.-M., Li, C. J., et al. (2013). ALKBH5 is a mammalian RNA demethylase that impacts RNA metabolism and mouse fertility. *Molecular Cell*, *49*, 18–29.

Zhong, S., Li, H., Bodi, Z., Button, J., Vespa, L., Herzog, M., et al. (2008). MTA is an Arabidopsis messenger RNA adenosine methylase and interacts with a homolog of a sex-specific splicing factor. *Plant Cell*, *20*, 1278–1288.

CHAPTER SEVEN

Transcriptome-Wide Mapping of N^6-Methyladenosine by m^{6}A-Seq

Dan Dominissini*,†, Sharon Moshitch-Moshkovitz‡,§, Ninette Amariglio‡,§, Gideon Rechavi‡,§,1

*Department of Chemistry, The University of Chicago, Chicago, Illinois, USA
†Institute for Biophysical Dynamics, The University of Chicago, Chicago, Illinois, USA
‡Cancer Research Center, Chaim Sheba Medical Center, Tel Hashomer, Israel
§Sackler School of Medicine, Tel Aviv University, Tel Aviv, Israel
[1]Corresponding author: e-mail address: gidi.rechavi@sheba.health.gov.il

Contents

Methods in Enzymology, Volume 560
ISSN 0076-6879
http://dx.doi.org/10.1016/bs.mie.2015.03.001

Abstract

A detailed protocol for isolation and sequencing of an enriched population of m^{6}A-methylated RNA fragments to create m^{6}A methylome maps is outlined. Our approach was developed to fill a void that existed because of a lack of methods for the detection of m^{6}A in RNA in an unbiased, high-throughput, and high-resolution manner. This method integrates immunoprecipitation of methylated, randomly fragmented RNA using a highly specific anti-m^{6}A antibody to obtain an enriched population of modified fragments and massively parallel sequencing, resulting in mapping of this modification throughout the transcriptome.

1. THEORY

When the constituents of messenger RNA were analyzed four decades ago, a variant of adenosine, methylated on the exocyclic nitrogen, was noted to be internally positioned and of sizeable abundance (Desrosiers, Friderici, & Rottman, 1974). Since its initial discovery and recognition as the most abundant internally modified nucleotide in RNA, research into N^6-methyladenosine (m^{6}A) progressed rather slowly, dictated by available methods at the time that posed a limit to the study of phenomena in transcripts of relatively low abundance (reviewed in Bokar, 2005). Chemical adducts on nucleotides are lost during *in vitro* PCR amplification and are therefore especially challenging to investigate. Consequently, abundant cellular species, like tRNAs and rRNAs, took center stage when it came to the study of RNA modifications (Cantara et al., 2011).

Lack of chemical derivatization approaches and the marginal effects that m^{6}A has on Watson–Crick base pairing as far as conventional polymerases are considered caused research into m^{6}A to lag behind that of other modified bases, with 5-methylcytosine in DNA (and its bisulfite sequencing) and inosine in RNA (read as guanosine by reverse transcriptases) being the most notable examples (Klose & Bird, 2006; Levanon et al., 2004).

Fitful efforts utilizing laborious techniques have uncovered only a handful of methylated positions in just a few transcripts but nonetheless firmly established—based on genetic manipulation and broad-spectrum methylation inhibitors—that absence of adenosine methylation was incompatible with life on both the organismal and cellular levels (Bokar, 2005). What makes m^{6}A so essential and how is it integrated into cellular metabolism in general and the flow of genetic information in particular? The gap between the plentiful existence of m^{6}A and its unquestionable importance remained unbridged.

As compelling as these initial observations seem today, wide acceptance of the presence and the impact of m^6A on mRNA were late to emerge. The recent introduction in 2012 of a novel method to localize m^6A sites throughout the entire transcriptome in high resolution, termed m^6A-seq, began to narrow this gap and provided a first glance into the function of this modification. It entails immunoprecipitation of RNA fragments using an m^6A-specific antibody followed by deep sequencing analyzed to identify regions of signal enrichment (Dominissini et al., 2012). Immunoprecipitated fragments originating from a single methylated locus may contain m^6A at any position, and when aligned to the genome overlap to form a peak whose midpoint, or summit, corresponds to an underlying methylated adenosine, characterized by the typical consensus sequence DR(m^6A)CH (D = A,G,U; R = G,A; H = A,C,U).

Determining the location of m^6A with single-nucleotide precision is still a challenge, though. DRACH motifs are prevalent and only a fraction of them are methylated *in vivo*. Furthermore, peaks can sometimes be broad or asymmetric, owing to the nonrandom nature of fragmentation or to close clustering of m^6A residues. Consequently, it may be difficult to unequivocally assign m^6A to a specific site. A recent improvement, photo-crosslinking-assisted m^6A-seq, relies on a mutational signature after UV-induced antibody–RNA crosslinking and reverse transcription to produce high-resolution maps that recapitulate the results of classic m^6A-seq and partially meet this shortcoming (Chen et al., 2015).

The second relative blind spot of m^6A-seq is stoichiometric quantification. Various lines of evidence suggest that m^6A is a dynamic modification, the result of active demethylation and changing substoichiometric methylation rates (Jia et al., 2011; Zheng et al., 2013). Positions may seem conserved or "static," however, the fraction of transcripts modified at these positions are subject to dynamic regulation. As m^6A-seq relies on physical enrichment of methylated RNA fragments, the extent of modification in a particular locus cannot be determined in absolute terms, and very strict criteria should be applied even when attempting to determine it in relative terms (Dominissini et al., 2012).

m^6A-seq allowed the first transcriptome-wide glimpse of the human and mouse methylomes that opened up the field to increasingly active investigation. It revealed that methylation nonrandomly punctuates thousands of positions along a large fraction of expressed transcripts. The newly created maps demonstrated that the whole is greater than the sum of its parts: the bird's eye view of the methylome proved pivotal to the identification of the evolutionarily conserved pattern to which m^6A invariably conforms: it is strongly enriched around stop codons, within unusually long internal exons and at transcription start sites.

The full potential of these maps was again realized when m^6A profiles were compared to other transcript attributes and processing events to expose m^6A's control over transcript stability (Wang et al., 2014) and splicing (Dominissini et al., 2012; Zhao et al., 2014), as well as the dynamic nature of its deposition under changing physiological conditions, and most recently its role in sustaining a fine balance essential to maintain pluripotent cells in a state poised to begin differentiation (Batista et al., 2014; Geula et al., 2015). The full potential of these maps has yet to be fully realized.

Here, we outline a protocol for mapping m^6A in RNA. Other similar protocols have been published (Meyer et al., 2012; Schwartz et al., 2013). For details on the bioinformatic analyses, please refer to our recent publication (Dominissini, Moshitch-Moshkovitz, Salmon-Divon, Amariglio, & Rechavi, 2013).

2. EQUIPMENT

- Refrigerated bench top microcentrifuge (capable of at least 16,000 × *g*)
- Weighing scale
- CO_2 incubator for mammalian cell culture
- Magnetic separation rack for 0.2-ml tubes
- NanoDrop spectrophotometer (NanoDrop Technologies, ND-1000, or equivalent)
- Agilent 2100 Bioanalyzer or equivalent
- Head-over-tail rotating wheel
- PCR thermal cycler
- Vortex mixer
- Microspin minicentrifuge

3. MATERIALS

3.1 Reagents and Chemicals

- Appropriate cell culture medium
- Phosphate-buffered saline (Sigma-Aldrich, cat. no. D8537)
- $ZnCl_2$ (Sigma-Aldrich, cat. no. 96468)

- Glycogen (5 mg/ml; Life Technologies, cat. no. AM9510)
- Ribonucleoside vanadyl complexes (RVC; 200 m*M*; Sigma-Aldrich, cat. no. R3380)
- RNasin Plus RNase inhibitor (Promega, cat. no. N2611)
- NaCl (5 *M*; Sigma-Aldrich, cat. no. S6546)
- Igepal CA-630 (Sigma-Aldrich, cat. no. I8896)
- Affinity-purified anti-m^6A rabbit polyclonal antibody (Synaptic Systems, cat. no. 202 003)
- N^6-Methyladenosine, 5′-monophosphate sodium salt (m^6A, Sigma-Aldrich, cat. no. M2780)
- Ethanol, absolute (Sigma-Aldrich, cat. no. E7023)
- Bead-immobilized recombinant protein A (Repligen, cat. no. IPA300)
- BSA (20 mg/ml; Sigma-Aldrich, cat. no. B8667)
- β-Mercaptoethanol (β-ME; Sigma-Aldrich, cat. no. M7522)
- RNaseKiller solution (5 PRIME, cat. no. 2900630)
- Ultrapure water (Biological Industries, cat. no. 01-866-1B)
- Agencourt AMPure XP Beads (Beckman Coulter)
- Sodium acetate (pH 5.2, 3 *M*; Sigma-Aldrich, cat. no. S7899)
- Tris–HCl (pH 7.0, 1 *M*; Sigma-Aldrich, cat. no. T2413)
- Tris–HCl (pH 7.4, 1 *M*; Sigma-Aldrich, cat. no. T2663)
- EDTA (pH 8.0, 0.5 *M*; Sigma-Aldrich, cat. no. 03690)

3.2 Solutions and Buffers Setup

3.2.1 Step 1: RNA Preparation and Fragmentation

Lysis solution supplemented with β-ME

Component	Stock	Amount	Final concentration
β-Mercaptoethanol	14.3 *M*	100 µl	0.143 *M*
PerfectPure RNA Cultured Cell Kit lysis solution	Stock	10 ml	

1 *M* $ZnCl_2$

Component	Amount
$ZnCl_2$	1.363 g
RNase-free water	10 ml

10 × Fragmentation buffer

Component	Stock	Amount	Final concentration
$ZnCl_2$	1 *M*	100 µl	0.1 *M*
Tris–HCl (pH 7.0)	1 *M*	100 µl	0.1 *M*
RNase-free water		800 µl	

75% (v/v) ethanol

Component	Stock	Amount	Final concentration
Ethanol	100%	7.5 ml	75%
RNase-free water		To 10 ml final volume	

Freshly prepare.
Note that 75% ethanol is used also in step 2.

3.2.2 Step 2: RNA Immunoprecipitation

10% Igepal CA-630

Component	Stock	Amount	Final concentration
Igepal CA-630	100%	1 ml	10%
RNase-free water		9 ml	

5 × IP buffer

Component	Stock	Amount	Final concentration
Tris–HCl (pH 7.4)	1 *M*	0.5 ml	0.05 *M*
NaCl	5 *M*	1.5 ml	0.75 *M*
Igepal CA-630	10%	0.5 ml	0.5%
RNase-free water		To 10 ml final volume	

Freshly prepare the buffer.
1 × IP buffer with RNase inhibitors

Component	Stock	Amount	Final concentration
5 × IP buffer	5 ×	1 ml	1 ×
RNasin Plus		50 µl	200–400 U
RVC		50 µl	2 m*M*
RNase-free water		To 5 ml final volume	

0.5 mg/ml anti-m^{6}A antibody stock solution

Component	Amount
Anti-m^{6}A antibody	50 µg
RNase-free water	100 µl

IP blocking buffer

Component	Stock	Amount	Final concentration
5 × IP buffer	5 ×	200 µl	1 ×
BSA	20 mg/ml	25 µl	0.5 mg/ml
RNase-free water		775 µl	

20 m*M* m^{6}A

Component	Amount
m^{6}A	10 mg
RNase-free water	1.3 ml

Elution buffer

Component	Stock	Amount	Final concentration
5 × IP buffer	5 ×	90 µl	1 ×
m^{6}A	20 m*M*	150 µl	6.7 m*M*
RNasin Plus		7 µl	200–400 U
RNase-free water		203 µl	

3.3 Kits

- PerfectPure RNA Cultured Cell Kit (5 PRIME, cat. no. 2302340)
- GenElute mRNA Miniprep Kit (Sigma–Aldrich)
- Ribo-Zero rRNA Removal Kit (Illumina, cat. no. MRZH116)
- NEBNext Ultra RNA Library Prep Kit for Illumina (New England Biolabs, cat. no. NEB #E7530)
- Agilent DNA 100 Kit (Agilent, cat. no. 5067-1504)
- TruSeq SR Cluster Kit v3 (Illumina, cat. no.15037929)
- TruSeq SBS Kit v3-HS (Illumina, cat. no. 15021662)
- TruSeq Multiplex Sequencing Primer Kit (Illumina, cat. no. 15017557)
- Quant-iT RNA Assay Kit (100 assays; Life Technologies, cat. no. Q32852)
- Agilent RNA 6000 Pico Kit (Agilent, cat. no. 5067-1513)
- Quant-iT™ dsDNA High-Sensitivity Assay Kit (Life Technologies, cat no. Q32854)

3.4 Consumables

- Microcentrifuge tubes (1.75 ml; Axygen, cat. no. MCT-175-C)
- Thin-walled PCR tubes with flat cap (0.2 ml; Axygen, cat. no. PCR-02-A)
- 15-ml RNAse-free tubes
- 50-ml RNAse-free tubes
- Pipette filter tips
- Weighing boats
- Cell scraper

4. PROTOCOL

4.1 Preparation

4.1.1 Cells

Expand cells in an appropriate growth medium supplemented with the appropriate selection antibiotics, if needed. It is recommended to lyse cells for RNA purification straight in the culture plate. However, it is possible to harvest the cells and flash-freeze cell pellet at −80 °C until further processing.

4.1.2 RNA

The number of identified m^6A locations is largely influenced from the methylation levels in the corresponding cells/organism, the starting amounts, the nature of the RNA sample, as well as from the sequencing coverage of the library. Since 18S rRNA is methylated (Machnicka et al., 2013), starting from total RNA would result in a high fraction of rRNA reads of the total sequenced reads. Mapping m^{6}A in mRNA would reduce significantly the rRNA reads and allow the use lower RNA quantities. The burden of rRNA can be reduced either by removal of rRNA with Ribo-Zero (Illumina) or Ribominus (Life Technologies), or by mRNA enrichment.

We successfully used a minimum of 250 μg total RNA and 5 μg mRNA (one-round polyA-selected RNA) for mapping thousands of m^{6}A sites and obtaining the global features of the m^{6}A methylome.

4.2 Tip

Note that m^{6}A methylation of a specific position is nonstoichiometric and may occur in only a fraction of transcripts. Thus, although the main m^{6}A peaks and the features of the methylome in terms of its nonrandom distribution and consensus motif can be identified with the minimal amounts indicated above, starting with larger RNA amounts will enable the identification of more peaks (due to their presence at a smaller fraction of transcripts or in lowly expressed transcripts).

4.3 Duration

Preparation	Cell expansion and growth	7 days
Protocol	RNA preparation and immunoprecipitation	3 days
	Library preparation	2 days
	Deep sequencing	6 days

5. STEP 1: RNA PREPARATION AND FRAGMENTATION

5.1 Overview

In this step, RNA is purified and fragmented to generate the starting material for mapping. RNA purification can be carried out using various protocols and reagents. Most RNA purification procedures are suitable for m^{6}A mapping, providing that purified RNA does not contain EDTA or salts that may interfere with RNA fragmentation. Typically, we prefer column-based purification kits for their high yield, RNA quality, and their optional genomic DNA removal step, which may interfere with downstream analysis.

DNA removal is of great significance in organisms known to contain m^6A in their DNA, since it is also recognized by anti-m^6A antibodies. The size of the RNA fragments affects the mapping resolution. Larger RNA fragments reduce the peak resolution.

5.2 Duration

RNA purification	3 h
RNA fragmentation	3 h
Ethanol precipitation	2–16 h (overnight)
Reconstitution and validation of fragmented RNA	2 h

5.3 Procedure

1. Remove growth media completely and wash cells once with cold PBS.
2. Add PerfectPure RNA Cultured Cell Kit lysis solution supplemented with 143 m*M* β-ME directly onto the cells.
3. Scrape lysed cells with a cell scraper, collect and transfer to a 50-ml tube.
4. Vortex cell lysates for 2 min at room temperature or until the sample is homogenous.

   Tip: When purifying RNA from fresh or frozen tissues, samples should be thoroughly homogenized in lysis solution using a homogenizer (either blade- or bead-based homogenizer). Lysates should be then passed four to eight times through a 18–21 gauge syringe needle, to complete tissue lysis and ensure genomic DNA shearing.
5. Purify total RNA using PerfectPure RNA Cultured Cell Kit according to the manufacturer's instructions, including the optional on-column DNase I treatment, to avoid genomic DNA contamination.
6. Elute RNA in RNase-free water. Refrain from using the supplied elution buffers as it may interfere with RNA fragmentation.

   Tip: Elution volume should be as small as possible to avoid working in low RNA concentrations.
7. Measure the concentration of the RNA using a NanoDrop spectrophotometer (Thermo Scientific).
8. Estimate the integrity of the RNA by Agilent 2100 Bioanalyzer or by agarose gel electrophoresis.

Tip: Since subsequent RNA fragmentation may reduce that size of RNA degradation products even further up to the point where they will escape downstream processing and analysis, maintaining the integrity of the RNA is important, especially if comparison between samples is warranted.

9. Optional: Enrich for polyadenylated RNA by at least one round of oligo-dT selection using GenElute mRNA Miniprep Kit (Sigma–Aldrich). Alternatively, rRNA can be depleted by using Ribo-Zero rRNA Removal Kit (Epicetre).

   Tip: To ensure maximum mRNA yield elution volumes of the polyA+/rRNA depleted RNA should be large (according to the manual of the kit used), resulting in low RNA concentration which may affect fragmentation (see below).

10. Adjust RNA concentration to ~0.7–1.1 μg/μl with RNase-free water prior to fragmentation.

    Tip: Using lower RNA concentrations may result in over-fragmentation and generation of smaller fragments. When using mRNA for m^6A mapping, RNA concentrations after enrichment are low and require either concentrating the RNA (by ethanol precipitation) or adjustment of the fragmentation protocol.

11. Preheat a heated lid thermocycler to 94 °C.
12. Assemble the fragmentation mixture in a thin-walled 0.2-ml PCR tube:

Component	**Volume (μl)**	**Final**
RNA (at 1 μg/μl)	18	up to 18 μg
10 × fragmentation buffer	2	1 ×
Total volume	20	

13. Vortex and spin down.
14. Incubate the fragmentation mix at 94 °C for 5 min in the preheated thermal cycler.
15. Stop fragmentation by removing the tubes from block to 4 °C and immediately adding 1/10 volumes of 0.5 *M* EDTA.
16. Vortex and spin down, and place on ice.

    Tip: Working with multiple tubes at the volumes indicated above is highly recommended. Scaling often affects fragmentation efficiency and results a wider size distribution.

Tip: To avoid lengthening of the handling time when working with a large number of tubes, it is recommended to work in batches of five tubes.

17. Ethanol precipitate the fragmentation mix by adding 1/10 volumes of 3 *M* sodium acetate, pH 5.2, glycogen (100 μg/ml final), and 2.5 volumes of 100% ethanol. Mix well and incubate at −80 °C overnight.

Tip: To increase the yield of recovered RNA fragments, it is recommended to use glycogen as a carrier. Using nucleic acids carriers is strongly discouraged, since they may interfere with immunoprecipitation and sequencing later in the protocol. Alternatively, oligo cleanup kits or beads can also be used as long as care is taken to insure compatibility in terms of length cutoffs.

Tip: The precipitation mixture can be stored at −80 °C until further use.

18. Precipitate the fragmented RNA by centrifugation at 15,000 × *g* for 25 min, 4 °C. Discard the supernatant, without disturbing the pellet. Wash the pellet with 1 ml of 75% (v/v) ethanol and centrifuge again at 15,000 × *g* for 15 min, 4 °C.

19. Carefully remove the supernatant and air-dry the pellet.

20. Resuspend the RNA pellet in RNase-free water. Make sure the pellet is completely resuspended.

21. Measure the concentration of fragmented RNA using a NanoDrop spectrophotometer (NanoDrop Technologies) and evaluate the size distribution of the RNA fragments by running an aliquot on a Bioanalyzer (Agilent) with an Agilent RNA 6000 Pico Kit (Agilent) according to the manufacturer's instructions. Alternatively, RNA fragmentation can be validated by running 0.5 μg of fragmented RNA on 1.5–2% (w/v) agarose gel for approximately 30 min.

Tip: If needed, RNA can be stored at −80 °C until further use.

6. STEP 2: RNA IMMUNOPRECIPITATION

6.1 Overview

In this step, fragmented RNA is subjected to immunoprecipitation to obtain a pool of RNA fragments enriched in m^6A methylation. It is important to save a small amount of the original, input, RNA for determining the levels of enrichment in the immunoprecipitated sample.

This step is prone to RNA degradation; thus, it is strongly advised to perform it in the presence of RNase inhibitors. We recommend combining

RVC (Sigma-Aldrich) that inhibit most nucleases with RNasin Plus (Promega), which inhibits eukaryotic RNase A and RNase B.

Since the enrichment is compared to the input, it is highly important to reduce background levels as much as possible. Although the anti-m^6A antibody (Synaptic Systems) is highly specific, preblocking the beads with BSA will reduce nonspecific binding of RNA fragments, resulting in lower background noise. In addition, preferring competitive elution over solvent extraction (Schwartz et al., 2013) would result in a higher signal-to-noise ratio. Controls such as immunoprecipitation with a nonrelevant antibody or a bead-only control yield undetectable levels of competitively eluted RNA.

6.2 Duration

RNA immunoprecipitation	5 h
Competitive elution of RNA fragments	2.5 h
Ethanol precipitation	2–16 h (overnight)
Reconstitution eluted RNA	1 h

6.3 Procedure

1. Keep a small amount of the original fragmented RNA for input control.
2. Adjust the volume of fragmented RNA to 755 μl using RNase-free water.
3. Assemble the following reagents in a 1.7–2.0 ml low-binding microcentrifuge tube:

Reagents	**Volume (μl)**	**Final**
5 × IP buffer	200	1 ×
Anti-m^6A antibody	25	12.5 μg
RVC	10	2 m*M*
RNasin	10	200–400 U
Fragmented RNA	755	Varies (>5 μg mRNA, >300 μg total RNA)
Total volume	1000	

4. Vortex and spin down.

   Tip: It is recommended to set up a bead-only control containing the same immunoprecipitation reagents as indicated above excluding the antibody, to evaluate background levels and elution efficiency.
5. Incubate the immunoprecipitation mixtures and the optional bead-only control for 2 h at 4 °C on a rotating wheel to allow the formation of antibody–RNA complexes.
6. During the incubation, wash 0.2 ml of recombinant protein-A bead slurry (Repligen) in 1 ml of 1 × IP buffer, vortex, and centrifuge for 1 min at 1000 rpm. Repeat wash.
7. Resuspend beads in 1 ml of IP blocking buffer and incubate for 2 h on a rotating wheel.
8. Spin beads down, discard supernatant, and wash twice with 1 ml of 1 × IP buffer.
9. Resuspend beads in 1 ml of 1 × IP buffer supplemented RNasin and RVC and divide beads equally into two microcentrifuge tubes. Spin again to remove the IP buffer.

   Tip: Exceeding the quantity of beads may increase background levels.
10. Transfer the antibody–RNA complexes (formed in 5) into the tubes containing the blocked beads
11. Incubate for 2 h at 4 °C, on a rotating wheel.
12. Centrifuge the beads for 1 min at 1000 rpm.
13. Remove the supernatant carefully without disturbing the bead pellet.
14. Wash the beads with 1 ml of 1 × IP buffer, vortex, and centrifuge for 1 min at 1000 rpm. Discard the supernatant.
15. Repeat the wash for a total of three times.
16. Elute the antibody-bound RNA fragments by resuspending the beads in 0.1 ml of elution buffer.
17. Incubate for 1 h under continuous shaking at 4 °C.
18. Centrifuge the beads for 1 min at 1000 rpm, carefully collect the supernatant.
19. Repeat the elution (as described in 16–18) and collect the supernatant.
20. Ethanol-precipitate the RNA fragments from the eluate by adding 1/10 volumes of 3 *M* sodium acetate, pH 5.2, and 2.5 volumes of absolute ethanol.
21. Vortex and incubate 2 h to overnight at −80 °C.

    Tip: It is important to precipitate the RNA fragments without precipitating free m^6A, as it may interfere with subsequent measurements and enzymatic reactions. Thus it is important to avoid any carrier (i.e., glycogen) in the precipitation mixture. Alternatively, an oligo cleanup kit or beads with a low enough length cutoff can be used to purify eluted RNA fragments.

22. Centrifuge the precipitation mixture at 4 °C for 25 min, 15,000 × *g*.
23. Carefully discard the supernatant without disrupting the pellet at the bottom of the tube which may not be visible.
24. Wash the RNA pellet once with 1 ml of 75% ethanol and centrifuge again at 4 °C for 25 min, 15,000 × *g*.
25. Discard the supernatant and let the pellet air-dry.
26. Resuspend the RNA pellet in 9 μl of RNase-free water.
27. Take 1 μl from for measuring the concentration of your immunoprecipitated RNA using Quant-iT RNA Assay Kit (Life Technologies). Depending on the source of RNA and the amounts used, eluted RNA may be below the threshold of measurement sensitivity. We recommend to continue and prepare the libraries in this case.
28. RNA can be stored at −80 °C until further use.

7. STEP 3: LIBRARY PREPARATION AND DEEP SEQUENCING

7.1 Overview

In this step, the immunoprecipitated RNA fragments that are enriched for m^6A methylation are used as template for library preparation in parallel to a library prepared from an equivalent amount of the original fragmented RNA (which will serve as input control).

7.2 Duration

Library preparation	2 days
Deep sequencing	5 days

7.3 Procedure

1. Take the remaining 8 μl of immunoprecipitated fragmented RNA and comparable amounts of input control to first-strand cDNA synthesis using the NEBNext Ultra RNA Library Prep Kit for Illumina (New England Biolabs) starting from step (B) of the protocol and skip the incubation time at 94 °C, since RNA fragmentation is performed prior to immunoprecipitation step in this protocol.

2. Proceed to second-strand cDNA synthesis, DNA end-repair, adapter ligation, and PCR amplification using the NEBNext Ultra RNA Library Prep Kit for Illumina (New England Biolabs) according to the manufacturer's instructions.
3. Perform the optional size selection in the manual according to the manufacturer's instructions using Agencourt AMPure XP Beads (Beckman Coulter).
4. Measure DNA concentration of the libraries (using 1 μl from each library) with Quant-iT™ dsDNA High-Sensitivity Assay Kit.
5. Assess library quality using Agilent Technologies 2100 Bioanalyzer to validate size range to ensure the absence of adaptor-ligated DNA.

   Tip: If the optional bead-only control sample was performed, this sample should not produce library.
6. In case that adaptor-ligated DNA population is evident in your library perform size selection according to the NEBNext Ultra RNA Library Prep Kit for Illumina (New England Biolabs) manual.
7. Store libraries at −20 °C until sequencing.
8. Perform cluster generation for deep sequencing with Illumina HiSeq platform (or similar NGS machine) using TruSeq SR Cluster kit v3 (Illumina) according to the manufacturer's instructions.
9. Sequence clustered libraries for 50 cycles on Illumina HiSeq platform using the TruSeq SBS Kit v3-HS (Illumina).

   Tip: We advise sequencing three libraries per lane to obtain adequate coverage for peak calling.

ACKNOWLEDGMENTS

D.D. is supported by Human Frontier Science Program Long-Term Fellowship. G.R. is supported by grants from the Flight Attendant Medical Research Institute (FAMRI), Israel Science Foundation (Grant no. 1667/12), by the I-CORE Program and The Israel Science Foundation (Grant Nos. 41/11 and 1796/12), by Teva National Network of Excellence in Neuroscience (NNE), and by the Ernest and Bonnie Beutler Research Program. G.R. is a member of the Sagol Neuroscience Network and holds the Djerassi Chair for Oncology (Sackler Faculty of Medicine, Tel Aviv University, Israel).

REFERENCES

Batista, P. J., Molinie, B., Wang, J., Qu, K., Zhang, J., Li, L., et al. (2014). m(6)A RNA modification controls cell fate transition in mammalian embryonic stem cells. *Cell Stem Cell*, *15*, 707–719.

Bokar, J. (2005). The biosynthesis and functional roles of methylated nucleosides in eukaryotic mRNA. In H. Grosjean (Ed.), *Fine-tuning of RNA functions by modification and editing: Vol. 12* (pp. 141–177). Berlin/Heidelberg: Springer.

Cantara, W. A., Crain, P. F., Rozenski, J., McCloskey, J. A., Harris, K. A., Zhang, X., et al. (2011). The RNA Modification Database, RNAMDB: 2011 update. *Nucleic Acids Research*, *39*, D195–D201.

Chen, K., Lu, Z., Wang, X., Fu, Y., Luo, G. Z., Liu, N., et al. (2015). High-resolution N(6)-methyladenosine (m(6) A) map using photo-crosslinking-assisted m(6) A sequencing. *Angewandte Chemie, International Edition in English*, *54*, 1587–1590.

Desrosiers, R., Friderici, K., & Rottman, F. (1974). Identification of methylated nucleosides in messenger RNA from Novikoff hepatoma cells. *Proceedings of the National Academy of Sciences of the United States of America*, *71*, 3971–3975.

Dominissini, D., Moshitch-Moshkovitz, S., Salmon-Divon, M., Amariglio, N., & Rechavi, G. (2013). Transcriptome-wide mapping of N(6)-methyladenosine by m(6) A-seq based on immunocapturing and massively parallel sequencing. *Nature Protocols*, *8*, 176–189.

Dominissini, D., Moshitch-Moshkovitz, S., Schwartz, S., Salmon-Divon, M., Ungar, L., Osenberg, S., et al. (2012). Topology of the human and mouse m6A RNA methylomes revealed by m6A-seq. *Nature*, *485*, 201–206. http://dx.doi.org/10.1038/nature11112.

Geula, S., Moshitch-Moshkovitz, S., Dominissini, D., Mansour, A. A., Kol, N., Salmon-Divon, M., et al. (2015). m6A mRNA methylation facilitates resolution of naive pluripotency toward differentiation. *Science*, *347*, 1002–1006.

Jia, G., Fu, Y., Zhao, X., Dai, Q., Zheng, G., Yang, Y., et al. (2011). N6-methyladenosine in nuclear RNA is a major substrate of the obesity-associated FTO. *Nature Chemical Biology*, *7*, 885–887. http://dx.doi.org/10.1038/nchembio.1687.

Klose, R. J. A., & Bird, P. (2006). Genomic DNA methylation: The mark and its mediators. *Trends in Biochemical Sciences*, *31*, 89–97.

Levanon, E. Y., Eisenberg, E., Yelin, R., Nemzer, S., Hallegger, M., Shemesh, R., et al. (2004). Systematic identification of abundant A-to-I editing sites in the human transcriptome. *Nature Biotechnology*, *22*, 1001–1005.

Machnicka, M. A., Milanowska, K., Osman Oglou, O., Purta, E., Kurkowska, M., Olchowik, A., et al. (2013). MODOMICS: A database of RNA modification pathways—2013 update. *Nucleic Acids Research*, *41*, D262–D267.

Meyer, K. D., Saletore, Y., Zumbo, P., Elemento, O., Mason, C. E., & Jaffrey, S. R. (2012). Comprehensive analysis of mRNA methylation reveals enrichment in 3' UTRs and near stop codons. *Cell*, *149*, 1635–1646.

Schwartz, S., Agarwala, S. D., Mumbach, M. R., Jovanovic, M., Mertins, P., Shishkin, A., et al. (2013). High-resolution mapping reveals a conserved, widespread, dynamic mRNA methylation program in yeast meiosis. *Cell*, *155*, 1409–1421.

Wang, X., Lu, Z., Gomez, A., Hon, G. C., Yue, Y., Han, D., et al. (2014). N6-methyladenosine-dependent regulation of messenger RNA stability. *Nature*, *505*, 117–120.

Zhao, X., Yang, Y., Sun, B. F., Shi, Y., Yang, X., Xiao, W., et al. (2014). FTO-dependent demethylation of N6-methyladenosine regulates mRNA splicing and is required for adipogenesis. *Cell Research*, *24*, 1403–1419.

Zheng, G., Dahl, J. A., Niu, Y., Fedorcsak, P., Huang, C. M., Li, C. J., et al. (2013). ALKBH5 is a mammalian RNA demethylase that impacts RNA metabolism and mouse fertility. *Molecular Cell*, *49*, 18–29.

CHAPTER EIGHT

Probing RNA Modification Status at Single-Nucleotide Resolution in Total RNA

Nian Liu*, Tao Pan[†,‡,1]

*Department of Chemistry, University of Chicago, Chicago, Illinois, USA
[†]Department of Biochemistry and Molecular Biology, University of Chicago, Chicago, Illinois, USA
[‡]Institute of Biophysical Dynamics, University of Chicago, Chicago, Illinois, USA
[1]Corresponding author: e-mail address: taopan@uchicago.edu

Contents

Abstract

RNA modifications, with over one hundred known so far, are commonly proposed to fine-tune the structure and function of RNA. While modifications in rRNA and tRNA are used to modulate RNA folding and decoding properties, little is known about the function of internal modifications in mRNA/lncRNA, which includes N^6-methyl adenosine (m^6A), 5-methyl cytosine (m^5C), 2′-*O*-methylated nucleotides (Nm), pseudouridine (**Ψ**), and possible others. Functional studies of mRNA/lncRNA modifications have been hindered by the lack of methods for their identification at single-nucleotide resolution. Challenges for the determination of mRNA/lncRNA modifications at single-nucleotide resolution are mainly due to the low abundance of mRNA/lncRNA. Traditional deep sequencing methods cannot identify mRNA/lncRNA modifications, such as m^6A, m^5C, Nm, and **Ψ**, because reverse transcriptase is insensitive to their presence in cDNA synthesis. Antibody-based approach enables the identification of m^6A regions in mRNA/lncRNA, but currently at ~100 nucleotide resolution. Here, we describe a method that accurately identifies m^6A position and modification fraction in human mRNA and lncRNAs at single-nucleotide resolution, termed "Site-specific Cleavage And Radioactive-labeling followed by Ligation-assisted Extraction and Thin-layer chromatography (SCARLET)." This method combines two previously established techniques,

Methods in Enzymology, Volume 560
ISSN 0076-6879
http://dx.doi.org/10.1016/bs.mie.2015.03.005

site-specific cleavage and splint ligation, to probe the RNA modification status at any mRNA/lncRNA site in the total RNA pool. SCARLET can potentially analyze any nucleotide that maintains Watson–Crick base pairing in the transcriptome and determine whether it contains m^6A, m^5C, Nm, Ψ, or other modifications yet to be discovered. Precise determination of the position and modification fraction of RNA modifications reveals crucial parameters for functional investigation of RNA modifications.

1. INTRODUCTION

Discovered in the 1970s, m^6A is the most abundant internal mRNA/lncRNA modification in eukaryotes, present on average in over three sites per mRNA molecule in mammals (Bokar, 2005; Dai et al., 2007; Fu, Dominissini, Rechavi, & He, 2014; Horowitz, Horowitz, Nilsen, Munns, & Rottman, 1984; Liu & Pan, 2015; Pan, 2013). Recently developed m^6A/MeRIP-seq revealed m^6A enrichment within 3′ untranslated region and near the stop codons, and m^6A modifications occur in cell-type- and cell-state-dependent manner (Chen et al., 2014; Dominissini et al., 2012; Meyer et al., 2012; Saletore et al., 2012). So far, two mammalian m^6A methyltransferases (METTL3 and METTL14) (Liu, Yue, et al., 2014; Wang, Lu, et al., 2014) have been discovered to posttranscriptionally install the methyl group on the N^6 position of adenosine within the consensus sequence RRACH (R=A/G; H=A/C/U) (Harper, Miceli, Roberts, & Manley, 1990). Knockdown of the m^6A methyltransferases led to apoptosis in human cells, and significantly, impaired development in other species (Hongay & Orr-Weaver, 2011; Schwartz et al., 2013; Zhong et al., 2008). m^6A modification can be removed by two m^6A demethylases FTO and ALKBH5 (Jia et al., 2011; Zheng et al., 2013). Perturbations of m^6A modifications have been shown to affect cell apoptosis, circadian rhythm, meiosis, and stem cell development (Bokar, 2005; Fustin et al., 2013; Schwartz et al., 2013; Wang, Lu, et al., 2014). Recently, Wang et al. discovered that many m^6A-modified mRNAs are recognized by the cytoplasmic YTHDF2 protein resulting in the regulation of their stability (Wang, Li, et al., 2014). Two independent groups found that *METTL3* knockout mouse and human embryonic stem cells failed to exit from pluripotency toward differentiation (Batista et al., 2014; Geula et al., 2015). Liu et al. found that m^6A can function as a RNA structure remodeler and modulate RNA–protein interactions transcriptome wide (termed m^6A-switch) (Liu et al., 2015). Other suggested functions for m^6A modification, including effects on mRNA splicing, transport, stability, translation, and immune tolerance (Bokar, 2005; Kariko, Buckstein, Ni, & Weissman, 2005), need further exploration. Functional investigation of mRNA/lncRNA modifications requires a method for their identification at single-nucleotide resolution.

In this chapter, we describe a broadly applicable protocol to determine the precise location as well as the modification fraction of mRNA/lncRNA modifications at single-nucleotide resolution without the need to isolate target RNA. This method has been applied to reveal the m^6A status in human mRNA/lncRNAs, revealing important localization and structural implications (Liu et al., 2013). Further, we expect the method to be applicable in determining many other RNA modifications, thus advancing the field of "RNA epigenetics" or "epitranscriptome."

2. METHODS

Our method is termed "Site-specific Cleavage And Radioactive-labeling followed by Ligation-assisted Extraction and Thin-layer chromatography (SCARLET)" (Fig. 1). SCARLET is composed of four main steps: site-specific cleavage at the target nucleotide site, radioactive labeling of the target nucleotide, splint-assisted ligation followed by RNases T1/A digestion, and thin layer chromatography (TLC). Below, we describe the SCARLET method in these four steps in details.

Perform at room temperature (RT) unless specifically indicated. Prepare all solutions using RNase-free water (prepared by autoclaving deionized water). All Chimeric oligos and DNA oligos were ordered from Integrated DNA Technologies (IDT) and gel purified before use.

2.1 Site-Specific Cleavage at the Target Nucleotide Site

The site-specific cleavage method was developed from the previously reported one working on Ψ and Nm RNA modifications from purified/isolated small nuclear RNAs (Ma et al., 2005; Wu, Xiao, Yang, & Yu, 2011; Yu, Shu, & Steitz, 1997). Supposing we are interested in the modification status of the target nucleotide X along the target mRNA Y, the first step of site-specific cleavage at the target nucleotide site is to design chimeric oligos according to the sequence of the target mRNA Y, as previously reported (Liu et al., 2013; Ma et al., 2005; Wu et al., 2011; Yu et al., 1997). The chimeric oligos-guided RNase H cleavage enables site-specific cleavage at the target nucleotide site.

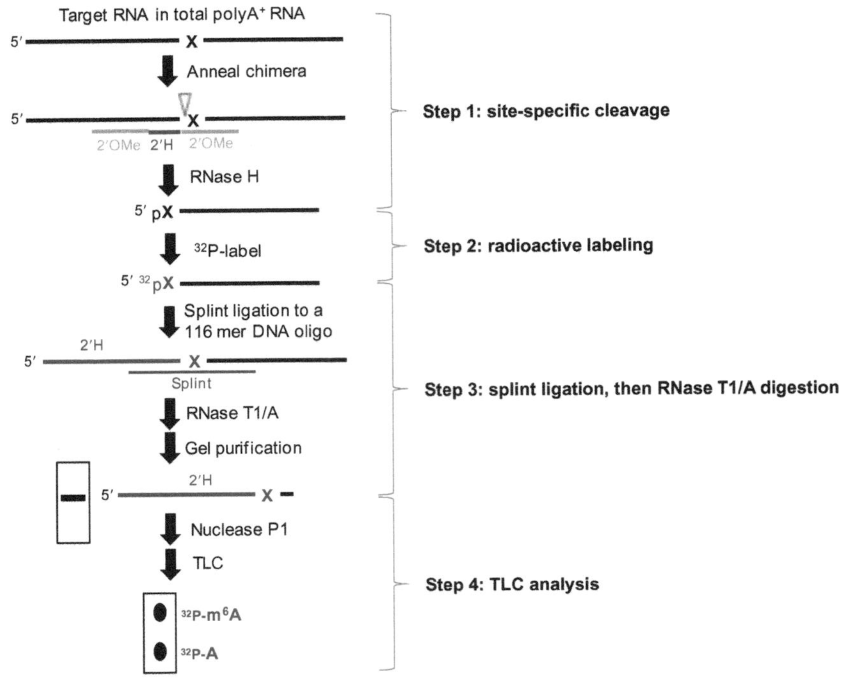

Figure 1 Schematic diagram of SCARLET. SCARLET consists of four steps: site-specific cleavage at the target nucleotide site, radioactive labeling of the target nucleotide, splint-assisted ligation followed by RNases T1/A digestion, and thin layer chromatography (TLC). (See the color plate.)

Detailed steps are as follows:

2.1.1. Isolate total RNA from HeLa cells or other cell lines using PerfectPure RNA cultured cell kit (# 2302340, 5 Prime) according to the manual.

2.1.2. Isolate the polyadenylated RNA (polyA$^+$ RNA) from the total RNA sample obtained in the 2.1.1 step via the GenElute mRNA miniprep kit (# MRN 10, Sigma-Aldrich) according to the manual.

2.1.3. Mix 1 μg polyA$^+$ RNA with (3 pmol) corresponding chimeric oligo in a total volume of 3 μl 30 m*M* Tris–HCl, pH 7.5. Anneal the oligo to RNA by heating at 95 °C for 1 min, followed by incubation at RT for 3 min before putting on ice for the next step.

2.1.4. Add 1 μl 5 × RNase H reaction mixture [2 × T4 polynucleotide kinase buffer (T4 PNK, USB), 1 U/μl RNase H (Epicentre #R0601K)] and 1 μl thermosensitive alkaline phosphatase (1 U/μl, TAP, Thermo-Scientific) to the annealed RNA sample. Incubate at 44 °C for 1 h for site-specific cleavage and dephosphorylation at the 5′-end of nucleic acids.

2.1.5. Terminate the reaction by heating the reaction mixture at 75 °C for 5 min followed by immediate incubation on ice to inactivate RNase H and TAP.

2.2 Radioactive Labeling of the Target Nucleotide

After the site-specific cleavage, the target nucleotide is at the 5′-end of the cut target mRNA Y with a phosphate group attached. In the dephosphorylation step of 2.1.4, the phosphate group is removed, and the target nucleotide is now present as a hydroxyl group.

For identification and quantification, we then label the target nucleotide with the radioactive phosphate through the T4 PNK catalyzed phosphorylation using γ-^{32}P-ATP (BLU002Z/NEG002Z, PerkinElmer).

Detailed steps are as follows:

2.2.1. Add 1 μl 6 × T4 PNK reaction mixture [1 × T4 PNK buffer, 6 U/μl T4 PNK (# 70031-X, USB), 28 μCi/μl [γ-^{32}P] ATP] to the mixture from the 2.1.5 step.

2.2.2. Incubate at 37 °C for 1 h.

2.2.3. Terminate the reaction by heating the mixture at 75 °C for 5 min followed by immediate incubation on ice to inactive the T4 PNK.

2.3 Splint-Assisted Ligation Followed by RNase T1/A Digestion

After Sections 2.1 and 2.2, the target nucleotide has been radioactively labeled. However, numerous other RNAs in this mixture have also been radioactively labeled. Identification of the modification status at the target nucleotide site requires removing all radioactive signals other than the radioactively labeled target sites. To do that, we perform the splint ligation followed by RNase T1/A digestion. We first ligate the cleaved target mRNA Y to a long single-stranded DNA (ssDNA) oligo (116 mer in our protocol), thus the radioactively labeled target nucleotide can be specifically connected to the ssDNA oligo. Therefore, the following RNase T1/A digestion can degrade all other labeled RNA into small pieces except for the short RNA region containing the radioactively labeled target nucleotide, which is covalently linked to the ssDNA oligo. In other words, the ssDNA oligo connects the radioactively labeled target nucleotide to its 3′-end and protects the target nucleotide from RNase T1/A digestion, which removes all other radioactively labeled nucleotides. Then, through denaturing gel purification, the ssDNA oligo with the radioactive target nucleotide can be isolated for subsequent steps.

It is noteworthy that the splint-ligation step was derived from previous established protocols (Maroney, Chamnongpol, Souret, & Nilsen, 2008). And the sequence design of splint oligos and 116 mer ssDNA oligos (ssDNA-116) was as previously reported (Liu et al., 2013; Maroney et al., 2008).

Detailed steps are as follows:

2.3.1. Add to the reaction mixture from above 1.5 μl the splint/ssDNA-116 oligo mixture (4 pmol splint oligos and 5 pmol ssDNA-116 oligos), mix well.

2.3.2. Anneal the RNA samples, splint oligos and ssDNA-116 oligos by heating the mixture at 75 °C for 3 min, followed by incubation at RT for 3 min before putting on ice for the next step.

2.3.3. Add 2.5 μl 4× ligation mixture [1.4× T4 PNK buffer, 0.27 m*M* ATP, 57% DMSO, 1.9 U/μl T4 DNA ligase (# EL0011, Thermo-Scientific)].

2.3.4. Incubate at 37 °C for 3.5 h for the splint ligation.

2.3.5. Terminate the reaction by mixing the reaction samples with equal volume of 2× RNA loading buffer (9 *M* urea, 100 m*M* EDTA, xylene, and bromophenol dyes).

2.3.6. Add 1 μl RNase T1/A mixture [160 U/μl RNase T1 (# EN0541, Thermo-Scientific) and 0.16 mg/ml RNase A (R6513-50MG, Sigma-Aldrich) in distilled water], mix well.

2.3.7. Incubate at 37 °C overnight (~16 h) to ensure complete RNase digestion.

2.3.8. Spin down the reaction mixture, load all samples to one preran 10% urea denaturing PAGE gels. Run the bromophenol dye to the bottom.

2.3.9. Disassembly the gel electrophoresis equipment, wrap the gel with plastic film, and expose the gel to one blanked phosphorimager screen. To get clear phosphorimaging figures with visible target bands, the exposure time varies from 10 to 30 min, depending on the radioactive signal strength (Fig. 2).

2.3.10. Visualize and print the Phosphorimager figure in the actual size. Put the printed figures under the gel for localization of the target bands. Cut the bands with flame-sterilized blades from the gel and transfer the cut gel slices into a clean 1.5-ml plastic tube. Add 0.4 ml crush–soak buffer (50 m*M* potassium acetate, 200 m*M* KCl, pH 7.0) to the tube, invert and rotate the tube for 4 h at RT.

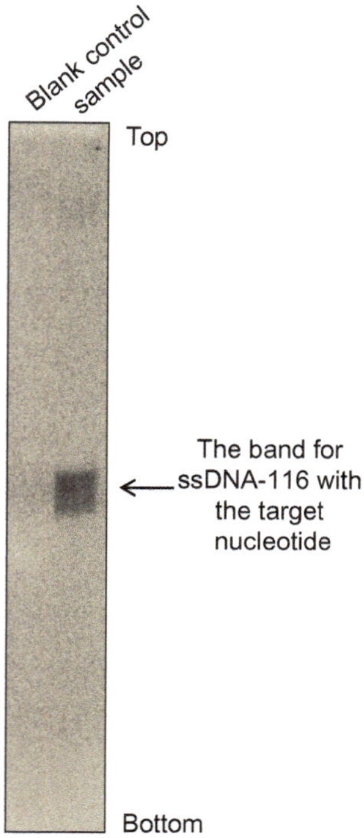

Figure 2 An example of radiolabeled product at the end of RNases T1/A digestion. Denaturing PAGE shows the band for ssDNA-116 oligos with ligated radioactive target nucleotide.

2.3.11. Transfer the crush–soak buffer containing the target oligos out to new tubes, add 2.7 × volume of pure ethanol. Mix well, freeze at −20 °C for at least 1 h, and then ethanol precipitate the oligos. Vacuum or air dry the pellet. Expect good signals for the next step when the radioactive signal is detectable by the Geiger counter.

2.4 TLC Reveals the Modification Status

By now, the alcohol-precipitated nucleic acids contain the radioactive signals, which belong to the target nucleotide. We then digest the nucleic acids into mononucleosides using nuclease P1, and run all the mononucleosides on a TLC plate. Since all radioactive signals on the TLC plates are labeled with the target nucleotides, we can obtain the following information from the TLC result: whether the target nucleotide is actually modified and what fraction the target nucleotide is modified.

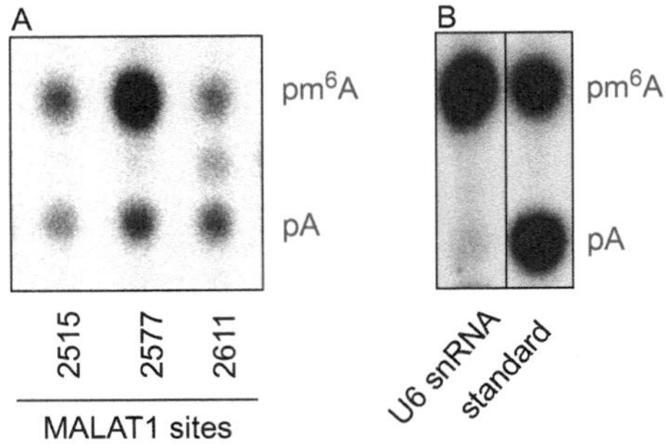

Figure 3 Examples of SCARLET results. (A) TLC result showing the modification status of the 2515-A, 2577-A, and 2611-A sites on the MALAT1 lncRNA (NR_002819). The methylation fraction varies from 40% to 85% among these three sites in HFF-1 cells. (B) TLC result showing the modification status of a known m^6A site in the HeLa U6 small nuclear RNA (snRNA). The methylation fraction is above 95%.

Detailed steps are as follows:

2.4.1. Resuspend and dissolve the alcohol-precipitated RNA pellet with 3 μl nuclease P1 mixture [0.33 U/μl nuclease P1 (N8630-1VL, Sigma-Aldrich) in 30 m*M* sodium acetate/acetic acid, pH 4.8].

2.4.2. Incubate at 37 °C for 2 h to allow complete digestion.

2.4.3. Spot the reaction mix 0.5–1 μl at a time on one clean TLC cellulose plastic sheet (20 × 20 cm; 1.05577.0001, Merck) as previously described (Liu et al., 2013). If multiple spotting is needed, wait for the TLC plate to dry completely before spotting the next time and ensure spotting at the same position on the TLC plate.

2.4.4. Develop the TLC plate in the developing tank (# 416180-0000, Thomas-Scientific) with 100 ml running buffer [isopropanol:HCl: water (70:15:15, v/v/v)]. This process takes ~14 h.

2.4.5. After that, dry the TLC plate at RT for 1 h, wrap the plate in plastic film, and expose it to a blanked phosphorimager screen. The exposure time varies from 1 to 20 h, depending on the strength of radioactive signals on the TLC plate.

2.4.6. Visualize and quantify the TLC result through phosphorimagering to get the modification status of the target nucleotide (Fig. 3).

3. NOTES

3.1. We normally perform ~10 SCARLET experiments in parallel to analyze 10 potential m^6A sites at once.

3.2. It is advisable to add a mixture of unmodified and m^6A-modified synthetic oligos as internal control, since synthetic oligos normally ensure significant signal of m^6A on the electrophoresis gel and TLC result, providing valuable information whether everything is going right and the correct position of the bands or dots containing the target nucleotides (Fig. 3B).

3.3. The m^6A RNA modification fraction in mRNA/lncRNA normally ranges between 5% and 98%. We set the threshold at 5%, so any modification signal with fraction less than 5% is considered to be background noise or unmodified.

3.4. RNA modifications on abundant RNA, such as ribosomal RNA and small nuclear RNA, normally generate strong SCARLET signal (Fig. 3B; Liu et al., 2013). RNA modifications on abundant mRNA/lncRNA tend to have stronger SCARLET signal and lower background noise on the TLC plate (Fig. 3A).

3.5. Since each SCARLET experiment works on only one candidate site, careful work on selecting candidate sites are needed to ensure successful detection of m^6A RNA modification. Potential candidate m^6A sites are evaluated through m^6A/MeRIP-seq results, RNA abundance through previous RNA-seq data or else, presence of RRACH consensus motif (Harper et al., 1990; Liu et al., 2013), structural motif embedded, species-conserved level, and so on.

ACKNOWLEDGMENTS

We thank Dr. Q. Dai for assistance with chemical synthesis, Dr. M. Parisien for bioinformatic analysis to choose potential target nucleotide sites in mRNA/lncRNA. This work was supported by a NIH grant (GM088599 to T.P. and C.H.).

REFERENCES

Batista, P. J., Molinie, B., Wang, J., Qu, K., Zhang, J., Li, L., et al. (2014). m^6A RNA modification controls cell fate transition in mammalian embryonic stem cells. *Cell Stem Cell*, *15*, 707–719.

Bokar, J. A. (2005). The biosynthesis and functional roles of methylated nucleosides in eukaryotic mRNA. In H. Grosjean (Ed.), *Fine-tuning of RNA functions by modification and editing* (pp. 141–178). Berlin, Heidelberg, New York: Springer-Verlag.

Chen, K., Lu, Z., Wang, X., Fu, Y., Luo, G. Z., Liu, N., et al. (2014). High-resolution N^6-methyladenosine (m^6A) map using photo-crosslinking-assisted m^6A sequencing. *Angewandte Chemie*, *53*, 1–5.

Dai, Q., Fong, R., Saikia, M., Stephenson, D., Yu, Y. T., Pan, T., et al. (2007). Identification of recognition residues for ligation-based detection and quantitation of pseudouridine and N^6-methyladenosine. *Nucleic Acids Research*, *35*, 6322–6329.

Dominissini, D., Moshitch-Moshkovitz, S., Schwartz, S., Salmon-Divon, M., Ungar, L., Osenberg, S., et al. (2012). Topology of the human and mouse m^6A RNA methylomes revealed by m^6A-seq. *Nature*, *485*, 201–206.

Fu, Y., Dominissini, D., Rechavi, G., & He, C. (2014). Gene expression regulation mediated through reversible m^6A RNA methylation. *Nature Reviews Genetics*, *15*, 293–306.

Fustin, J. M., Doi, M., Yamaguchi, Y., Hida, H., Nishimura, S., Yoshida, M., et al. (2013). RNA-methylation-dependent RNA processing controls the speed of the circadian clock. *Cell*, *155*, 793–806.

Geula, S., Moshitch-Moshkovitz, S., Dominissini, D., Mansour, A. A., Kol, N., Salmon-Divon, M., et al. (2015). m^6A mRNA methylation facilitates resolution of naïve pluripotency toward differentiation. *Science*, *347*, 1002–1006. http://dx.doi.org/10.1126/science.1261417.

Harper, J. E., Miceli, S. M., Roberts, R. J., & Manley, J. L. (1990). Sequence specificity of the human mRNA N^6-adenosine methylase in vitro. *Nucleic Acids Research*, *18*, 5735–5741.

Hongay, C. F., & Orr-Weaver, T. L. (2011). Drosophila Inducer of MEiosis 4 (IME4) is required for Notch signaling during oogenesis. *Proceedings of the National Academy of Sciences of the United States of America*, *108*, 14855–14860.

Horowitz, S., Horowitz, A., Nilsen, T. W., Munns, T. W., & Rottman, F. M. (1984). Mapping of N^6-methyladenosine residues in bovine prolactin mRNA. *Proceedings of the National Academy of Sciences of the United States of America*, *81*, 5667–5671.

Jia, G., Fu, Y., Zhao, X., Dai, Q., Zheng, G., Yang, Y., et al. (2011). N^6-methyladenosine in nuclear RNA is a major substrate of the obesity-associated FTO. *Nature Chemical Biology*, *7*, 885–887.

Kariko, K., Buckstein, M., Ni, H., & Weissman, D. (2005). Suppression of RNA recognition by Toll-like receptors: The impact of nucleoside modification and the evolutionary origin of RNA. *Immunity*, *23*, 165–175.

Liu, N., Dai, Q., Zheng, G., He, C., Parisien, M., & Pan, T. (2015). N^6-methyladenosine-dependent RNA structural switches regulate RNA-protein interactions. *Nature*, *518*, 560–564.

Liu, N., & Pan, T. (2015). RNA epigenetics. *Translational Research: The Journal of Laboratory and Clinical Medicine*, *165*, 28–35.

Liu, N., Parisien, M., Dai, Q., Zheng, G., He, C., & Pan, T. (2013). Probing N^6-methyladenosine RNA modification status at single nucleotide resolution in mRNA and long noncoding RNA. *RNA*, *19*, 1848–1856.

Liu, J., Yue, Y., Han, D., Wang, X., Fu, Y., Zhang, L., et al. (2014). A METTL3-METTL14 complex mediates mammalian nuclear RNA N^6-adenosine methylation. *Nature Chemical Biology*, *10*, 93–95.

Ma, X., Yang, C., Alexandrov, A., Grayhack, E. J., Behm-Ansmant, I., & Yu, Y. T. (2005). Pseudouridylation of yeast U2 snRNA is catalyzed by either an RNA-guided or RNA-independent mechanism. *The EMBO Journal*, *13*, 2403–2413.

Maroney, P. A., Chamnongpol, S., Souret, F., & Nilsen, T. W. (2008). Direct detection of small RNAs using splinted ligation. *Nature Protocols*, *3*, 279–287.

Meyer, K. D., Saletore, Y., Zumbo, P., Elemento, O., Mason, C. E., & Jaffrey, S. R. (2012). Comprehensive analysis of mRNA methylation reveals enrichment in 3′ UTRs and near stop codons. *Cell*, *149*, 1635–1646.

Pan, T. (2013). N^6-methyl-adenosine modification in messenger and long non-coding RNA. *Trends in Biochemical Sciences*, *38*, 204–209.

Saletore, Y., Meyer, K., Korlach, J., Vilfan, I. D., Jaffrey, S., & Mason, C. E. (2012). The birth of the Epitranscriptome: Deciphering the function of RNA modifications. *Genome Biology*, *13*, 175.

Schwartz, S., Agarwala, S. D., Mumbach, M. R., Jovanovic, M., Mertins, P., Shishkin, A., et al. (2013). High-resolution mapping reveals a conserved, widespread, dynamic mRNA methylation program in yeast meiosis. *Cell*, *155*, 1409–1421.

Wang, Y., Li, Y., Toth, J. I., Petroski, M. D., Zhang, Z., & Zhao, J. C. (2014). N^6-methyladenosine modification destabilizes developmental regulators in embryonic stem cells. *Nature Cell Biology*, *16*, 191–198.

Wang, X., Lu, Z., Gomez, A., Hon, G. C., Yue, Y., Han, D., et al. (2014). N^6-methyladenosine-dependent regulation of messenger RNA stability. *Nature*, *505*, 117–120.

Wu, G., Xiao, M., Yang, C., & Yu, Y. T. (2011). U2 snRNA is inducibly pseudouridylated at novel sites by Pus7p and snR81 RNP. *The EMBO Journal*, *30*, 79–89.

Yu, Y. T., Shu, M. D., & Steitz, J. A. (1997). A new method for detecting sites of 2′-*O*-methylation in RNA molecules. *RNA*, *3*, 324–331.

Zheng, G., Dahl, J. A., Niu, Y., Fedorcsak, P., Huang, C. M., Li, C. J., et al. (2013). ALKBH5 is a mammalian RNA demethylase that impacts RNA metabolism and mouse fertility. *Molecular Cell*, *49*, 18–29.

Zhong, S., Li, H., Bodi, Z., Button, J., Vespa, L., Herzog, M., et al. (2008). MTA is an Arabidopsis messenger RNA adenosine methylase and interacts with a homolog of a sex-specific splicing factor. *Plant Cell*, *20*, 1278–1288.

CHAPTER NINE

High-Resolution Mapping of N^6-Methyladenosine in Transcriptome and Genome Using a Photo-Crosslinking-Assisted Strategy

Kai Chen, Guan-Zheng Luo, Chuan He[1]
Department of Chemistry, Department of Biochemistry and Molecular Biology, Institute for Biophysical Dynamics, The University of Chicago, Chicago, Illinois, USA
Howard Hughes Medical Institute, The University of Chicago, Chicago, Illinois, USA
[1]Corresponding author: e-mail address: chuanhe@uchicago.edu

Contents

Methods in Enzymology, Volume 560
ISSN 0076-6879
http://dx.doi.org/10.1016/bs.mie.2015.03.012

Abstract

A photo-crosslinking-assisted strategy has been developed and applied to the high-resolution mapping of N^6-methylation of adenosine (N^6-methyladenosine or m^{6}A/6mA, we chose to use m^{6}A to represent the RNA methylation, and 6mA to represent the DNA methylation in order to be consistent with DNA 5mC methylation nomenclature) in transcriptome and genome. The new approach introduces a covalent interaction between the anti-m^{6}A antibody and the RNA/DNA molecule, followed by multiple washing steps which reduce nonspecific binding in immunoprecipitation and nuclease digestion in order to significantly increase the resolution by removing unprotected polynucleotides. By using this protocol, a high-resolution transcriptome-wide human m^{6}A map and genome-wide *Chlamydomonas* 6mA map were obtained, providing new insights into the distribution and biological functions of the m^{6}A/6mA. This well-established strategy is reproducible and widely applicable to other biological systems for high-throughput sequencing investigations of other RNA/DNA modifications.

1. INTRODUCTION

Chemical modifications of nucleosides in RNA and DNA have been studied for decades. Among these modifications, methylation is the most abundant and plays important roles in multiple biological functions in both RNA- and DNA-related pathways in prokaryotes and eukaryotes (Cedar, 1988; Desrosiers, Friderici, & Rottman, 1974, 1975; Fu, Dominissini, Rechavi, & He, 2014; Jia, Fu, & He, 2013; Robertson, 2005; Sasaki & Matsui, 2008).

RNA modifications could be reversible and may possess epigenetic features that directly impact gene expression (He, 2010). Particularly, N^6-methyladenosine (m^{6}A) is a widely present modification found within eukaryotic messenger RNA and various nuclear noncoding RNAs (Desrosiers et al., 1974, 1975; Fu et al., 2014; Jia et al., 2013; Meyer & Jaffrey, 2014; Pan, 2013). Recent discoveries have revealed that methylation of adenosine in mRNA is a dynamic and reversible process. The m^{6}A formation in the nucleus is catalyzed by a complex that contains methyltransferase-like 3 (METTL3), methyltransferase-like 14 (METTL14), and Wilms' tumor 1-associating protein (WTAP) (Bokar, 2005; Bokar, Rath-Shambaugh, Ludwiczak, Narayan, & Rottman, 1994; Bokar, Shambaugh, Polayes, Matera, & Rottman, 1997; Liu et al., 2014; Ping et al., 2014; Wang, Li, et al., 2014). Two human AlkB family proteins, the fat mass- and obesity-associated protein (FTO) and ALKBH5, serve as RNA demethylases which reverse m^6A methylation in mammalian poly(A)-tailed RNA (Jia et al., 2011; Zheng et al., 2013). The "reader" protein YTHDF2, a protein specifically binds to m^6A and is shown to interact with thousands of mRNA targets, mediates a methylation-dependent mRNA decay of its substrates, demonstrating a significant role of methylation in mRNA metabolism (Dominissini et al., 2012; Wang, Lu, et al., 2014). The latest work on the *Mettl3*-knockout mouse embryonic stem cells has revealed critical roles of m^6A in stem cell differentiation (Batista et al., 2014; Geula et al., 2015). The formation of m^6A on mRNA could also be affected by microRNA (Chen, Hao, et al., 2015).

The same methylation of adenosine, 6mA, is also present in the genomic DNA of viruses, bacteria, protists, fungi, algae, plants, and mosquitoes (Ratel, Ravanat, Berger, & Wion, 2006). In bacteria, 6mA is a crucial marker that plays regulatory roles in DNA repair, replication, and cell defense (Campbell & Kleckner, 1990; Collier, McAdams, & Shapiro, 2007; Low, Weyand, & Mahan, 2001; Lu, Campbell, Boye, & Kleckner, 1994; Messer & Noyer-Weidner, 1988; Ogden, Pratt, & Schaechter, 1988). Another important feature of bacterial 6mA is that 6mA serves as a component in restriction–modification (R–M) systems, in which methylation of adenosine is recognized by restriction enzyme partners in order to ensure digestion of unmethylated foreign DNA (Murray, 2002). The deletion of 6mA methyltransferase in pathogenic *Escherichia coli* results in global transcription changes, indicating the likelihood of 6mA not only serving as a host genetic marker but also possessing regulatory functions (Fang et al., 2012). Besides bacteria, certain unicellular eukaryotes also have relatively abundant 6mA in their genomes, for instance, the protozoan *Tetrahymena* (Harrison, Findly, & Karrer, 1986; Hattman, Kenny, Berger, & Pratt, 1978), *Oxytricha fallax* (Rae & Spear, 1978), *Paramecium aurelia* (Cummings, Tait, & Goddard, 1974), and green algae *Chlamydomonas reinhardtii* (Hattman et al., 1978). However, the lack of methylation-sensitive restriction endonucleases suggests that 6mA possesses specific regulatory functions in these unicellular eukaryotes (Ehrlich & Zhang, 1990). Moreover, the existence of 6mA in plants, insects, and mammals has also been uncovered, indicating that 6mA is likely an epigenetic mark in eukaryotes (Ratel et al., 2006).

Knowledge of the precise location and distribution of m^6A/6mA is essential for understanding exact biological functions. The progress of high-throughput sequencing technology facilitates the transcriptome-wide and/or genome-wide identification and detection of specific base derivatives. The recently developed high-throughput sequencing approach to localize m^6A sites in mammalian transcriptomes, termed m^6A-seq (or MeRIP-seq, m^6A-specific methylated RNA immunoprecipitation coupled with next-generation sequencing), has provided insightful distribution information on m^6A. This approach employs specific anti-m^6A antibody and immunoprecipitation of m^6A-containing RNA fragments followed by deep sequencing (Dominissini, Moshitch-Moshkovitz, Salmon-Divon, Amariglio, & Rechavi, 2013; Dominissini et al., 2012; Meyer et al., 2012). This modification was shown to accumulate at 3′-UTR around stop codons and within exons and localize in well-known consensus sequence R_1R_2-m^6A-CH (R=G or A; H=A or C or U; R_2: G>A) as previously reported (Canaani, Kahana, Lavi, & Groner, 1979; Dimock & Stoltzfus, 1977; Dominissini et al., 2012, 2013; Harper, Miceli, Roberts, & Manley, 1990; Horowitz, Horowitz, Nilsen, Munns, & Rottman, 1984; Kane & Beemon, 1985; Meyer et al., 2012; Schibler, Kelley, & Perry, 1977; Wei & Moss, 1977). On the other hand, the gene-specific detection of 6mA in bacterial genome conducted using methylation-sensitive restriction enzyme digestion has been the approach widely used for decades (Roberts & Halford, 1993; Roberts, Vincze, Posfai, & Macelis, 2010). More recently, the genome-wide maps of 6mA in pathogenic *E. coli* and other bacteria strains have been reported by using single-molecule real-time (SMRT) sequencing, which could uncover strand-specific and quantitative information for methylation sites and sequence motifs recognized by the methyltransferase (Fang et al., 2012; Murray et al., 2012).

Although high-throughput approaches have provided insights into the properties of m^6A/6mA in both the transcriptome and the genome, these methods can be improved to achieve higher resolution. For m^6A-seq, the resolution of these maps is around 200 nt and therefore may not differentiate single m^6A sites in RNA fragments (Dominissini et al., 2012; Meyer et al., 2012). A yeast m^6A methylome map reached a higher resolution by using an improved m^6A-seq approach, which involved fragmenting the mRNA into shorter pieces (Schwartz et al., 2013). However, the precise identification of yeast m^6A sites was still based on consensus sequences under Immunoprecipitation (IP) peaks. A ligation-based detection method named SCARLET (site-specific cleavage and radioactive labeling followed by ligation-assisted extraction and TLC), developed in order to detect methylation sites with single-nucleotide resolution, is precise but time consuming, and not yet compatible with the requirement of high-throughput applications (Dai et al., 2007; Liu et al., 2013).

SMRT sequencing is able to achieve single-nucleoside resolution; however, the relatively low signal-to-noise ratio and the relatively high error rate and cost make this technique too expensive and time consuming when dealing with eukaryotic genomic DNA. Hence, Fu et al. applied the m^6A-seq immunoprecipitation principle to *Chlamydomonas* genomic DNA (6mA-IP), thus generating the first 6mA map in a eukaryotic system. The high-quality dataset sheds light on the distribution, localization, and the potential biological function of 6mA. Yet similar to problems with m^6A-seq, 6mA-IP is unable to afford higher resolution assignments of 6mA sites, which are critical in order to reveal detailed features of 6mA distribution in genomic DNA.

The strategy of covalent crosslinking between protein and polynucleotide followed by immunoprecipitation and nuclease digestion has been widely employed in order to significantly increase high-throughput sequencing resolution and reduce nonspecific binding. Recently, two cross-linking-based approaches, photoactivatable ribonucleoside-enhanced cross-linking and immunoprecipitation (PAR-CLIP) to study the binding behavior of RNA-binding proteins, and chromatin immunoprecipitation (ChIP) combined with lambda exonuclease digestion followed by high-throughput sequencing (ChIP-exo) to identify genomic locations of DNA-binding proteins at near single-nucleotide accuracy, were developed and have been broadly utilized (Hafner et al., 2010; Rhee & Pugh, 2011). Inspired by these two methodologies, we have designed and developed two similar approaches, named photo-crosslinking-assisted m^6A sequencing (PA-m^6A-seq) (Fig. 1; Chen, Lu, et al., 2015) and 6mA cross-linking and immunoprecipitation followed by exonuclease digestion of unprotected DNA and high-throughput sequencing (6mA-CLIP-exo) (Fig. 2), for transcriptome-wide and genome-wide mapping of m^6A and 6mA, respectively. These methods achieve a near single-nucleotide resolution for mammalian transcriptome-wide m^6A mapping and *Chlamydomonas* genome-wide 6mA mapping, respectively.

Here, we outline detailed protocols for PA-m^6A-seq and 6mA-CLIP-exo. High-throughput sequencing data analyses are mentioned briefly; for details, please refer to recent publications (Chen, Lu, et al., 2015; Fu et al., 2015).

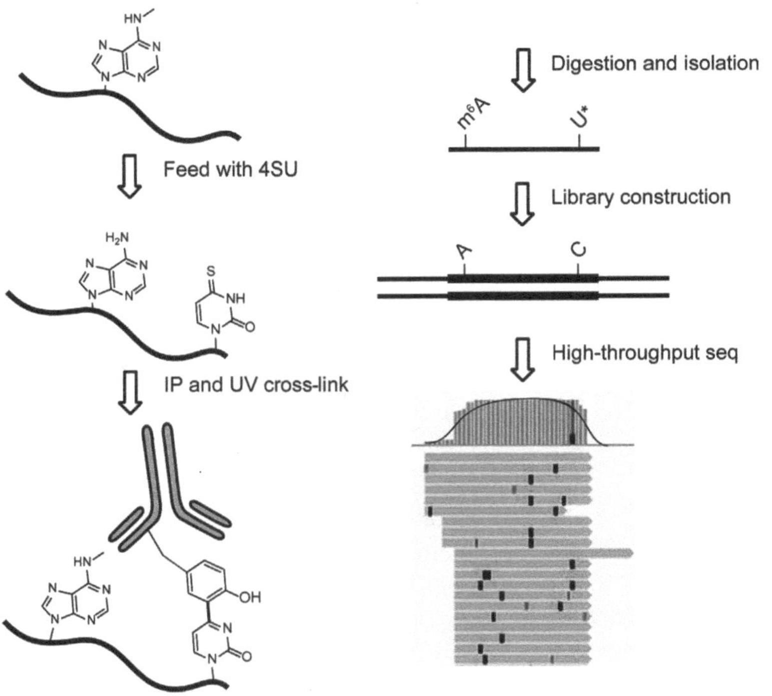

Figure 1 Scheme of PA-m^{6}A-seq. *Adapted from Scheme 1 in Chen, Lu, et al. (2015).*

2. EQUIPMENT

- Refrigerated bench-top microcentrifuge (capable of at least 16,000 × *g*)
- Weighing scale
- CO_2 incubator for mammalian cell culture
- Magnetic separation rack for 0.2 mL tubes
- Magnetic separation rack for 1.7 mL tubes
- Qubit fluorometer (Life Technologies, 2.0 system or equivalent)
- Agilent 2100 Bioanalyzer or equivalent
- Head-over-tail rotating wheel
- PCR thermal cycler
- Vortex mixer
- Microspin minicentrifuge
- UV Stratalinker 2400 (Stratagene, UV 254 nm, or equivalent)
- CL-1000 Ultraviolet Crosslinker (UVP, UV 365 nm, or equivalent)
- Multi-therm Heat Shake (Benchmark, or equivalent)
- Bioruptor pico (Diagenode, or equivalent)

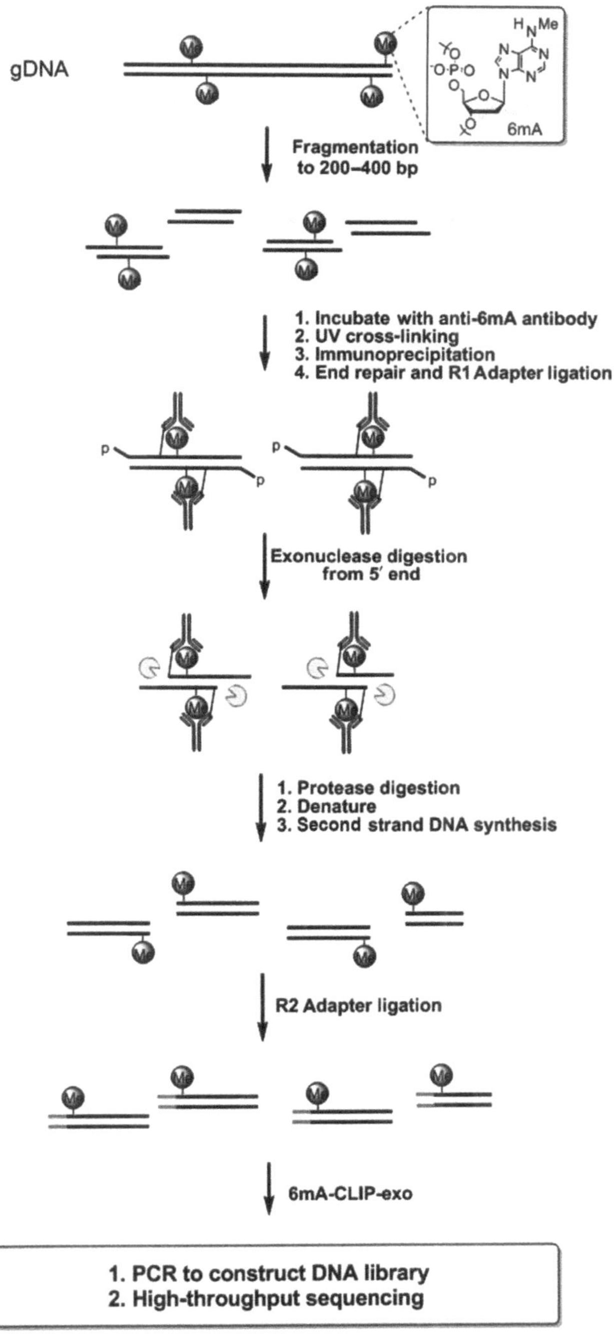

Figure 2 Scheme of 6mA-CLIP-exo. *Adapted from Figure 2 in Fu et al., 2015.* (See the color plate.)

3. MATERIALS

3.1 Reagents, Chemicals, and Enzymes

- Appropriate cell and algae culture medium
- 4-Thiouridine (Sigma-Aldrich)
- Glycogen (5 mg/mL; Life Technologies)
- RNasin Plus RNase inhibitor (Promega)
- NaCl (5 *M*; Sigma-Aldrich)
- Igepal CA-630 (Sigma-Aldrich)
- Affinity-purified anti-m^6A rabbit polyclonal antibody (Synaptic Systems)
- 200 proof Ethanol (Decon Labs)
- Dynabeads® Protein A (Life Technologies)
- BSA (20 mg/mL; Sigma-Aldrich)
- RNaseKiller solution (5 PRIME)
- Sterile water for RNA work (Fisher Scientific)
- Agencourt AMPure XP beads (Beckman Coulter)
- Sodium acetate (pH 5.2, 3 *M*; Ambion)
- HEPES (GenScript)
- Tris (Sigma-Aldrich)
- UltraPure EDTA (pH 8.0, 0.5 *M*; Life Technologies)
- Lithium chloride solution (8 *M*; Sigma-Aldrich)
- Triton X-100 (Fisher Scientific)
- Potassium chloride (Fisher Scientific)
- Magnesium chloride hexahydrate (Sigma-Aldrich)
- Sodium deoxycholate (Sigma-Aldrich)
- SDS (Fisher Scientific)
- 25:24:1 phenol/chloroform/isoamyl alcohol (Sigma-Aldrich)
- Acid:phenol:chloroform, pH 4.5 (with isoamyl alcohol (IAA), 125:24:1, Ambion)
- dNTP mix (10 m*M*/each, NEB)
- RNase T1 (1000 U/μL, Thermo Fisher)
- Antarctic phosphatase (5 U/μL, NEB)
- Proteinase K (~20 mg/mL, Thermo Fisher)
- T4 DNA polymerase (3 U/μL, NEB)
- T4 polynucleotide kinase (10 U/μL, NEB)
- T4 DNA ligase (500 U/μL, NEB)
- Phi29 DNA polymerase (10 U/μL, NEB)
- Lambda exonuclease (5 U/μL, NEB)
- $RecJ_f$ exonuclease (30 U/μL, NEB)

- Phusion® High-Fidelity PCR Master Mix with HF Buffer (2×, NEB)
- Ethylene glycol-bis(β-aminoethyl ether)-*N,N,N′,N′*-tetraacetic acid tetrasodium salt (Sigma-Aldrich)
- TRIzol reagent (Ambion)
- Chloroform (Fisher Scientific)
- Isopropanol (Fisher Scientific)

3.2 Solutions and Buffers Setup

3.2.1 Protocol 1: Photo-Crosslinking-Assisted m6A Sequencing

- 5× IP Buffer

Component	Final Concentration
Tris–HCl (pH 7.4)	50 m*M*
NaCl	750 m*M*
Igepal CA-630	0.5% (v/v)

- 4-Thiouridine stock solution (1 *M* in DMSO)
- IP Wash Buffer

Component	Final Concentration
HEPES–KOH (pH 7.4)	50 m*M*
KCl	300 m*M*
Igepal CA-630	0.05% (v/v)

- High Salt Wash Buffer

Component	Final Concentration
HEPES–KOH (pH 7.4)	50 m*M*
KCl	500 m*M*
Igepal CA-630	0.05% (v/v)

- Phosphatase Wash Buffer

Component	Final Concentration
Tris–HCl (pH 7.4)	50 m*M*
EGTA	20 m*M*
Igepal CA-630	0.5%

- T4 PNK Wash Buffer

Component	Final Concentration
Tris–HCl (pH 7.4)	50 m*M*
NaCl	50 m*M*
$MgCl_2$	10 m*M*

- 2 × Proteinase K Buffer

Component	Final Concentration
Tris–HCl (pH 7.4)	100 m*M*
NaCl	150 m*M*
EDTA	12.5 m*M*
SDS	2% (w/v)

3.2.2 Protocol 2: 6mA-CLIP-exo Sequencing

- 5 × IP Buffer

Component	Final Concentration
Tris–HCl (pH 7.4)	50 m*M*
NaCl	750 m*M*
Igepal CA-630	0.5% (v/v)

- Exo High Salt Wash Buffer

Component	Final Concentration
HEPES–KOH (pH 7.4)	50 m*M*
NaCl	1 *M*
EDTA	2 m*M*
Triton X-100	1% (v/v)
Sodium deoxycholate	0.1% (w/v)

- Exo Wash Buffer 2

Component	Final Concentration
HEPES–KOH (pH 7.4)	50 m*M*
NaCl	0.5 *M*
EDTA	2 m*M*
Igepal CA-630	1% (v/v)
Sodium deoxycholate	0.1% (w/v)

- Exo Wash Buffer 3

Component	Final Concentration
Tris–HCl (pH 7.9)	10 m*M*
LiCl	25 m*M*
EDTA	2 m*M*
Triton X-100	1% (v/v)
Sodium deoxycholate	1% (w/v)

- Tris-EDTA (TE) Buffer

Component	Final Concentration
Tris–HCl (pH 8.0)	10 m*M*
EDTA	1 m*M*

- ChIP Elution Buffer

Component	Final Concentration
Tris–HCl (pH 8.0)	25 m*M*
NaCl	200 m*M*
SDS	0.5% (w/v)

3.3 Kits

- FastTrack MAG mRNA Isolation Kit (Life Technologies)
- NEBNext Ultra RNA Library Prep Kit for Illumina (NEB)
- Quick-gDNA MicroPrep Kit (Zymo Research)
- RNA Clean & Concentrator Kit (Zymo Research)
- QIAquick Gel Extraction Kit (QIAGEN)
- TruSeq Small RNA Sample Prep Kit (Illumina)

3.4 Oligonucleotides

- *R1 Adapter Upper:*
  5′ OH-**TGGAATTCTCGGGTGCC**-OH 3′
- *R1 Adapter Lower:*
  5′ Phos-**CCTTGGCACCCGAGAATTCCA**-OH 3′
- *R1 Primer Extension:*
  5′ OH-**GCCTTGGCACCCGAGAATTCCA**-OH 3′
- *R2 Adapter Upper:*
  5′ OH-**GTTCAGAGTTCTACAGTCCGACGATC**-OH 3′
- *R2 Adapter Lower:*
  5′ OH-**GATCGTCGGACTGTAGAACTCTGAACCACCA**-OH 3′

  Note: Oligonucleotides are compatible with Illumina TruSeq Small RNA Sample Prep Kit and Illumina high-throughput sequencer. All can be ordered from IDT (Integrated DNA Technologies, Inc.) or Eurofins MWG Operon.

3.5 Consumables

- Low-adhesive microcentrifuge tubes
- Thin-walled PCR tubes with flat cap
- 15 mL RNAse free tubes
- 50 mL RNAse free tubes
- Pipette filter tips
- Weighing boats
- Cell scraper
- 96-well cell culture plate (free of DNase and RNase)

4. PROTOCOL 1: PHOTO-CROSSLINKING-ASSISTED m^6A SEQUENCING

4.1 Preparation of 4-Thiouridine-Incorporated polyA$^+$ RNA

1. Culture cells in an appropriate growth medium supplemented with appropriate supplies, if needed.
2. When the confluence of cells is around 80%, add 1 *M* 4-Thiouridine DMSO stock to cell culture medium directly to a final concentration of 200 μ*M*. Shake the plates gently in order to expand the chemical.
3. Incubate for 16 h in a cell incubator.

   Note: 4-Thiouridine is light sensitive. Avoid strong light source.
4. Add 5 mL of TRIzol reagent directly to the cells after removing culture media. Homogenize adhesive cells in TRIzol reagent and pool the lysate in a 50-mL tube.

   Note: Collected TRIzol cell lysate can be kept for at least 1 month at −80 °C.
5. Extract total RNA following the TRIzol reagent protocol by adding chloroform for phase separation and then adding isopropanol for RNA precipitation.
6. Dissolve total RNA in RNase-free water. Isolate polyA$^+$ RNA from total RNA by using FastTrack mRNA isolation kit.

   Note: The yield of mRNA from HeLa total RNA is around 2–3%. If the yield is too high, there might be rRNA contamination.

4.2 Anti-m^6A Immunoprecipitation and UV Crosslinking

7. Set up 600 μL anti-m^6A immunoprecipitation reaction as following:

4-Thiouridine incorporated polyA$^+$ RNA	12 μg
Anti-m^6A polyclonal antibody	10 μg (0.5 mg/mL)
5 × IP Buffer	120 μL
RNase inhibitor (40 ×)	15 μL
RNase-free H_2O	to 600 μL

8. Mix thoroughly but gently. Transfer to head-over-tail rotating wheel to incubate for 2 h at 4 °C.
9. During the reaction incubation, pre-block Dynabeads protein A magnetic slurry with 1 × IP Buffer supplement with 0.5 mg/mL BSA and 1 U/μL RNase inhibitor.
10. Split the reaction mix into 8–10 wells in 96-well cell culture plate. Keep the plate on ice and expose in UV 365 nm three times with energy dosage of 0.15 J/cm^2.

    Note: Position the plate close to UV light source in order to increase the cross-linking efficiency. Expose three times with a 30-s interval.

4.3 Enzymatic Treatment on Beads

11. Pool the cross-linked fractions together, add RNase T1 to a final concentration of 0.2 U/μL for the first round RNase T1 digestion. Incubate the reaction at 22 °C for 10 min. Quench the reaction on ice for 5 min.

    Note: The final concentration of RNase T1 depends on the concentration of polyA^+ RNA. If the concentration of RNA changes, adjust the concentration of RNase T1 accordingly.
12. Transfer each reaction to 80 μL pre-blocked protein A slurry, mix gently, and then incubate on head-over-tail rotating wheel for 1.5 h at 4 °C.
13. Spin down the beads briefly and then place the tube on magnetic rack for 30 s until beads are captured. Discard the supernatant and wash the beads with 500 μL ice-cold IP Wash Buffer three times by gentle vortex.

    Note: Transferring the protein A beads to a new low-adhesive RNase-free tube is recommended in order to reduce nonspecific binding.
14. Resuspend the beads in one volume of IP Wash Buffer and then conduct second round RNase-T1 digestion with a final concentration of 20 U/μ L. Incubate for 15 min at 22 °C. Quench the reaction on ice for 5 min.
15. Spin down the beads briefly and then place the tube on magnetic rack for 30 s until beads are captured. Discard the supernatant and wash the beads with 500 μL ice-cold High Salt Wash Buffer by gentle vortex.

    Note: Transferring the protein A beads to a new low-adhesive RNase-free tube is recommended in order to reduce nonspecific binding.

16. Resuspend the beads in one volume of 1 × Antarctic Phosphatase Buffer (NEB) containing 1 U/μL Antarctic phosphatase. Incubate for 20 min at 37 °C to dephosphorylate the 5′-terminal of the RNA fragments.
17. Spin down the beads briefly and then place the tube on a magnetic rack for 30 s until beads are captured. Discard the supernatant and wash the beads with 500 μL ice-cold Phosphatase Wash Buffer twice.

    Note: Transferring the protein A beads to a new low-adhesive RNase-free tube is recommended in order to reduce nonspecific binding.
18. Repeat Step 17 with 500 μL ice-cold PNK Wash Buffer instead of Phosphatase Wash Buffer.

    Note: Transferring the protein A beads to a new low-adhesive RNase-free tube is recommended in order to reduce nonspecific binding.
19. Resuspend the beads in one volume of 1 × T4 PNK Buffer (NEB) containing 1 U/μL T4 PNK. Incubate for 30 min at 37 °C first to remove 3′-terminal phosphor group, add ATP to a final concentration of 1 m*M* and incubate for another 20 min at 37 °C.
20. Spin down the beads briefly and then place the tube on a magnetic rack for 30 s until beads are captured. Discard the supernatant and wash the beads with 500 μL ice-cold PNK Wash Buffer three times.

    Note: Transferring the protein A beads to a new low-adhesive RNase-free tube is recommended in order to reduce nonspecific binding.
21. Resuspend the beads in one volume of 1 × Proteinase K Buffer containing 2 mg/mL Proteinase K. Incubate for 30 min at 55 °C.
22. Spin down the beads briefly and then place the tube on magnetic rack for 30 sec until beads are captured. Transfer the supernatant (which contains RNA fragments) in a new tube.
23. Repeat Steps 21 and 22 once more and combine the supernatant together.

4.4 RNA Isolation and Purification

24. Use equal volume of phenol:chloroform to extract RNA fragment in supernatant from Step 23.
25. Wash the aqueous phase once more with equal volume of chloroform.

26. Add 1/10 volume of 3 *M* Sodium Acetate (pH 5.3) and 4 μL Glycogen. Then precipitate RNA fragments with 2.5 volume of cold pure ethanol. Store the mixture at −80 °C overnight.
Note: Pure ethanol is prechilled at −80 °C.

27. Spin down the RNA fragments, wash the pellet with 75% (v/v) ethanol, and dissolve the pellet into 50 μL RNase-free H_2O. Clean up by Zymo Research RNA Clean & Concentrator and elute in 10 μL RNase-free H_2O.

4.5 Library Construction, High-Throughput Sequencing and Data Analysis

28. Apply RNA from Step 27 to small RNA library construction. The library is purified by agarose gel size selection and extraction.

29. Library quality control by Bioanalyzer and qPCR.

30. Align single-end 50 bp reads to reference genome by Bowtie (Langmead, Trapnell, Pop, & Salzberg, 2009). Call peaks by using PARalyzer (Corcoran et al., 2011).

5. PROTOCOL 2: 6mA-CLIP-EXO SEQUENCING

5.1 Preparation of *Chlamydomonas* Genomic DNA

1. Culture the algae under constant light to exponential growth stage in Erlenmeyer flask in 22 °C shaker.

2. Collect the algae and extract genomic DNA (gDNA) by Zymo Quick-gDNA MicroPrep Kit

3. Sonicate gDNA into ~150 bp fragment by using Bioruptor Pico (following user manual)

5.2 Anti-6mA Immunoprecipitation and UV Cross-linking

4. Set up 500 μL anti-6mA immunoprecipitation reaction as following:

Algae gDNA fragment	10 μg
Anti-m^6A polyclonal antibody	8 μg (0.5 mg/mL)
5 × IP Buffer	100 μL
DNase-free H_2O	to 500 μL

5. Mix thoroughly but gently. Transfer to head-over-tail rotating wheel to incubate at 4 °C overnight.

6. Pre-block Dynabeads protein A magnetic slurry with 1 × IP Buffer supplement with 0.5 mg/mL BSA for around 1 h.
7. Split the reaction mix into 8–10 wells in 96-well cell culture plate. Keep the plate on ice and expose in UV 254 nm six times with energy dosage of 0.15 J/cm^2.

   Note: Position the plate close to UV light source in order to increase the cross-linking efficiency. Expose six times with a 30-s interval.

5.3 Enzymatic Treatment on Beads

8. Pool the cross-linked fractions together and then transfer each reaction to 80 μL pre-blocked protein A slurry. Mix gently and then incubate on head-over-tail rotating wheel for 1.5 h at 4 °C.
9. Spin down the beads briefly and then place the tube on a magnetic rack for 30 s until beads are captured. Discard the supernatant and wash the beads with 500 μL ice-cold following buffers once as order: 1 × IP Buffer, Exo High Salt Wash Buffer, Exo Wash Buffer 2, Exo Wash Buffer 3, TE Buffer, and 10 m*M* Tris–HCl (pH 7.9) by gentle vortex.

   Note: Transferring the protein A beads to a new low-adhesive tube is recommended in order to reduce nonspecific binding.
10. *Polishing reaction:* resuspend the beads in the following reaction system:

10 m*M* Tris–HCl (pH 8.0)	80 μL
10 × NEBuffer 2	10 μL
20 × BSA (2 mg/mL)	5 μL
10 m*M* dNTP mix	3 μL
T4 DNA polymerase	2 μL

   Incubate the reaction for 30 min at 25 °C.
11. Repeat Step 9 once.

   Note: Transferring the protein A beads to a new low-adhesive tube is recommended in order to reduce nonspecific binding.
12. *Kinase reaction:* resuspend the beads in following reaction system:

10 m*M* Tris–HCl (pH 7.5)	87 μL
10 × T4 DNA Ligase Buffer	10 μL
T4 PNK	3 μL

   Incubate for 30 min at 37 °C.

13. Repeat Step 9 once.

Note: Transferring the protein A beads to a new low-adhesive tube is recommended in order to reduce nonspecific binding.

14. *R1 Adapter ligation reaction:* resuspend the beads in following reaction system:

10 m*M* Tris–HCl (pH 7.5)	82 μL
10 × T4 DNA Ligase Buffer	10 μL
R1 Adapter (40 μ*M*)	5 μL
T4 DNA polymerase	3 μL

Incubate for 2 h at 25 °C.

15. Repeat Step 9 once.

Note: Transferring the protein A beads to a new low-adhesive tube is recommended in order to reduce nonspecific binding.

16. *Filling-in reaction:* resuspend the beads in the following reaction system:

10 m*M* Tris–HCl (pH 7.5)	75 μL
10 × Phi29 DNA Polymerase Buffer	10 μL
20 × BSA (2 mg/mL)	5 μL
10 m*M* dNTP mix	3 μL
Phi29 DNA polymerase	2 μL

Incubate for 20 min at 30 °C.

17. Repeat Step 9 once.

Note: Transferring the protein A beads to a new low-adhesive tube is recommended in order to reduce nonspecific binding.

18. *Optional—Phosphorylation reaction*: resuspend the beads in the following reaction system:

10 m*M* Tris–HCl (pH 7.5)	87 μL
10 × T4 DNA Ligase Buffer	10 μL
T4 PNK	3 μL

Incubate for 30 min at 37 °C.

Note: Even when phosphorylated adapter is used, the optional phosphorylation reaction is still recommended in order to increase efficiency and specificity of lambda exonuclease digestion.

19. Repeat Step 9 once.
    Note: Transferring the protein A beads to a new low-adhesive tube is recommended in order to reduce nonspecific binding.
20. *Lambda exonuclease digestion:* resuspend the beads in the following reaction system:

10 m*M* Tris–HCl (pH 7.5)	86 μL
10 × Lambda Exonuclease Buffer	10 μL
Lambda exonuclease	4 μL

Incubate for 30 min at 37 °C.

21. Repeat Step 9 once.
    Note: Transferring the protein A beads to a new low-adhesive tube is recommended in order to reduce nonspecific binding.
22. *$RecJ_f$ exonuclease digestion*: resuspend the beads in the following reaction system:

10 m*M* Tris–HCl (pH 7.9)	87 μL
10 × $RecJ_f$ Exonuclease Buffer	10 μL
$RecJ_f$ exonuclease	3 μL

Incubate for 30 min at 37 °C.

23. Repeat Step 9 once.
    Note: Transferring the protein A beads to a new low-adhesive tube is recommended in order to reduce nonspecific binding.
24. *ChIP elution*: add 150 μL ChIP elution buffer, mix thoroughly, and incubate for 20 min at 65 °C on Heat Shake with 1000 rpm speed. Then, spin down the beads briefly and place the tube on magnetic rack for 30 s until beads are captured and transfer the supernatant (which contains ChIP complex) in a new tube.
25. Repeat Step 24 once and combine the supernatant together.
26. *Reverse cross-linking*: add 10 μL Proteinase K to the supernatant and incubate overnight at 65 °C on Heat Shake with 600 rpm speed.
27. *DNA fragment isolation*: spin down the solution briefly. Extract DNA fragment by adding equal volume 25:24:1 phenol/chloroform/isoamyl alcohol. Then wash aqueous phase once with chloroform and isolate the DNA fragment by ethanol precipitation. Store the mixture at −80 °C for 3 h.

28. *DNA fragment clean-up*: spin down DNA fragment precipitation, wash the pellet with 75% (v/v) ethanol, and dissolve the pellet into 50 μL DNase-free H_2O. Clean up by Zymo Research DNA Clean & Concentrator and elute in 12 μL TE Buffer.

5.4 Enzymatic Treatment in Solution

29. *Denaturing and primer extension*: prepare the following reaction mix:

DNA fragment in TE Buffer	12 μL
10× Phi29 DNA Polymerase Buffer	2 μL
20× BSA (2 mg/mL)	2 μL
10 m*M* dNTP mix	1 μL
R1 primer extension (20 μ*M*)	1 μL

Set up a program on the thermal cycler:

95 °C Denature	5 min
60 °C Anneal	5 min
30 °C Extension	30 min
65 °C Deactivate	10 min

Incubate the above reaction mixture at 95 °C for denaturing and 60 °C for primer annealing. Leave the reaction to cool down to room temperature. Then add 2 μL Phi29 DNA polymerase and continue the thermal cycler program.

30. *R2 Adapter ligation reaction*: prepare the following reaction mixture:

dsDNA fragment from Step 29	20 μL
10× T4 DNA Ligase Buffer	3 μL
T4 DNA ligase	2 μL
R2 Adapter (30 μ*M*)	1 μL
TE Buffer	4 μL

Incubate for 2 h at 25 °C.

5.5 dsDNA Isolation and Purification

31. *dsDNA fragment purification*: purify dsDNA from Step 30 by using 52 μL well-mixed AMPure beads.

5.6 Library Construction, High-Throughput Sequencing, and Data Analysis

32. *dsDNA elution and PCR amplification*: elute dsDNA fragment in 30 μL TE Buffer from completely dried AMPure beads and prepare the PCR reaction mix:

Purified dsDNA fragment from Step 31	15 μL
2× Phusion High-Fidelity PCR Master Mix	25 μL
PCR Primer (Illumina Small RNA)	2 μL
PCR Index Primer (Illumina Small RNA)	2 μL
H_2O	6 μL

Set up a program on the thermal cycler:

98 °C	30 s	1 step
98 °C	10 s	16 cycles
60 °C	30 s	
72 °C	45 s	
72 °C	10 min	1 step
4 °C	–	Hold

Note: All purified dsDNA fragment from Step 31 can be used for amplification; however, using half is recommended in order to avoid over-amplification.

33. *Library purification*: purify library from Step 32 by using 50 μL well-mixed AMPure beads or agarose gel size selection and purification.

    Note: Either way is compatible; however, to avoid adapter dimer contamination, follow the size selection recommendations.

34. *Data analysis*: align single-end 50 bp reads to reference genome by Bowtie (Langmead et al., 2009). Call peaks by using MACE (*M*odel-based *A*nalysis of ChIP-*e*xo) (Wang, Chen, et al., 2014).

6. CONCLUSION

Photo-crosslinking-assisted strategy followed by high-throughput sequencing is able to significantly improve the resolution of mapping m^6A on transcriptome and 6mA in genomic DNA. The combination of covalent cross-linking and nuclease digestion is compatible with investigating other chemical modifications on nucleosides, suggesting potential widespread applications.

ACKNOWLEDGMENTS

This work was supported by National Institutes of Health (HG006827 to C.H.). S.F. Reichard contributed editing.

REFERENCES

Batista, Pedro J., Molinie, B., Wang, J., Qu, K., Zhang, J., Li, L., et al. (2014). m^6A RNA modification controls cell fate transition in mammalian embryonic stem cells. *Cell Stem Cell*, *15*, 707–719.

Bokar, J. A. (2005). The biosynthesis and functional roles of methylated nucleosides in eukaryotic mRNA. In H. Grosjean (Ed.), *Fine-tuning of RNA functions by modification and editing* (pp. 141–177). Berlin Heidelberg: Springer-Verlag.

Bokar, J. A., Rath-Shambaugh, M. E., Ludwiczak, R., Narayan, P., & Rottman, F. (1994). Characterization and partial purification of mRNA N^6-adenosine methyltransferase from HeLa cell nuclei. Internal mRNA methylation requires a multisubunit complex. *Journal of Biological Chemistry*, *269*, 17697–17704.

Bokar, J. A., Shambaugh, M. E., Polayes, D., Matera, A. G., & Rottman, F. M. (1997). Purification and cDNA cloning of the AdoMet-binding subunit of the human mRNA (N6-adenosine)-methyltransferase. *RNA*, *3*, 1233–1247.

Campbell, J. L., & Kleckner, N. (1990). *E. coli oriC* and the *dnaA* gene promoter are sequestered from dam methyltransferase following the passage of the chromosomal replication fork. *Cell*, *62*, 967–979.

Canaani, D., Kahana, C., Lavi, S., & Groner, Y. (1979). Identification and mapping of N^6-methyladenosine containing sequences in Simian Virus 40 RNA. *Nucleic Acids Research*, *6*, 2879–2899.

Cedar, H. (1988). DNA methylation and gene activity. *Cell*, *53*, 3–4.

Chen, T., Hao, Y.-J., Zhang, Y., Li, M.-M., Wang, M., Han, W., et al. (2015). m^6A RNA methylation is regulated by microRNAs and promotes reprogramming to pluripotency. *Cell Stem Cell*, *16*, 1–13.

Chen, K., Lu, Z., Wang, X., Fu, Y., Luo, G.-Z., Liu, N., et al. (2015). High-resolution N^6-methyladenosine (m^6A) map using photo-crosslinking-assisted m^6A sequencing. *Angewandte Chemie International Edition*, *54*, 1587–1590.

Collier, J., McAdams, H. H., & Shapiro, L. (2007). A DNA methylation ratchet governs progression through a bacterial cell cycle. *Proceedings of the National Academy of Sciences of the United States of America*, *104*, 17111–17116.

Corcoran, D. L., Georgiev, S., Mukherjee, N., Gottwein, E., Skalsky, R. L., Keene, J. D., et al. (2011). PARalyzer: Definition of RNA binding sites from PAR-CLIP short-read sequence data. *Genome Biology*, *12*, R79.

Cummings, D. J., Tait, A., & Goddard, J. M. (1974). Methylated bases in DNA from *Paramecium aurelia*. *Biochimica et Biophysica Acta*, *374*, 1–11.

Dai, Q., Fong, R., Saikia, M., Stephenson, D., Yu, Y. T., Pan, T., et al. (2007). Identification of recognition residues for ligation-based detection and quantitation of pseudouridine and N^6-methyladenosine. *Nucleic Acids Research*, *35*, 6322–6329.

Desrosiers, R., Friderici, K., & Rottman, F. (1974). Identification of methylated nucleosides in messenger RNA from Novikoff hepatoma cells. *Proceedings of the National Academy of Sciences of the United States of America*, *71*, 3971–3975.

Desrosiers, R. C., Friderici, K. H., & Rottman, F. M. (1975). Characterization of Novikoff hepatoma mRNA methylation and heterogeneity in the methylated 5′ terminus. *Biochemistry*, *14*, 4367–4374.

Dimock, K., & Stoltzfus, C. M. (1977). Sequence specificity of internal methylation in B77 avian sarcoma virus RNA subunits. *Biochemistry*, *16*, 471–478.

Dominissini, D., Moshitch-Moshkovitz, S., Salmon-Divon, M., Amariglio, N., & Rechavi, G. (2013). Transcriptome-wide mapping of N^6-methyladenosine by m^6A-seq based on immunocapturing and massively parallel sequencing. *Nature Protocols*, *8*, 176–189.

Dominissini, D., Moshitch-Moshkovitz, S., Schwartz, S., Salmon-Divon, M., Ungar, L., Osenberg, S., et al. (2012). Topology of the human and mouse m^6A RNA methylomes revealed by m^6A-seq. *Nature*, *485*, 201–206.

Ehrlich, M., & Zhang, X. Y. (1990). Chapter 10 Naturally occurring modified nucleosides in DNA. In W. G. Charles & C. T. K. Kenneth (Eds.), *Chromatography and Modification of Nucleosides: Biological roles and function of modification. Journal of chromatography library: Vol. 45B* (pp. B327–B362). Elsevier.

Fang, G., Munera, D., Friedman, D. I., Mandlik, A., Chao, M. C., Banerjee, O., et al. (2012). Genome-wide mapping of methylated adenine residues in pathogenic *Escherichia coli* using single-molecule real-time sequencing. *Nature Biotechnology*, *30*, 1232–1239.

Fu, Y., Dominissini, D., Rechavi, G., & He, C. (2014). Gene expression regulation mediated through reversible m^6A RNA methylation. *Nature Reviews. Genetics*, *15*, 293–306.

Fu, Y., Luo, G.-Z., Chen, K., Deng, X., Yu, M., Han, D., et al. (2015). N^6-Methyldeoxyadenosine Marks Active Transcription Start Sites in Chlamydomonas. *Cell*, *161*, 879–892.

Geula, S., Moshitch-Moshkovitz, S., Dominissini, D., Mansour, A. A., Kol, N., Salmon-Divon, M., et al. (2015). m^6A mRNA methylation facilitates resolution of naïve pluripotency toward differentiation. *Science*, *347*, 1002–1006.

Hafner, M., Landthaler, M., Burger, L., Khorshid, M., Hausser, J., Berninger, P., et al. (2010). Transcriptome-wide identification of RNA-binding protein and microRNA target sites by PAR-CLIP. *Cell*, *141*, 129–141.

Harper, J. E., Miceli, S. M., Roberts, R. J., & Manley, J. L. (1990). Sequence specificity of the human mRNA N6-adenosine methylase in vitro. *Nucleic Acids Research*, *18*, 5735–5741.

Harrison, G. S., Findly, R. C., & Karrer, K. M. (1986). Site-specific methylation of adenine in the nuclear genome of a eucaryote, *Tetrahymena thermophila*. *Molecular and Cellular Biology*, *6*, 2364–2370.

Hattman, S., Kenny, C., Berger, L., & Pratt, K. (1978). Comparative study of DNA methylation in three unicellular eucaryotes. *Journal of Bacteriology*, *135*, 1156–1157.

He, C. (2010). Grand challenge commentary: RNA epigenetics? *Nature Chemical Biology*, *6*, 863–865.

Horowitz, S., Horowitz, A., Nilsen, T. W., Munns, T. W., & Rottman, F. M. (1984). Mapping of N^6-methyladenosine residues in bovine prolactin mRNA. *Proceedings of the National Academy of Sciences of the United States of America*, *81*, 5667–5671.

Jia, G., Fu, Y., & He, C. (2013). Reversible RNA adenosine methylation in biological regulation. *Trends in Genetics*, *29*, 108–115.

Jia, G., Fu, Y., Zhao, X., Dai, Q., Zheng, G., Yang, Y., et al. (2011). N^6-methyladenosine in nuclear RNA is a major substrate of the obesity-associated FTO. *Nature Chemical Biology*, *7*, 885–887.

Kane, S. E., & Beemon, K. (1985). Precise localization of m^6A in Rous sarcoma virus RNA reveals clustering of methylation sites: Implications for RNA processing. *Molecular and Cellular Biology*, *5*, 2298–2306.

Langmead, B., Trapnell, C., Pop, M., & Salzberg, S. L. (2009). Ultrafast and memory-efficient alignment of short DNA sequences to the human genome. *Genome Biology*, *10*, R25.

Liu, N., Parisien, M., Dai, Q., Zheng, G., He, C., Pan, T., et al. (2013). Probing N^6-methyladenosine RNA modification status at single nucleotide resolution in mRNA and long noncoding RNA. *RNA*, *19*, 1848–1856.

Liu, J., Yue, Y., Han, D., Wang, X., Fu, Y., Zhang, L., et al. (2014). A METTL3-METTL14 complex mediates mammalian nuclear RNA N^6-adenosine methylation. *Nature Chemical Biology*, *10*, 93–95.

Low, D. A., Weyand, N. J., & Mahan, M. J. (2001). Roles of DNA adenine methylation in regulating bacterial gene expression and virulence. *Infection and Immunity*, *69*, 7197–7204.

Lu, M., Campbell, J. L., Boye, E., & Kleckner, N. (1994). *SeqA*: A negative modulator of replication initiation in *E. coli*. *Cell*, *77*, 413–426.

Messer, W., & Noyer-Weidner, M. (1988). Timing and targeting: The biological functions of Dam methylation in *E. coli*. *Cell*, *54*, 735–737.

Meyer, K. D., & Jaffrey, S. R. (2014). The dynamic epitranscriptome: N^6-Methyladenosine and gene expression control. *Nature Reviews Molecular Cell Biology*, *15*, 313–326.

Meyer, K., Saletore, Y., Zumbo, P., Elemento, O., Mason, C., Jaffrey, S. R., et al. (2012). Comprehensive analysis of mRNA methylation reveals enrichment in 3′ UTRs and near stop codons. *Cell*, *149*, 1635–1646.

Murray, N. E. (2002). 2001 Fred Griffith review lecture. Immigration control of DNA in bacteria: Self versus non-self. *Microbiology*, *148*, 3–20.

Murray, I. A., Clark, T. A., Morgan, R. D., Boitano, M., Anton, B. P., Luong, K., et al. (2012). The methylomes of six bacteria. *Nucleic Acids Research*, *40*, 11450–11462.

Ogden, G. B., Pratt, M. J., & Schaechter, M. (1988). The replicative origin of the *Escherichia coli* chromosome binds to cell membranes only when hemimethylated. *Cell*, *54*, 127–135.

Pan, T. (2013). N^6-Methyl-adenosine modification in messenger and long non-coding RNA. *Trends in Biochemical Sciences*, *38*, 204–209.

Ping, X.-L., Sun, B.-F., Wang, L., Xiao, W., Yang, X., Wang, W. J., et al. (2014). Mammalian WTAP is a regulatory subunit of the RNA N^6-methyladenosine methyltransferase. *Cell Research*, *24*, 177–189.

Rae, P. M., & Spear, B. B. (1978). Macronuclear DNA of the hypotrichous ciliate *Oxytricha fallax*. *Proceedings of the National Academy of Sciences of the United States of America*, *75*, 4992–4996.

Ratel, D., Ravanat, J.-L., Berger, F., & Wion, D. (2006). N^6-Methyladenine: The other methylated base of DNA. *Bioessays*, *28*, 309–315.

Rhee, Ho S., & Pugh, B. F. (2011). Comprehensive genome-wide protein-DNA interactions detected at single-nucleotide resolution. *Cell*, *147*, 1408–1419.

Roberts, R. J., & Halford, S. E. (1993). 2 Type II Restriction Endonucleases. *Cold Spring Harbor Monograph Archive*, *25*, 35–88.

Roberts, R. J., Vincze, T., Posfai, J., & Macelis, D. (2010). REBASE—A database for DNA restriction and modification: Enzymes, genes and genomes. *Nucleic Acids Research*, *38*, D234–D236.

Robertson, K. D. (2005). DNA methylation and human disease. *Nature Reviews. Genetics*, *6*, 597–610.

Sasaki, H., & Matsui, Y. (2008). Epigenetic events in mammalian germ-cell development: Reprogramming and beyond. *Nature Reviews. Genetics*, *9*, 129–140.

Schibler, U., Kelley, D. E., & Perry, R. P. (1977). Comparison of methylated sequences in messenger RNA and heterogeneous nuclear RNA from mouse L cells. *Journal of Molecular Biology*, *115*, 695–714.

Schwartz, S., Agarwala, Sudeep D., Mumbach, Maxwell R., Jovanovic, M., Mertins, P., Shishkin, A., et al. (2013). High-resolution mapping reveals a conserved, widespread, dynamic mRNA methylation program in yeast meiosis. *Cell*, *155*, 1409–1421.

Wang, L., Chen, J., Wang, C., Uusküla-Reimand, L., Chen, K., Medina-Rivera, A., et al. (2014). MACE: Model based analysis of ChIP-exo. *Nucleic Acids Research*, *42*, e156.

Wang, Y., Li, Y., Toth, J. I., Petroski, M. D., Zhang, Z., Zhao, J. C., et al. (2014). N^6-Methyladenosine modification destabilizes developmental regulators in embryonic stem cells. *Nature Cell Biology*, *16*, 191–198.

Wang, X., Lu, Z., Gomez, A., Hon, G. C., Yue, Y., Han, D., et al. (2014). N^6-Methyladenosine-dependent regulation of messenger RNA stability. *Nature*, *505*, 117–120.

Wei, C.-M., & Moss, B. (1977). Nucleotide sequences at the N^6-methyladenosine sites of HeLa cell messenger ribonucleic acid. *Biochemistry*, *16*, 1672–1676.

Zheng, G., Dahl, John A., Niu, Y., Fedorcsak, P., Huang, C.-M., Li, C. J., et al. (2013). ALKBH5 Is a mammalian RNA demethylase that impacts RNA metabolism and mouse fertility. *Molecular Cell*, *49*, 18–29.

CHAPTER TEN

Pseudouridine in mRNA: Incorporation, Detection, and Recoding

Guowei Wu*, Chao Huang†, Yi-Tao Yu‡,1

*Department of Cellular and Molecular Medicine, University of California, San Diego, La Jolla, California, USA

†Process Science Downstream, Bristol-Myers Squibb Company, Syracuse, New York, USA

‡Center for RNA Biology, Department of Biochemistry and Biophysics, University of Rochester Medical Center, Rochester, New York, USA

[1]Corresponding author: e-mail address: yitao_yu@urmc.rochester.edu

Contents

Methods in Enzymology, Volume 560
ISSN 0076-6879
http://dx.doi.org/10.1016/bs.mie.2015.03.009

Abstract

It has long been known that pseudouridine (Ψ) is the most abundant modified nucleotide in stable RNAs, including tRNA, rRNA, and snRNA. Recent studies using massive parallel sequencing have uncovered the presence of hundreds of Ψs in mRNAs as well. In eukaryotes and archaea, RNA pseudouridylation is introduced predominantly by box H/ACA RNPs, RNA–protein complexes each consisting of a single RNA moiety and four core proteins. It has been well established that Ψ plays an essential role in regulating the structure and function of stable RNAs in several model organisms, including yeast, *Xenopus laevis*, and humans. However, the functional role of Ψ in mRNA remains to be elucidated. One possibility (and true for stop/termination codons) is that Ψ influences decoding during translation. It is imperative, therefore, to establish a system, in which one can site-specifically introduce pseudouridylation into target mRNA and biochemically test the impact of mRNA pseudouridylation on protein translation. Here, we present a method for (1) site-specific conversion of uridine into Ψ in mRNA by designer box H/ACA RNP, (2) detection of Ψ in target mRNA using site-specific labeling followed by nuclease digestion and thin layer chromatography, and (3) analysis of recoding of pseudouridylated premature termination codon in mRNA during translation.

1. INTRODUCTION

Once dubbed the fifth nucleotide, pseudouridine (Ψ) is the most abundant posttranscriptionally modified nucleotide, constituting ~5% of total RNA ribonucleotides (Davis & Allen, 1957). The conversion of uridine to Ψ (pseudouridylation) requires two distinct chemical reactions: the breaking of the C1′—N1 glycosidic bond and the making of a new carbon–carbon (C1′—C5) bond that relinks the base to the sugar (Wu, Yu, Kantartzis, & Yu, 2011). Pseudouridylation is a true isomerization reaction, which creates an extra hydrogen bond donor.

Earlier biophysics studies using NMR revealed that Ψ, when present in RNA, favors the formation of 3′-*endo* conformation and promotes base stacking, thus stabilizing RNA structures (Davis, 1995; Lee & Tinoco, 1980). In line with this, computational simulation studies have independently established the role for Ψ in stabilizing the structure of the anticodon stem loop of tRNA (Bilbille et al., 2009; Durant & Davis, 1999). In addition, it has been reported that loss of a single Ψ in the tRNA anticodon stem loop results in a decrease in the efficiency of recoding (Lecointe et al., 2002). Furthermore, experimental results indicate that Ψs in U2 snRNA contribute to the formation of active U2 snRNP and the spliceosome, and are thus important for splicing in *Xenopus* oocytes (Yu, Shu, & Steitz, 1998). Likewise, Ψs in rRNAs are concentrated in functional important regions, such as peptidyl transferase center and decoding center, and play important roles in protein synthesis. Specifically, abolishing these Ψs severely affects ribosome functions, resulting in reduced rate of translation and increased stop-codon readthrough and frame shifting (King, Liu, McCully, & Fournier, 2003; Liang, Liu, & Fournier, 2007).

In eukaryotes and archaea, pseudouridylation is introduced predominantly by box H/ACA RNPs, each of which contains a unique small RNA (box H/ACA RNA) and four core proteins (NAP57/dyskerin/Cbf5, Nhp2, Nop10, and Gar1) (Karijolich & Yu, 2008). The RNA component serves as a guide that specifies, through base-paring interaction with its substrate RNA, the target uridine for pseudouridylation (Ge & Yu, 2013). One of the four core proteins, NAP57/dyskerin/Cbf5, catalyzes the chemical reactions, converting the target uridine to Ψ (Huang, Karijolich, & Yu, 2011).

Based on this guide-substrate base-pairing scheme, Karijolich and Yu designed an artificial box H/ACA RNA to introduce Ψ into mRNA at a premature termination codon (PTC). They demonstrated that Ψ was indeed incorporated into *TRM4* mRNA at the PTC, and remarkably, pseudouridylated PTC promotes nonsense suppression by altering ribosome decoding (Fernandez et al., 2013; Karijolich & Yu, 2011). Recently, three groups developed Ψ-seq and identified hundreds of naturally occurring Ψs in both yeast and human mRNAs (Carlile et al., 2014; Lovejoy, Riordan, & Brown, 2014; Schwartz et al., 2014). Many of these novel Ψs reside in the coding regions, and the majority of them respond to environmental stress, indicating functional significance (Carlile et al., 2014).

Here, we present a protocol that details how to introduce and detect Ψs in mRNA, and how to analyze the protein products generated by Ψ-mediated recoding (nonsense suppression). Starting with the plasmid bearing the C-terminally tagged *TRM4* gene, we create a PTC in this gene using site-directed mutagenesis. We then construct a second plasmid containing an artificial box H/ACA RNA gene (based on *snR81*) (Ma et al., 2005) designed to target the uridine of the PTC in *TRM4* mRNA for pseudouridylation. Transformation of yeast cells with these two plasmids allows site-specific pseudouridylation at the PTC, leading to the production of a full-length PTC-readthrough protein, which can readily be detected by Western blotting analysis.

2. INCORPORATION OF PSEUDOURIDINE IN *TRM4* mRNA

2.1 Overview

All naturally occurring ORF stop codons, UAA, UGA, and UAG, contain a uridine in the first position. To study Ψ-mediated nonsense suppression, we choose a plasmid containing a C-terminally tagged (3C-HA-HIS6X-Protein A) reporter gene, *TRM4*, which, when transformed into yeast cells, is highly expressed (Chernyakov, Whipple, Kotelawala, Grayhack, & Phizicky, 2008). Using PCR-based site-directed mutagenesis, we then introduce a PTC into the ORF of *TRM4* at codon 602 (TTT-to-TAA change), as shown in Fig. 1.

To introduce Ψs into PTC, we generate an artificial box H/ACA guide RNA (based on yeast *snR81* box H/ACA RNA) (Ma et al., 2005), in which the guide sequences are designed to target the uridine of the PTC codon for pseudouridylation. The artificial box H/ACA RNA is amplified by PCR with four overlapping primers. The first primer (*snR81-TRM4-TAA*-F1) is a sense forward primer and the other three (*snR81-TRM4-TAA*-R1, *snR81-TRM4-TAA*-R2, and *snR81-TRM4-TAA*-R3) are antisense reverse primers (Figs. 2 and 3). Within each primer, there is a tract of nucleotides that represents the guide sequence (indicated as guide1, guide1′, guide2, and guide2′ in Figs. 2 and 3). Guide1 and guide1′ form the first

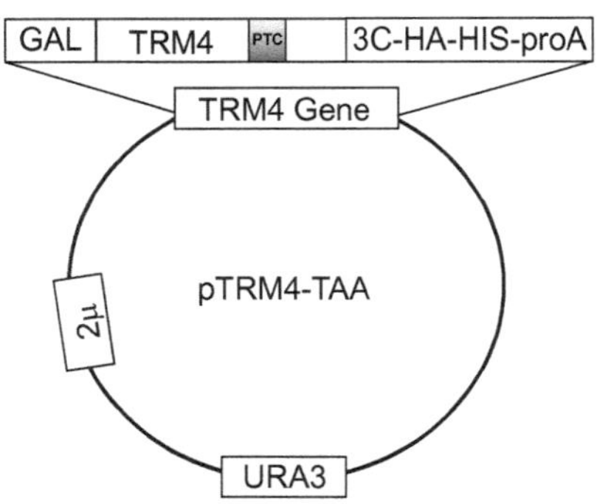

Figure 1 Schematic representation of p*TRM4-TAA* expression plasmid. The wild-type *TRM4* gene is mutated at F602 position (TTT converted to TAA) to generate a premature termination codon (PTC) (indicated). The plasmid-borne *TRM4* gene is fused with a multicomponent C-terminal tag to facilitate the detection and purification of full-length Trm4 product. The tag contains a 3C protease cleavage site that can be used to release the purified protein from the tag. The expression of *TRM4* gene is under the control of Gal promoter, which is galactose-inducible. *URA3* is an auxotroph selective marker in yeast. This is a high copy 2 μ plasmid.

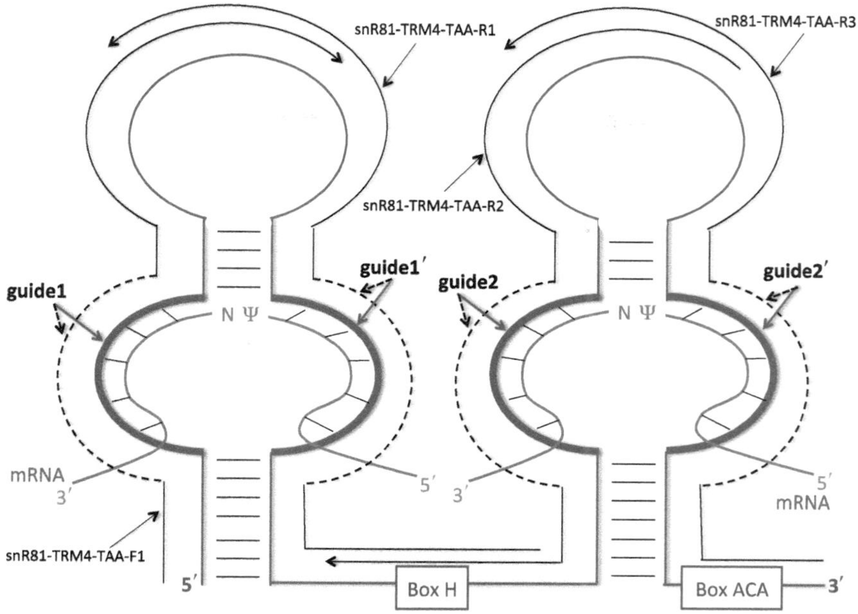

Figure 2 Schematic representation of box H/ACA guide RNA and its interaction with its substrate RNAs. The guide RNA (red) folds into a hairpin-hinge-hairpin-tail structure. The pseudouridylation pockets responsible for target site specification are the internal loops (thick red lines) within the hairpin structures. The H and ACA boxes and the guide sequences (guide1, guide1′, guide2, and guide2′) are indicated. The mRNA substrates (blue) are paired with the guide sequences of the guide RNA, as indicated by short lines between strands. The pseudouridylated uridine (Ψ) and its downstream residue (N) in the mRNA substrates are left unpaired. The four overlapping primers, used for guide RNA construction (see text), are also shown (black lines). The dotted black lines denote specific guide sequences. (See the color plate.)

(5′) pseudouridylation pocket, and guide2 and guide2′ form the second (3′) pseudouridylation pocket in the guide RNA (Fig. 2). The guide sequences in both pseudouridylation pockets of the artificial guide RNA (*snR81-TRM4-TAA*) are designed to target the PTC site in *TRM4-TAA*. Similar to *snR81-TRM4-TAA* construct, a control *snR81* guide RNA construct is generated by PCR with four overlapping primers, *snR81*-Control-F1, *snR81*-Control-R1, *snR81*-Control-R2, and *snR81*-Control-R3. The guide sequences in *snR81*-Control do not target the PTC of *TRM4-TAA* and have no targets in yeast genome.

Finally, yeast BY4741 strain is cotransformed with the *TRM4-TAA* reporter and one of the guide RNA constructs (control or PTC-specific).

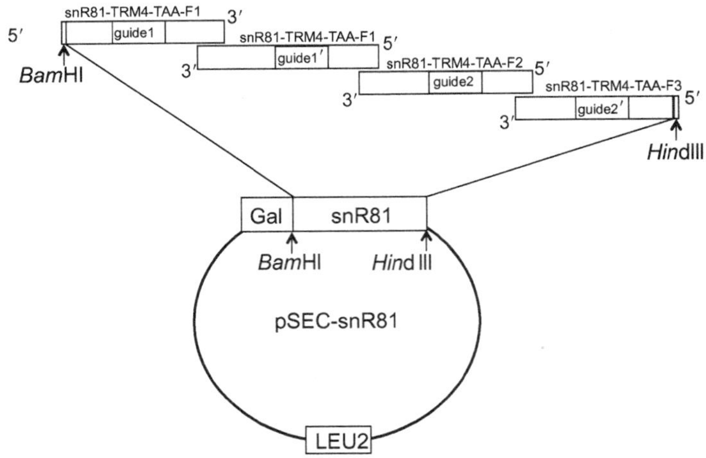

Figure 3 The strategy for cloning artificial *snR81* using four primers overlapping PCR (also see Fig. 2). The construction is based on naturally occurring yeast *snR81* box H/ACA RNA. The 5′-most primer is sense-strand sequence (forward), while the other primers have antisense sequences (reverse). The guide sequences (guide1, guide1′, guide2, and guide2′) are changed to base pair with substrate mRNAs, while the rest of the *snR81* sequence remain unchanged. The amplified artificial *snR81* is digested with *Bam*HI and *Hind*III and subsequently cloned into pSEC plasmid under the control of the Gal promoter. LEU2 is an auxotroph selective marker in yeast.

It is expected that only the PTC-specific guide RNA (*snR81-TRM4-TAA*) will direct *TRM4-TAA* mRNA pseudouridylation at the PTC codon.

2.2 Construction of p*TRM4-TAA* Expression Plasmid (PCR-Based Mutagenesis)

2.2.1 Buffers, Reagent, and Solutions

p*TRM4-WT*: a plasmid bearing the wild-type *TRM4* sequence with C-terminally tag (gift from Eric Phizicky at University of Rochester Medical Center) (Huang, Wu, & Yu, 2012)

TRM4-TAA-F1 primer: 5′-AAA TTA AGC TCT GGT TGC GCC **TAA** ATT GAT GTG TCA AGA-3′

TRM4-TAA-R1 primer: 5′-TCT TGA CAC ATC AAT **TTA** GGC GCA ACC AGA GCT TAA TTT-3′

10 m*M* dNTPs (Fermentas, cat. no. R0181)

6 × DNA loading dye (Fermentas, cat. no. R0611)

Pfu DNA polymerase (5 U/μL) and 10 × buffer (Stratagene, cat. no. 600250)

*Dpn*I restriction endonuclease (Fermentas, cat. no. ER1705)

XL1-blue competent cells (Stratagene, cat. no. 200521)

100 mg/mL Ampicillin (IBI Scientific, cat. no. IB02040)
5 × TBE buffer: 445 m*M* Tris, 445 m*M* boric acid, 16 m*M* EDTA
0.5 × TBE buffer: mix 100 mL of 5 × TBE buffer with 900 mL of ddH_2O
LB liquid medium: 10 g of NaCl, 10 g of peptone, 5 g of yeast extract, fill to 1 L with ddH_2O and autoclave
LB-ampicillin solid medium: 20 g of agar, 10 g of NaCl, 10 g of peptone, 5 g of yeast extract, fill to 1 L with ddH_2O and autoclave. Allow to cool to ~55 °C, and add 1 mL of 100 mg/mL ampicillin. Gently mix and pour about 25 mL of medium in Petri dishes to achieve a bed height of ~0.5 cm
70% Ethanol: mix 700 mL of ethanol with 300 mL of ddH_2O

2.2.2 Protocol

1. Purify p*TRM4-WT* plasmid using miniprep kit from Qiagen. Prepare the plasmid stock solution at 1 μg/μL, and dilute with ddH_2O to 10 ng/μL working concentration.
2. Prepare DNA oligonucleotide primers with ddH_2O to 10 μ*M* working concentration.
3. Mix the following components in a 0.2-mL PCR tube:

5 μL	10 × *Pfu* DNA polymerase buffer
2 μL	10 m*M* dNTPs
2 μL	10 ng/μL p*TRM4-WT* plasmid
2 μL	10 μ*M* *TRM4-TAA*-F1 primer
2 μL	10 μ*M* *TRM4-TAA*-R1 primer
1 μL	5 U/μL *Pfu* DNA polymerase
Add ddH_2O to a final reaction volume of 50 μL.	

4. In a separate 0.2-mL PCR tube, prepare a control PCR with ddH_2O substituting the primers.
5. Perform the PCR cycles as follows:
   A: 95 °C for 2 min (1 cycle)
   B: 95 °C for 30 s
   55 °C for 1 min
   68 °C for 9 min
   (Repeat B for 28 cycles)
   C: 72 °C for 2 min (1 cycle)
   4 °C for indefinite time

6. Take 10 μL of the PCR product and add 2 μL of 6 × DNA loading dye before loading the samples onto 1% agarose gel containing 0.05% (v/v) ethidium bromide.
7. Carry out electrophoresis in 0.5 × TBE buffer for 30 min (at 10 V/cm gel length). Visualize the PCR products under UV light. The control PCR should have no visible bands.
8. Once the successful amplification is confirmed, add 2 μL (10 U/μL) of *Dpn*I (specific for methylated sites) to the PCR to linearize (remove) the wild-type plasmid template (p*TRM4-WT*) (the wild-type plasmid template is methylated, whereas the PCR product is not) and incubate at 37 °C for 2 h.
9. Precipitate the PCR product by adding 200 μL of 100% ethanol and 5 μL of 3 *M* NaOAc, followed by centrifugation at the maximum speed (~14,000 × *g*) in a bench-top centrifuge for 10 min.
10. Wash the pellet with 70% ethanol and air-dry. Carefully dissolve the PCR product in 10 μL of ddH_2O.
11. Under sterile conditions, add 2 μL of PCR products to 100 μL of XL-blue competent cells (prealiquoted in 1.5-mL tube). Incubate on ice for 20 min.
12. Heat shock the sample for 90 s at 42 °C and immediately return to ice for 2 min.
13. Add 900 μL of autoclaved LB liquid medium to the sample and shake at 200 rpm for 1 h at 37 °C.
14. Plate the sample onto a LB-ampicillin solid medium and incubate at 37 °C for 16 h.
15. When colonies appear, pick 5–10 individual colonies and prepare plasmid DNA from each colony. Sequence the candidate plasmids using a proper primer. Choose one plasmid with correct sequence (codon TTT converted to TAA at F602 position) and label it as p*TRM4-TAA*. Adjust the concentration of the plasmid to 1 μg/μL.

2.3 Construction of *TRM4-TAA*-Targeting *snR81* Expression Plasmid

2.3.1 Buffers, Reagent, and Solutions

pSEC plasmid: plasmid with SnoRNA expression cassette. A high copy *E. coli*–yeast shuttle plasmid with LEU2 as the auxotroph selective marker in yeast.

snR81-TRM4-TAA-F1: 5′-GCG GGA TCC GGG ACT GCA TCA ATT GCG GCG AGG CAG CCC ACA TCA AGT GGA ACT ACA C-3′
snR81-TRM4-TAA-R1: 5′-TGT TAG GAT TGC TCT TGG GAC CGT TGC GCC GCG ACA AGG AAG TCT GTG TAG TTC CAC TTG-3′
snR81-TRM4-TAA-R2: 5′-CCA TCC GTG GAC TGT ACA GGT TCA GCG GGG GAA TTG ATG TTT GCT TGT TAG GAT TGC TCT-3′
snR81-TRM4-TAA-R3: 5′-GCG AAG CTT AGA TGT GAA AAA GCG CGC CCC CCC GAA TCA TAT AAC TTC TGC ACC ATC CGT GGA CTG T-3′
snR81-Control-F1: 5′-GCG GGA TCC GGG ACT GCT CGA TTA GCG GCG AGG CAG CCC ACA TCA AGT GGA ACT ACA C-3′
snR81-Control-R1: 5′-TGT TAG GAT TGC TCT TGG GAC CGA TTA CTC GCG ACA AGG AAG TCT GTG TAG TTC CAC TTG-3′
snR81-Control-R2: 5′-CCA TCC GTG GAC TGT ACA GGT TCA GCG GGG GTA ATC GAT TTT GCT TGT TAG GAT TGC TCT-3′
snR81-Control-R3: 5′-GCG AAG CTT AGA TGT GAA AAA GCG ACT CCC CCC GAA TCA TAT AAC TTC TGC ACC ATC CGT GGA CTG T-3′
Taq DNA polymerase (5 U/μL) (Fermentas, cat. no. EP0401)
10 × *Taq* DNA polymerase buffer (Fermentas, cat. no. B34)
10 m*M* dNTPs (Fermentas, cat. no. R0181)
6 × DNA loading dye (Fermentas, cat. no. R0611)
*Bam*HI restriction endonuclease (Fermentas, cat. no. ER0051)
*Hin*dIII restriction endonuclease (Fermentas, cat. no. ER0501)
T4 DNA ligase (1 U/μL) (Fermentas, cat. no. EL0011) and 5 × buffer (cat. no. B69)
DH5α competent cells (Invitrogen, cat. no. 18265-017)
Phenol (Alfa Aesar, cat. no. A15760-0E)
Chloroform (J.T. Baker, cat. no. 9182-01)
Isoamyl alcohol (Sigma-Aldrich, cat. no. I9392)
Ampicillin (IBI Scientific, cat. no. IB02040)
Glycogen (Sigma-Aldrich, cat. no. G0885)

5× TBE buffer: 445 m*M* Tris, 445 m*M* boric acid, 16 m*M* EDTA
0.5× TBE buffer: mix 100 mL of 5× TBE buffer with 900 mL of ddH_2O
Qiagen Gel Extraction kit (Qiagen, cat. no. 28704)
LB liquid medium: 10 g of NaCl, 10 g of peptone, 5 g of yeast extract, fill to 1 L with ddH_2O and autoclave
LB-ampicillin solid medium: 20 g of agar, 10 g of NaCl, 10 g of peptone, 5 g of yeast extract, fill to 1 L with ddH_2O and autoclave. Allow to cool, add 1 mL of 100 mg/mL of ampicillin, and mix well. Pour about 20–25 mL of medium in Petri dishes to achieve a bed height of ~0.5 cm
PCA (phenol/chloroform/isoamyl alcohol=25/24/1 [v/v/v]): saturated (1:1 [v/v]) with 20 m*M* Tris–HCl, pH 8.0
G50 buffer: 20 m*M* Tris–HCl at pH 7.5, 300 m*M* sodium acetate, 2 m*M* EDTA, 0.2% SDS
70% Ethanol: mix 700 mL of ethanol with 300 mL of ddH_2O

2.3.2 Protocol

1. Prepare following DNA oligonucleotide primers with ddH_2O to 10 μ*M* working concentration: *snR81-TRM4-TAA*-F1, *snR81-TRM4-TAA*-R3, *snR81*-Control-F1, and *snR81*-Control-R3.
2. Prepare following DNA oligonucleotide primers with ddH_2O to 1 μ*M* working concentration: *snR81-TRM4-TAA*-R1, *snR81-TRM4-TAA*-R2, *snR81*-Control-R1, and *snR81*-Control-R2.
3. Mix the following buffers and reagents in a 0.2-mL PCR tube (to generate *snR81-TRM4-TAA*). The ratio of four primers is optimized to yield the PCR product at expected size.

5 μL	10× *Taq* DNA polymerase buffer
2 μL	10 m*M* dNTPs
2 μL	10 μ*M* *snR81-TRM4-TAA*-F1 primer
1 μL	1 μ*M* *snR81-TRM4-TAA*-R1 primer
1 μL	1 μ*M* *snR81-TRM4-TAA*-R2 primer
2 μL	10 μ*M* *snR81-TRM4-TAA*-R3 primer
1 μL	5 U/μL *Taq* DNA polymerase
Add ddH_2O to a final reaction volume of 50 μL.	

4. In a separate 0.2-mL PCR tube, prepare the following buffers and reagents (to generate *snR81*-Control):

5 μL	10× *Taq* DNA polymerase buffer
2 μL	10 m*M* dNTPs
2 μL	10 μ*M* *snR81*-Control-F1 primer
1 μL	1 μ*M* *snR81*-Control-R1 primer
1 μL	1 μ*M* *snR81*-Control-R2 primer
2 μL	10 μ*M* *snR81*-Control-R3 primer
1 μL	5 U/μL *Taq* DNA polymerase
Add ddH_2O to a final reaction volume 50 μL.	

5. Perform the PCR as follows:
   A: 95 °C for 2 min (1 cycle)
   B: 95 °C for 30 s
   42 °C for 30 s
   72 °C for 30 s
   (Repeat B for 35 cycles)
   C: 72 °C for 2 min (1 cycle)
   4 °C for indefinite time
6. Take 10 μL of the PCR product and add 2 μL of 6 × DNA loading dye before loading the samples on 1% agarose gel containing 0.05% (v/v) of ethidium bromide (to check the PCR products).
7. Conduct electrophoresis in 0.5 × TBE buffer for 30 min (at 10 V/cm gel length). Visualize the PCR products under UV light.
8. Once the success of PCR amplification is confirmed, precipitate the remaining PCR products by adding 200 μL of 100% ethanol and 4 μL of 3 *M* NaOAc to each tube, and spinning at maximum speed (~14,000 × *g*) in a bench-top centrifuge for 10 min.
9. Wash the pellet with 70% ethanol and air-dry. Carefully dissolve each PCR product in 10 μL of ddH_2O.
10. The dissolved PCR products are then mixed with 2 μL of 6 × DNA loading dye and loaded on a 2% agarose gel (with 0.05% (v/v) ethidium bromide). The bands at expected position (size) are excised and purified using Qiagen Gel Extraction kit (refer to kit manual for detailed procedure).

11. The PCR products (*snR81-TRM4-TAA* and *snR81*-Control) are each eluted with 40 μL of ddH_2O. The final concentration of the DNA is adjusted to ~1 μg/μL using ddH_2O.
12. Digest the PCR products with *Bam*HI by mixing the following buffer and reagents in a 1.5-mL tube:

5 μL	1 μg/μL PCR product (*snR81-TRM4-TAA* or *snR81*-Control)
5 μL	10 × *Bam*HI digestion buffer
2.5 μL	*Bam*HI (10 U/μL)
Add ddH_2O to a final reaction volume of 50 μL.	

13. In a separate 1.5-mL tube, digest, in parallel, the pSEC plasmid with *Bam*HI:

5 μL	1 μg/μL pSEC plasmid
5 μL	10 × *Bam*HI digestion buffer
2.5 μL	*Bam*HI (10 U/μL)
Add ddH_2O to a final reaction volume of 50 μL.	

14. Seal the caps of both tubes with parafilm and incubate for 2 h in a 37 °C water bath.
15. Add 450 μL of G50 buffer and 500 μL of PCA to each tube. Vigorously vortex for 30 s.
16. Spin the tubes at the maximum speed (~14,000 × *g*) in a bench-top centrifuge for 1 min.
17. Carefully transfer the upper aqueous phase to a new 1.5-mL tube. Add 1 mL of 100% ethanol and 1 μL of 10 mg/mL glycogen (as carrier).
18. Place the tube in dry ice for 5 min.
19. Spin the tubes at the maximum speed (~14,000 × *g*) in a bench-top centrifuge for 10 min.
20. Wash the pellet with 70% ethanol and air-dry. Carefully dissolve the pellet in 10 μL of ddH_2O.
21. Digest the *Bam*HI-digested PCR product with *Hin*dIII by mixing the following buffer and reagents in a new 1.5-mL tube:

10 μL	1 μg/μL *Bam*HI-digested PCR product (*snR81-TRM4-TAA* or *snR81*-Control)
5 μL	10 × *Hin*dIII digestion buffer
2.5 μL	*Hin*dIII (10 U/μL)
Add ddH_2O to a final reaction volume of 50 μL.	

22. In parallel, digest the *Bam*HI-digested pSEC with *Hin*dIII in a separate 1.5-mL tube:

10 μL	1 μg/μL *Bam*HI-digested pSEC plasmid
5 μL	10 × *Hin*dIII digestion buffer
2.5 μL	*Hin*dIII (10 U/μL)
Add ddH_2O to a final reaction volume of 50 μL.	

23. Seal the caps of both tubes with parafilm and incubate for 2 h in 37 °C water bath.
24. Add 450 μL of G50 buffer and 500 μL of PCA to each tube. Vigorously vortex for 30 s.
25. Spin the tubes at the maximum speed (~14,000 × *g*) in a bench-top centrifuge for 5 min.
26. Carefully transfer the upper aqueous phase to a new 1.5-mL tube. Add 1 mL of 100% ethanol and 1 μL of 10 mg/mL glycogen (as carrier).
27. Place the tube in dry ice for 5 min.
28. Spin the tubes at the maximum speed (~14,000 × *g*) in a bench-top centrifuge for 10 min.
29. Wash the pellet with 70% ethanol and air-dry. Carefully dissolve the pellet in 10 μL of ddH_2O.
30. Prepare the ligation reaction in a new 1.5-mL tube:

5 μL	Digested PCR product (*snR81-TRM4-TAA* or *snR81*-Control)
1 μL	Digested pSEC plasmid
2 μL	5 × T4 DNA ligase buffer
2 μL	T4 DNA ligase (1 U/μL)

31. Seal the tube cap with parafilm and incubate for overnight in a 16 °C water bath.
32. Under sterile conditions, add 5 μL of the ligation reaction to 100 μL of DH5α competent cells (prealiquoted in 1.5-mL tube), mix, and lay on ice for 10 min.
33. Heat shock the cells at 42 °C for 45 s and immediately return to ice for 2 min.
34. Add 900 μL of autoclaved ampicillin-free LB liquid medium to the tube and shake at 200 rpm for 1 h at 37 °C.
35. Spin the tube at ~4000 × *g* in a bench-top centrifuge for 5 min to pellet the cells. Remove supernatant.
36. Resuspend the pellet in 100 μL of ampicillin-containing LB liquid medium and plate the cells onto a LB-ampicillin solid medium. Incubate for overnight at 37 °C.
37. Pick 5–10 individual colonies for each sample (*snR81-TRM4-TAA* and *snR81*-Control) and prepare plasmid DNA from each colony. Sequence the candidate plasmids using a proper primer. Choose one correct plasmid for *snR81-TRM4-TAA* and one for *snR81*-Control, and label them as p*snR81-TRM4-TAA* and p*snR81*-Control, respectively. The final concentration of the plasmid is adjusted with ddH_2O to 1 μg/μL (working concentration).

2.4 Cotransformation of Yeast with p*TRM4-TAA* and p*snR81-TRM4-TAA*

2.4.1 Buffers, Reagent, and Solutions

S. cerevisiae strain BY4741 (Open Biosystems, cat. no. YSC1048)
Yeast extract (BD Diagnostics, cat. no. 90000-444)
Peptone (BD Diagnostics, cat. no. 90000-264)
Yeast nitrogen base (AMRESCO, cat. no. 97064-322)
Ammonium sulfate (EMD Millipore, cat. no. EM-AX1385-1)
L-Isoleucine (Sigma-Aldrich, cat. no. I2752)
L-Valine (Sigma-Aldrich, cat. no. V0500)
Adenine hemisulfate salt (Sigma-Aldrich, cat. no. A9126)
L-Arginine monohydrochloride (Sigma-Aldrich, cat. no. A5131)
L-Histidine monohydrochloride (Sigma-Aldrich, cat. no. H8125)
L-Lysine monohydrochloride (Sigma-Aldrich, cat. no. L5626)
L-Methionine (Sigma-Aldrich, cat. no. M9625)
L-Phenylalanine (Sigma-Aldrich, cat. no. P2126)
L-Tryptophan (Sigma-Aldrich, cat. no. T0254)
L-Tyrosine (Sigma-Aldrich, cat. no. T3754)

Galactose (J.T. Baker, cat. no. M672-07)
Lithium acetate (Alfa Aesar, cat. no. AA13417-30)
Polyethylene glycol PEG-3350 (J.T. Baker, cat. no. JTU221-8)
Synthetic leucine/uracil double drop-out powder:

Yeast nitrogen base	25.1 g/15 L
Ammonium sulfate	75.4 g/15 L
Isoleucine	450 mg/15 L
Valine	2.25 g/15 L
Adenine	300 mg/15 L
Arginine	300 mg/15 L
Histidine	300 mg/15 L
Leucine	0 mg/15 L
Lysine	450 mg/15 L
Methionine	300 mg/15 L
Phenylalanine	750 mg/15 L
Tryptophan	300 mg/15 L
Tyrosine	450 mg/15 L
Uracil	0 mg/15 L

YPD liquid medium: 10 g of yeast extract, 20 g of peptone, and 20 g of dextrose, fill with ddH_2O to 1 L and autoclave
One-step-transformation buffer: 100 m*M* lithium acetate, 50% (w/v) PEG-3350 solution
SGal-LEU-URA double drop-out liquid media: 7.5 g of synthetic leucine/uracil double drop-out powder, 20 g of galactose, fill to 1 L with ddH_2O and autoclave
SGal-LEU-URA double drop-out solid media: 7.5 g of synthetic leucine/uracil double drop-out powder, 20 g of galactose, fill to 1 L with ddH_2O and autoclave. Pour about 20–25 mL of medium in Petri dishes to achieve a bed height of ~0.5 cm

2.4.2 Protocol

1. Under sterile conditions, pick a single yeast colony to grow in 5 mL of YPD liquid medium. Shake at 200 rpm for overnight at 30 °C.

2. The cells are diluted to ~0.5 OD (600 nm) with 5 mL of fresh YPD liquid medium and continue propagation.
3. When the OD_{600nm} reaches 2.0, precipitate the cells by centrifugation at 2500 × *g* at 4 °C for 5 min in an SH3000 rotor (Sorvall RC-5C Plus centrifuge).
4. Carefully remove all supernatant YPD medium and resuspend the cell pellets in 200 μL of one-step-transformation buffer.
5. In a new 1.5-mL tube, prepare the transformation mixture:

50 μL	Resuspended cells
1 μL	1 μg/μL p*TRM4-TAA* plasmid
1 μL	1 μg/μL p*snR81-TRM4-TAA* plasmid

6. Seal the tube cap with parafilm and incubate in a 42 °C water bath for 30 min.
7. Supply 300 μL of fresh YPD media to the mixture and shake at 200 rpm in a 30 °C-shaker for 1 h.
8. Pellet the cells by centrifugation at 2500 × *g* for 5 min in a bench-top centrifuge.
9. Carefully remove all supernatant YPD medium, and resuspend the cells in 100 μL of ddH_2O.
10. Spread the cells on SGal-LEU-URA solid media and incubate at 30 °C for 2–3 days in order to see colonies.

3. DETECTION OF PSEUDOURIDINE IN THE *TRM4* mRNA

3.1 Overview

After the *TRM4-PTC* mRNA and corresponding guide RNA are coexpressed in yeast, *TRM4* mRNA pseudouridylation at the PTC site is assessed. Here we present a protocol for purification, site-specific cleavage, isotope-labeling, and nuclease digestion of *TRM4* mRNA, and analysis of U-to-Ψ conversion at the PTC by thin layer chromatography (TLC). [Note, the conventional approach—CMCT modification coupled with primer extension—is not sensitive enough to allow clear detection and accurate quantification of Ψ in low abundance mRNA.]

The first step in this protocol is to extract a large amount of total RNA from yeast cells. From this pool of RNA, *TRM4* mRNA is further enriched by oligonucleotide affinity chromatography, where *TRM4* mRNA, when hybridized to a biotinylated antisense *TRM4* oligonucleotide, is pulled-down by streptavidin-coupled agarose beads.

The second step is to cleave the enriched *TRM4* mRNA specifically at the site 5′ of the PTC uridine. This is achieved by site-specific RNase H cleavage directed by 2′-*O*-methyl RNA–DNA chimera. Briefly, an antisense *TRM4* 2′-*O*-methyl RNA–DNA chimera is designed, which contains, from 5′ to 3′, two 2′-*O*-methyl RNA nucleotides (complementary to the first two nucleotides of the PTC), four deoxynucleotides (complementary to the four nucleotides immediately upstream of the PTC), and thirteen 2′-*O*-methyl RNA nucleotides (complementary to the 13 nucleotide sequence that is 5 nucleotides upstream of the PTC) (Fig. 4; Yu, 1999). This chimeric oligo, when annealed to *TRM4* mRNA, is able to direct RNase H to cleave the phosphodiester bond 3′ of the substrate nucleotide (immediately preceding the uridine of PTC) that base pairs with the 5′-most deoxynucleotide of the chimera. The resulting 3′-half fragment of the mRNA begins with the PTC codon (UAA) and carries a 5′ phosphate (pU) (Fig. 4).

The third step is to radiolabel the U residue of the PTC codon of the 3′-half fragment. To this end, the 3′-half fragment is dephosphorylated by calf intestine phosphatase (CIP) and rephosphorylated by T4 polynucleotide kinase (PNK) in the presence of [γ^{32}P]ATP (Huang & Yu, 2013).

The last step is to purify the radiolabeled 3′-half fragment of *TRM4* mRNA by gel electrophoresis, digest the purified RNA with nuclease P1, and analyze Ψ/U ratio using TLC (Fig. 5). In order to precisely locate the Ψ on the TLC plate, the nuclease P1 digested sample is spiked with 5′-^{32}P-adenosine-5′-monophosphate, 5′-^{32}P-cytidine-5′-monophosphate,

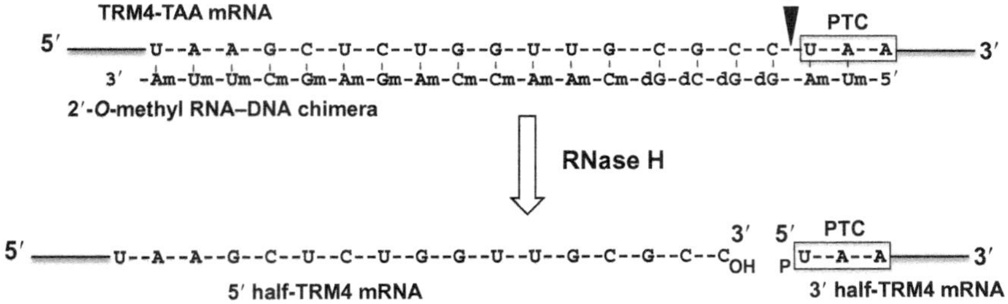

Figure 4 Schematic representation of site-specific RNase H cleavage of *TRM4* mRNA directed by 2′-*O*-methyl RNA–DNA chimera. The target sequence of *TRM4* mRNA is shown and the PTC (UAA) is indicated. The antisense *TRM4* 2′-*O*-methyl RNA–DNA chimera is also shown. d represents deoxy, and m stands for 2′-*O*-methyl. The arrow indicates the RNase H cleavage site.

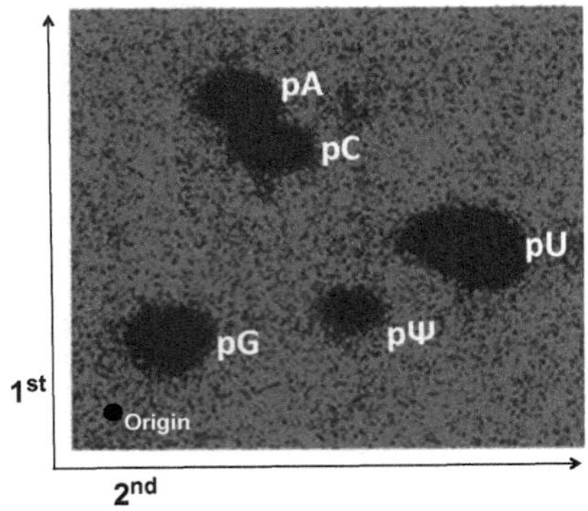

Figure 5 TLC analysis of uridine-to-Ψ conversion at the PTC within the *TRM4* mRNA transcript. The PTC-containing *TRM4* mRNA, coexpressed with a PTC-specific guide RNA, is purified by oligonucleotide affinity chromatography. The RNA is cleaved by RNase H (directed by a specific 2′-*O*-methyl RNA–DNA chimera) at the site 5′ of the PTC (UAA). The resulting 3′-half fragment is 5′ radiolabeled with ^{32}P through dephosphorylation and rephosphorylation (see text). The labeled RNA is digested with nuclease P1 to completion. The digested sample is mixed with an equal amount of 5′-^{32}P-adenosine-monophosphate, 5′-^{32}P-cytidine-monophosphate, and 5′-^{32}P-guanosine-monophosphate, and analyzed by two-dimensional TLC. The first and second dimensions are indicated. The origin and the positions of each 5′-phosphorylated nucleotide are also indicated.

and 5′-^{32}P-guanosine-5′-monophosphate before being spotted on the TLC plate. The percentage of pseudouridylation at the PTC is calculated using the formula: $\Psi/(\Psi+U)$ (Fig. 5).

3.2 Isolation of Total RNA from *S. cerevisiae*

3.2.1 Buffers, Reagent, and Solutions

1 × RIB buffer: 0.2 *M* Tris–HCl pH 7.5, 0.5 *M* NaCl, 0.01 *M* EDTA, 1% SDS

PCI-RIB (phenol/chloroform/isoamyl alcohol = 50/49/1 [v/v/v]): saturated (1:1 [v/v]) with 1 × RIB buffer

Sterile acid-washed glass beads

3.2.2 Protocol

1. Pick and grow a cotransformed yeast colony in 100 mL of SGal-LEU-URA liquid medium. Shake at 200 rpm at 30 °C and monitor OD_{600nm}.
2. When the OD_{600nm} reaches 2.0, precipitate cells by centrifugation at 2500 × *g* at 4 °C for 5 min in an SH3000 rotor (Sorvall RC-5C Plus centrifuge).

3. Resuspend cell pellet (~200 μL) in 400 μL of 1 × RIB buffer and transfer to a new 2-mL screw-cap tube.
4. Add 400 μL of sterile acid-washed glass beads and 400 μL of PCI-RIB and place on ice. Seal the tube cap with parafilm to avoid leakage during vortexing.
5. Vortex vigorously for 40 s and then immediately place the tube on ice for 40 s.
6. Repeat step 5 five times.
7. Spin the 2-mL screw-cap tube at the maximum speed (~14,000 × *g*) in a bench-top centrifuge for 10 min and transfer the supernatant to a new 1.5-mL tube.
8. Add 400 μL of PCI-RIB and vortex for 30 s.
9. Spin the 1.5-mL tube at maximum speed (~14,000 × *g*) in a bench-top centrifuge for 5 min and transfer the supernatant to a new 1.5-mL tube.
10. Repeat steps 8 and 9 one more time.
11. Carefully remove the upper aqueous phase to a new 1.5-mL tube and add 2 × volume of 100% ethanol and 1 μL of 10 mg/mL glycogen (as carrier).
12. Place the tube in dry ice for 5 min.
13. Precipitate the total RNA by centrifugation at the maximum speed (~14,000 × *g*) in a bench-top centrifuge for 10 min.
14. Wash the RNA pellet with 70% ethanol and air-dry. Dissolve the RNA pellet in ddH_2O to a final concentration of 1 μg/μL.

3.3 Purification of *TRM4* mRNA Using Biotinylated-*TRM4* Antisense Oligo and Streptavidin Agarose Beads

3.3.1 *Buffers, Reagent, and Solutions*

Biotin-streptavidin binding buffer: 0.1 *M* phosphate, 0.15 *M* NaCl, 0.1% SDS, 1% NP-40, pH 7.2

Biotinylated-*TRM4* antisense oligo (IDT): biotin-5′-CCACTCTTGTTGGTTCACCAGTGGC-3′

Streptavidin agarose bead (Pierce)

Elution buffer: 0.1 *M* glycine–HCl pH 7.0

3.3.2 *Protocol*

1. Mix 10 μL of 10 μ*M* biotinylated-*TRM4* antisense oligo with 100 μL of 1 μg/μL purified total RNA in a 1.5-mL tube and place in a 65 °C heating block for 10 min.

2. Transfer the tube to room temperature (place it on bench top) for 30 min to let the mRNA anneal to biotinylated-*TRM4* antisense oligo.
3. In the mean time, prepare 50 μL bed volume of streptavidin agarose beads in a new 1.5-mL tube.
4. Spin the tube at medium speed (~3000 × *g*) in a bench-top centrifuge for 1 min and carefully discard supernatant.
5. Add 200 μL of biotin-streptavidin binding buffer to the beads and mix by inverting the tube.
6. Spin the tube at the medium speed (~3000 × *g*) in a bench-top centrifuge for 1 min and carefully discard supernatant.
7. Repeat steps 5 and 6 one more time.
8. Add the annealed mRNA/biotinylated-*TRM4* antisense oligo mixture and 100 μL of biotin-streptavidin binding buffer to the streptavidin agarose beads and nutate at room temperature for 2 h.
9. Spin the tube at medium speed (~3000 × *g*) in a bench-top centrifuge for 1 min and carefully discard supernatant.
10. Add 200 μL of biotin-streptavidin binding buffer to the beads and mix by inverting the tube.
11. Repeat steps 9 and 10 three more times.
12. Add 200 μL of elution buffer to the beads and heat in a 65 °C heating block for 10 min. Mix by inverting the tube several times during the 10 min heating period.
13. Immediately spin the tube at the medium speed (~3000 × *g*) in a bench-top centrifuge for 1 min and transfer the supernatant to a new 1.5-mL tube.
14. Wash and elute the beads again by repeating steps 12 and 13 one more time.
15. Combine the supernatant from steps 13 and 14 and add 800 μL of 100% ethanol and 1 μL of 10 mg/mL glycogen (as carrier).
16. Place the tube in dry ice for 5 min.
17. Precipitate the *TRM4* mRNA by centrifugation at the maximum speed (~14,000 × *g*) in a bench-top centrifuge for 10 min.
18. Wash the pellet with 70% ethanol and air-dry. Dissolve the pellet in a small volume of ddH_2O.

3.4 Site-Specific RNase H Cleavage of *TRM4* mRNA Directed by 2′-*O*-Methyl RNA–DNA Chimeras

3.4.1 Buffers, Reagent, and Solutions

RNase H (Amersham)

10 × Buffer for RNase H: 200 m*M* Tris–HCl (pH 7.5), 100 m*M* $MgCl_2$, 1 *M* KCl, 250 m*M* DTT, 50% sucrose
2′-*O*-Methyl RNA–DNA chimera oligo (IDT): 5′-Um-Am-dG-dG-dC-dG-Cm-Am-Am-Cm-Cm-Am-Gm-Am-Gm-Cm-Um-Um-Am-3′
RNase inhibitor (Fermentas, cat. no. EO0381)
Urea (Sigma, cat. no. U5378)
40% Acrylamide (acrylamide:bis acrylamide = 19:1) (Fisher Scientific, cat. no. BP1406-1)
10% Ammonium persulfate (APS) (Fisher Scientific, cat. no. M2300)
N,N,N′,N′-Tetramethylethylenediamine (TEMED) (Sigma, cat. no. T9281)
GeneRuler Low Range (size markers) (Thermo, cat. no. SM1193)
5 × TBE buffer: 445 m*M* Tris, 445 m*M* boric acid, 16 m*M* EDTA
0.5 × TBE buffer: 44.5 m*M* Tris, 44.5 m*M* boric acid, 1.6 m*M* EDTA (prepare it by making a 1/10 dilution of 5 × TBE buffer with ddH_2O)
2 × RNA sample buffer: 95% formamide, 0.025% SDS, 0.025% bromophenol blue, 0.025% xylene cyanol, 0.5 m*M* EDTA
G50 buffer: 20 m*M* Tris–HCl at pH 7.5, 300 m*M* sodium acetate, 2 m*M* EDTA, 0.2% SDS
PCA (phenol/chloroform/isoamyl alcohol = 25/24/1 [v/v/v]): saturated (1:1 [v/v]) with 20 m*M* Tris–HCl, pH 8.0
2 × RNA loading dye: 95% formamide, 0.5 m*M* EDTA, 0.01% bromophenol blue, 0.005% xylene cyanol

3.4.2 Protocol

1. Mix 6 μL of *TRM4* mRNA (~100 pmol) with 1 μL of complementary 2′-*O*-methyl RNA–DNA chimera (~200 pmol) in a 1.5-mL tube and place in a 95 °C heating block for 3 min.
2. Place the tube on bench top for 10 min to let the *TRM4* mRNA anneal to the 2′-*O*-methyl RNA–DNA chimera.
3. Add 1 μL of 10 × RNase H buffer, 1 μL of RNase inhibitor (40 U), and 1 μL of RNase H (2 U) to the annealed *TRM4* mRNA/2′-*O*-methyl RNA–DNA chimera mixture.
4. Seal the tube cap with parafilm and incubate at 37 °C for 1 h.
5. Add 490 μL of G50 buffer and 500 μL of PCA and mix by vortexing for 30 s.
6. Spin the tube at maximum speed (~14,000 × *g*) in a bench-top centrifuge for 3 min.

7. Carefully transfer the upper aqueous phase to a new 1.5-mL tube. Add 1 mL of 100% ethanol and 1 μL of 10 mg/mL glycogen (as carrier).
8. Place the tube in dry ice for 5 min.
9. Spin the tube at maximum speed (~14,000 × *g*) in a bench-top centrifuge for 10 min.
10. Wash the pellet with 70% ethanol and air-dry. Carefully dissolve the pellet in 5 μL of ddH_2O.
11. To facilitate the next dephosphorylation step, the 3′ mRNA fragments can be separated through gel electrophoresis (Adachi & Yu, 2014). Prepare a 6% acrylamide gel solution following the recipe below:

14.41 g	Urea
3 mL	5 × TBE
4.5 mL	40% Acrylamide (acrylamide:bis acrylamide = 19:1)
Bring up to 30 mL with ddH_2O.	

12. Add 300 μL of freshly made 10% APS and 20 μL of TEMED to the solution and mix gently.
13. Immediately pour the gel solution in between two 20 × 30 cm glass plates (tape-sealed) with 0.42 mm thick spacers. And insert a comb into the gel (from the top).
14. Allow the gel to polymerize at room temperature for at least 20 min.
15. Remove the comb and clean up the wells with 0.5 × TBE. Remove the tape on the bottom of the glass plates and prerun the gel in 0.5 × TBE buffer at 20 W (constant power) for 30 min.
16. Add 5 μL of RNA loading dye to the above RNase H-treated RNA sample (5 μL), and heat it at 95 °C for 3 min and immediately chill on ice.
17. Load the RNA sample into a well of the gel and load a size marker into an adjacent well.
18. Carry out electrophoresis at 20 W for about 60 min.
19. Turn off the power supply when the bromophenol blue dye reaches the bottom of the gel.
20. Expose the gel to a phosphor imaging screen and scan the PhosphorImager.
21. Locate the 3′ half RNA on the gel according to the size marker and cut out the RNA band from the gel with a clean scalpel.

22. Place the gel slice in a 1.5-mL tube and add 400 μL of G50 buffer, and then place the tube in dry ice for 5 min.
23. Place the tube on a rotator and rotate at room temperature overnight.
24. Add 500 μL of PCA to the tube and mix by vortexing for 30 s.
25. Spin the tube at maximum speed (~14,000 × *g*) in a bench-top centrifuge for 5 min.
26. Carefully transfer the upper aqueous phase to a new 1.5-mL tube. Add 1 mL of 100% ethanol and 1 μL of 10 mg/mL glycogen (as carrier).
27. Place the tube in dry ice for 5 min.
28. Spin the tube at maximum speed (~14,000 × *g*) in a bench-top centrifuge for 10 min.
29. Wash the pellet with 70% ethanol and air-dry. Carefully dissolve the pellet in 10 μL of ddH_2O.

3.5 Dephosphorylation of the 3′ mRNA Fragment of *TRM4* to Generate a 5′-Hydroxyl End

3.5.1 *Buffers, Reagent, and Solutions*

CIP (Thermo Scientific)
10× Buffer for CIP

3.5.2 *Protocol*

1. Prepare the dephosphorylation reaction with the following buffer and reagents in a 1.5-mL tube:

2 μL	10× Buffer for CIP
10 μL	Cleaved 3′ RNA fragment
1 μL	RNase inhibitor (40 U)
2 μL	CIP (2 U)
Add ddH_2O to a final reaction volume of 20 μL.	

2. Seal the tube cap with parafilm and incubate at 37 °C for 30 min.
3. Add 480 μL of G50 buffer and 500 μL of PCA and mix by vortexing for 30 s.
4. Spin the tube at maximum speed (~14,000 × *g*) in a bench-top centrifuge for 5 min.
5. Carefully transfer the upper aqueous phase to a new 1.5-mL tube. Add 1 mL of 100% ethanol and 1 μL of 10 mg/mL glycogen (as carrier).

6. Place the tube in dry ice for 5 min.
7. Spin the tube at maximum speed (~14,000 × *g*) in a bench-top centrifuge for 10 min.
8. Wash the pellet with 70% ethanol and air-dry. Carefully dissolve the pellet in 10 μL of ddH_2O.

3.6 Rephosphorylation of the 3′ *TRM4* mRNA Fragment with [γ-^{32}P]ATP

3.6.1 *Buffers, Reagent, and Solutions*

T4 PNK (Thermo Scientific)

10 × Buffer for T4 PNK forward reaction

[γ^{32}P]-ATP (3000 Ci/mmol, 10 μCi/μL) (Perkin Elmer)

3.6.2 *Protocol*

1. Set up a rephosphorylation reaction in a 1.5-mL tube using the following buffer and reagents:

2 μL	10 × Buffer for T4 PNK forward reaction
10 μL	Dephosphorylated 3′ RNA fragment
2 μL	[γ^{32}P]-ATP (3000 Ci/mmol, 10 μCi/μL)
1 μL	RNase inhibitor (40 U)
2 μL	T4 PNK (20 U)
Add ddH_2O to a final reaction volume of 20 μL.	

2. Seal the tube cap with parafilm and incubate for 30 min in a 37 °C water bath.
3. Add 480 μL of G50 buffer and 500 μL of PCA and mix by vortexing for 30 s.
4. Spin the tube at maximum speed (~14,000 × *g*) in a bench-top centrifuge for 3 min.
5. Carefully transfer the upper aqueous phase to a new 1.5-mL tube. Add 1 mL of 100% ethanol and 1 μL of 10 mg/mL glycogen (as carrier).
6. Place the tube in dry ice for 5 min.
7. Spin at the maximum speed (~14,000 × *g*) in a bench-top centrifuge for 10 min.

8. Wash the pellet with 70% ethanol and air-dry. Carefully dissolve the pellet in 10 μL of ddH_2O.
9. For higher purity, the 3′ mRNA fragment is separated on a 6% urea-PAGE gel as described in Section 3.4.2, steps 11–29.

3.7 Nuclease P1 Digestion and TLC Analysis of Pseudouridine in *TRM4* mRNA

3.7.1 Buffers, Reagent, and Solutions

10 × Nuclease P1 digestion buffer: 200 m*M* acetic acid, 10 m*M* $CaCl_2$, pH 6.0

Nuclease P1 (0.2 μg/μL): 0.1 μg of nuclease P1 is sufficient to digest 10 pmol of RNA substrate (at 37 °C for 1 h)

TLC PEI cellulose F plate (EMD)

First dimension developing buffer: isobutyric acid/ddH_2O/NH_4OH (66:33:1) [v/v/v]

Second dimension developing buffer: isopropanol/HCl/ddH_2O (70:15:15) [v/v/v]

3.7.2 Protocol

1. Set up the following nuclease P1 digestion reaction in a 1.5-mL tube:

1 μL	10 × Nuclease P1 digestion buffer
1 μL	Rephosphorylated 3′ RNA fragment
1 μL	Nuclease P1 (0.2 μg/μL)
Add ddH_2O to a final reaction volume of 10 μL.	

2. Seal the tube cap with parafilm and incubate at 37 °C for 2 h.
3. The nuclease P1 digestion reaction is then spiked with 0.1 μL of each of the following controls: 5′-^{32}P-adenosine-monophosphate, 5′-^{32}P-cytidine-monophosphate, and 5′-^{32}P-guanosine-monophosphate.
4. Spot 2 μL of spiked P1 reaction at the origin (one of the corners) of the TLC PEI cellulose F plate (20 × 20 cm^2) and place the plate in a chemical hood until it is completely dry.
5. Place the plate (with the origin at the bottom left) in a chromatography tank containing 75 mL of first dimension developing buffer and allow the solvent to ascend.

6. Remove the plate from the tank, when the solvent front is about 1 cm from the top of the plate (it takes ~10 h).
7. Leave the plate in a chemical hood and let it air-dry for overnight.
8. Rotate the plate by 90° (counterclockwise) and place it (with the origin at the bottom right) in the chromatography tank with 75 mL of second dimension developing buffer. Allow the solvent to ascend.
9. Remove the plate from the tank, when the solvent front is about 1 cm from the top of the plate (it takes ~10 h).
10. Let the plate air-dry in a chemical hood for overnight.
11. Cover the plate with Saran Wrap and expose it to a PhosphorImager (Molecular Dynamics) for overnight. Visualize the ^{32}PΨ and ^{32}PU spots, and quantify the U-to-Ψ conversion efficiency.

4. RECODING OF PSEUDOURIDYLATED PTC CODON

4.1 Overview

To study Ψ-mediated stop-codon recoding, we take advantage of multiple tags at the C-terminus of Trm4. If the Ψ-containing PTC behaves similarly to an unmodified stop codon—terminating ribosome translation, no full-length Trm4 protein (including the C-terminal tags) will be generated. Consequently, no signal will appear when probed with anti-C-terminal tag antibodies (Western blot). If, however, pseudouridylation of the PTC codon leads to nonsense suppression, a full-length readthrough product will be generated and detected by Western blot (anti-C-terminal tag antibodies). The C-terminal tags can also be used to purify (immunoprecipitate) the readthrough product, which can then be analyzed by mass spectrometry to determine the amino acid(s) incorporated at the pseudouridylated stop codon (This is not discussed here).

The first step is to prepare the cell extract for SDS-PAGE gel electrophoresis. Since yeast has resilient cell wall, we need to lyse cells by mixing glass beads with cells and vortex at high frequency. The glass bead-beating technique is ideal for small-scale preparation (discussed below). For medium and large scale, homogenizer or French press should be used (not discussed here).

The second step is to run SDS-PAGE and Western blotting. Since the C-terminal tag contains Protein A, monoclonal anti-Protein A antibody is used to probe the full-length readthrough Trm4 protein (Fig. 6). Any housekeeping gene products in *S. cerevisiae* could be used as loading controls. GAPDH is used in this study (Fig. 6).

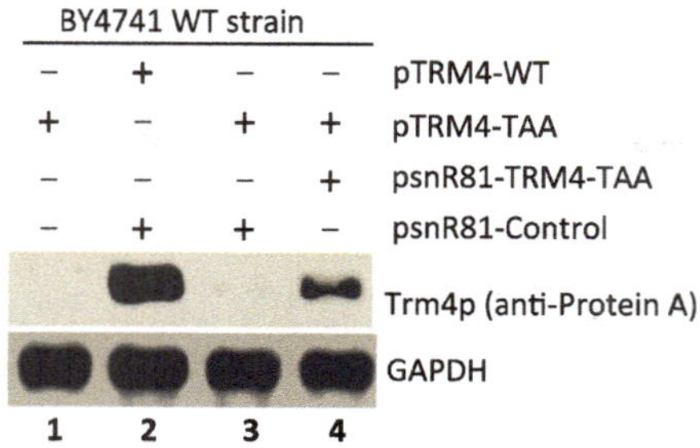

Figure 6 Ψ-Mediated nonsense suppression detected by Western blotting. Equal amounts of total protein are loaded (lanes 1–4), and anti-Protein A and anti-GAPDH (loading control) are used for blotting. Readthrough of PTC is visualized (lane 4), where a cognate (PTC-specific) guide RNA designed to target the PTC for pseudouridylation is coexpressed. In contrast, no significant suppression is observed when no guide RNA (lane 1) or a guide RNA-containing random guide sequences is coexpressed (lane 3). Lane 2 is a positive control where the wild-type *TRM4* mRNA (with no PTC) is expressed.

4.2 Preparation of Cell Extract from *S. cerevisiae*

4.2.1 Buffers, Reagent, and Solutions

Acid-washed glass beads, 0.5 mm (BioSpec Products Inc., cat. no. 11079105)

2× Cell extract preparation buffer: 100 m*M* Tris–HCl at pH 7.5, 300 m*M* KCl, 3 m*M* $MgCl_2$, 1 m*M* EDTA, 2 m*M* DTT (add before use), 1 m*M* PMSF (add before use), and 20% glycerol

4.2.2 Protocol

1. Pick a cotransformed yeast colony and grow in 100 mL of SGal-LEU-URA liquid medium. Shake at 200 rpm at 30 °C and monitor OD_{600nm}.
2. When OD_{600nm} reaches 2.0, pellet the cells by centrifugation at 2500 × *g* at 4 °C for 5 min in an SH3000 rotor (Sorvall RC-5C Plus centrifuge).
3. Transfer the cell pellet (~200 μL) to a new 2-mL screw-cap tube and add 200 μL of 2× cell extract preparation buffer and 200 μL of glass beads. Seal the tube cap with parafilm to avoid leakage.
4. Vortex vigorously for 1 min and then immediately place the tube on ice for 30 s.
5. Repeat step 4 for seven more times.
6. Spin the tube at the maximum speed (~14,000 × g) in a bench-top centrifuge for 10 min and transfer the supernatant to a new 1.5-mL tube.

7. Repeat step 6 for four more times until there is no cell pellet accumulated at the bottom of the new tube after spin.
8. Aliquot the supernatant into 1.5-mL tubes (10 μL each). Use one tube at a time for Western blotting.

4.3 Detection of Full-Length Readthrough Trm4 Protein by Western Blotting

4.3.1 *Buffers, Reagent, and Solutions*

2 × Loading dye: 0.5 *M* Tris–HCl (pH 6.8), 4.4% (w/v) SDS, 20% (v/v) glycerol, 2% (v/v) 2-mercaptoethanol, and 2% (v/v) bromophenol blue
4–15% Tris–HCl Ready Gels (Bio-Rad, cat. no. 161-1104)
10 × Tris–glycine–SDS buffer (Bio-Rad, cat. no. 161-0732)
10 × Tris–glycine buffer (Bio-Rad, cat. no. 161-0734)
Monoclonal anti-Protein A, mouse (Sigma-Aldrich, cat. no. P2921)
Antibody against GAPDH (Pierce, cat. no. MA5-15738)
Goat anti-mouse IgG (H + L)-AP conjugate (Bio-Rad, cat. no. 170-6520)
1-Step NBT/BCIP (Pierce, cat. no. 34042)
Western blot running buffer: mix 100 mL of 10 × Tris–glycine–SDS buffer with 900 mL of ddH_2O
Western blot transferring buffer: mix 100 mL of 10 × Tris–glycine buffer with 200 mL of methanol and 700 mL of ddH_2O
Western blot washing buffer: mix 100 mL of 10 × Tris–glycine buffer with 100 μL of Tween-20 and 899.9 mL of ddH_2O
Western blot blocking buffer: dissolve 5 g of BSA in 100 mL of ddH_2O, pass through 0.2-μm filter

4.3.2 *Protocol*

1. Leave the following buffers and equipment at 4 °C for overnight: 1 L of Western blot running buffer, 1 L of Western blot transferring buffer, gel electrophoresis apparatus, and gel transfer box.
2. Prepare the following loading mixture in a 1.5-mL tube:

10 μL	2 × loading dye
1 μL	Cell extract (∼10 μg)
9 μL	ddH_2O

3. Seal the tube cap with parafilm and place in a 95 °C heating block for 5 min.
4. Briefly centrifuge the tube in a bench-top centrifuge to collect condensation.
5. Load 8 μL of sample on the 4–15% Tris–HCl Ready Gel and carry out electrophoresis at 110 V at 4 °C until the bromophenol blue dye reaches the bottom of the gel.
6. In a container with Western blot transferring buffer, place the gel onto a piece of Whatman nitrocellulose membrane (pore size: 0.1 μm) and sandwich the gel/membrane with two pieces of Whatman filter paper.
7. Place the gel/membrane sandwich into a transfer cassette, with gel facing the anode and membrane facing the cathode.
8. Transfer the protein from gel to membrane at constant 100 V at 4 °C for 2 h.
9. Separate the membrane from the gel, place the membrane (with the protein side up) in a small plastic box containing 20 mL of 5% BSA solution, and shake it (low rpm) at room temperature for 1 h.
10. Transfer and soak the membrane in 20 mL of Western blot washing buffer and shake at room temperature for 10 min.
11. Take 5 μL of monoclonal anti-Protein A and mix with 10 mL of primary hybridization buffer. Soak the membrane in this primary hybridization buffer and shake at 4 °C for 16 h.
12. Transfer and soak the membrane in 20 mL of Western blot washing buffer (fresh) and shake at room temperature for 10 min.
13. Repeat step 12 for three more times.
14. Mix 1 μL of goat anti-mouse (H + L)-AP conjugated IgG with 10 mL of freshly made secondary hybridization buffer. Soak the membrane in this secondary hybridization buffer and shake at room temperature for 1 h.
15. Transfer and soak the membrane in 20 mL of freshly made Western blot washing buffer and shake at room temperature for 10 min.
16. Repeat step 15 for three more times.
17. Transfer and soak the membrane in 10 mL of 1-Step NBT/BCIP solution and shake while shielded from light at room temperature for 5–15 min to visualize protein bands.
18. Terminate the developing reaction by transferring and soaking the membrane in 20 mL of ddH_2O.
19. Air-dry the membrane in darkness before scanning.

ACKNOWLEDGMENTS

We thank the members of the Yu laboratory for inspiring discussions. This work was supported by grants GM104077 and AG039559 (to Y.-T.Y.) from the National Institutes of Health.

REFERENCES

Adachi, H., & Yu, Y. T. (2014). Purification of radiolabeled RNA products using denaturing gel electrophoresis. *Current Protocols in Molecular Biology. 105*. http://dx.doi.org/10.1002/0471142727.mb0420s105, 4.20.1–4.20.13.

Bilbille, Y., Vendeix, F. A., Guenther, R., Malkiewicz, A., Ariza, X., Vilarrasa, J., et al. (2009). The structure of the human tRNALys3 anticodon bound to the HIV genome is stabilized by modified nucleosides and adjacent mismatch base pairs. *Nucleic Acids Research, 37*(10), 3342–3353. http://dx.doi.org/10.1093/nar/gkp187.

Carlile, T. M., Rojas-Duran, M. F., Zinshteyn, B., Shin, H., Bartoli, K. M., & Gilbert, W. V. (2014). Pseudouridine profiling reveals regulated mRNA pseudouridylation in yeast and human cells. *Nature, 515*(7525), 143–146. http://dx.doi.org/10.1038/nature13802.

Chernyakov, I., Whipple, J. M., Kotelawala, L., Grayhack, E. J., & Phizicky, E. M. (2008). Degradation of several hypomodified mature tRNA species in Saccharomyces cerevisiae is mediated by Met22 and the 5′-3′ exonucleases Rat1 and Xrn1. *Genes & Development, 22*(10), 1369–1380. http://dx.doi.org/10.1101/gad.1654308.

Davis, D. R. (1995). Stabilization of RNA stacking by pseudouridine. *Nucleic Acids Research, 23*(24), 5020–5026.

Davis, F. F., & Allen, F. W. (1957). Ribonucleic acids from yeast which contain a fifth nucleotide. *The Journal of Biological Chemistry, 227*(2), 907–915.

Durant, P. C., & Davis, D. R. (1999). Stabilization of the anticodon stem-loop of tRNALys,3 by an A+ − C base-pair and by pseudouridine. *Journal of Molecular Biology, 285*(1), 115–131. http://dx.doi.org/10.1006/jmbi.1998.2297.

Fernandez, I. S., Ng, C. L., Kelley, A. C., Wu, G., Yu, Y. T., & Ramakrishnan, V. (2013). Unusual base pairing during the decoding of a stop codon by the ribosome. *Nature, 500*(7460), 107–110. http://dx.doi.org/10.1038/nature12302.

Ge, J., & Yu, Y. T. (2013). RNA pseudouridylation: New insights into an old modification. *Trends in Biochemical Sciences, 38*(4), 210–218. http://dx.doi.org/10.1016/j.tibs.2013.01.002.

Huang, C., Karijolich, J., & Yu, Y. T. (2011). Post-transcriptional modification of RNAs by artificial box H/ACA and box C/D RNPs. *Methods in Molecular Biology, 718*, 227–244. http://dx.doi.org/10.1007/978-1-61779-018-8_14.

Huang, C., Wu, G., & Yu, Y. T. (2012). Inducing nonsense suppression by targeted pseudouridylation. *Nature Protocols, 7*(4), 789–800. http://dx.doi.org/10.1038/nprot.2012.029.

Huang, C., & Yu, Y. T. (2013). Synthesis and labeling of RNA *in vitro*. *Current Protocols in Molecular Biology, 4*, Unit4.15. http://dx.doi.org/10.1002/0471142727.mb0415s102.

Karijolich, J., & Yu, Y. T. (2008). Insight into the protein components of the box H/ACA RNP. *Current Proteomics, 5*(2), 129–137. http://dx.doi.org/10.2174/157016408784911936.

Karijolich, J., & Yu, Y. T. (2011). Converting nonsense codons into sense codons by targeted pseudouridylation. *Nature, 474*(7351), 395–398. http://dx.doi.org/10.1038/nature10165.

King, T. H., Liu, B., McCully, R. R., & Fournier, M. J. (2003). Ribosome structure and activity are altered in cells lacking snoRNPs that form pseudouridines in the peptidyl transferase center. *Molecular Cell, 11*(2), 425–435.

Lecointe, F., Namy, O., Hatin, I., Simos, G., Rousset, J. P., & Grosjean, H. (2002). Lack of pseudouridine 38/39 in the anticodon arm of yeast cytoplasmic tRNA decreases *in vivo* recoding efficiency. *The Journal of Biological Chemistry*, *277*(34), 30445–30453. http://dx.doi.org/10.1074/jbc.M203456200.

Lee, C. H., & Tinoco, I., Jr. (1980). Conformation studies of 13 trinucleoside diphosphates by 360 MHz PMR spectroscopy. A bulged base conformation. I. Base protons and H1′ protons. *Biophysical Chemistry*, *11*(2), 283–294, http://dx.doi.org/0301-4622(80)80031-7 [pii].

Liang, X. H., Liu, Q., & Fournier, M. J. (2007). rRNA modifications in an intersubunit bridge of the ribosome strongly affect both ribosome biogenesis and activity. *Molecular Cell*, *28*(6), 965–977. http://dx.doi.org/10.1016/j.molcel.2007.10.012.

Lovejoy, A. F., Riordan, D. P., & Brown, P. O. (2014). Transcriptome-wide mapping of pseudouridines: Pseudouridine synthases modify specific mRNAs in S. cerevisiae. *PLoS One*, *9*(10), e110799. http://dx.doi.org/10.1371/journal.pone.0110799.

Ma, X., Yang, C., Alexandrov, A., Grayhack, E. J., Behm-Ansmant, I., & Yu, Y. T. (2005). Pseudouridylation of yeast U2 snRNA is catalyzed by either an RNA-guided or RNA-independent mechanism. *The EMBO Journal*, *24*(13), 2403–2413.

Schwartz, S., Bernstein, D. A., Mumbach, M. R., Jovanovic, M., Herbst, R. H., Leon-Ricardo, B. X., et al. (2014). Transcriptome-wide mapping reveals widespread dynamic-regulated pseudouridylation of ncRNA and mRNA. *Cell*, *159*(1), 148–162. http://dx.doi.org/10.1016/j.cell.2014.08.028.

Wu, G., Yu, A. T., Kantartzis, A., & Yu, Y. T. (2011). Functions and mechanisms of spliceosomal small nuclear RNA pseudouridylation. *Wiley Interdisciplinary Reviews RNA*, *2*(4), 571–581. http://dx.doi.org/10.1002/wrna.77.

Yu, Y. T. (1999). Construction of 4-thiouridine site-specifically substituted RNAs for cross-linking studies. *Methods*, *18*(1), 13–21. http://dx.doi.org/10.1006/meth.1999.0752.

Yu, Y. T., Shu, M. D., & Steitz, J. A. (1998). Modifications of U2 snRNA are required for snRNP assembly and pre-mRNA splicing. *The EMBO Journal*, *17*(19), 5783–5795.

CHAPTER ELEVEN

Pseudo-Seq: Genome-Wide Detection of Pseudouridine Modifications in RNA

Thomas M. Carlile, Maria F. Rojas-Duran, Wendy V. Gilbert[1]
Department of Biology, Massachusetts Institute of Technology, Cambridge, Massachusetts, USA
[1]Corresponding author: e-mail address: wgilbert@mit.edu

Contents

Methods in Enzymology, Volume 560
ISSN 0076-6879
http://dx.doi.org/10.1016/bs.mie.2015.03.011

Abstract

RNA molecules contain a variety of chemically diverse, posttranscriptionally modified bases. The most abundant modified base found in cellular RNAs, pseudouridine (Ψ), has recently been mapped to hundreds of sites in mRNAs, many of which are dynamically regulated. Though the pseudouridine landscape has been determined in only a few cell types and growth conditions, the enzymes responsible for mRNA pseudouridylation are universally conserved, suggesting many novel pseudouridylated sites remain to be discovered. Here, we present Pseudo-seq, a technique that allows the identification of sites of pseudouridylation genome-wide with single-nucleotide resolution. In this chapter, we provide a detailed description of Pseudo-seq. We include protocols for RNA isolation from *Saccharomyces cerevisiae*, Pseudo-seq library preparation, and data analysis, including descriptions of processing and mapping of sequencing reads, computational identification of sites of pseudouridylation, and assignment of sites to specific pseudouridine synthases. The approach presented here is readily adaptable to any cell or tissue type from which high-quality mRNA can be isolated. Identification of novel pseudouridylation sites is an important first step in elucidating the regulation and functions of these modifications.

1. INTRODUCTION

1.1 RNA Modifications

In addition to the four canonical bases, RNA molecules contain a wide array of posttranscriptional modifications. Across species, more than 100 chemically diverse modified bases have been identified within RNAs, primarily in stable noncoding RNAs (ncRNAs, e.g., tRNA and rRNA) (Cantara et al., 2011). Until recently, only three modified bases were known to occur within mRNAs—inosine, N^6-methyladenosine (m^6A), and 5-methylcytosine (m^5C). The most abundant modified base, pseudouridine, is found at multiple positions in both tRNAs and rRNAs (Ge & Yu, 2013), but was not discovered in mRNAs until new genome-wide methods for pseudouridine profiling enabled detection of this modification in low-abundance transcripts (Carlile et al., 2014; Lovejoy, Riordan, & Brown, 2014; Schwartz et al., 2014).

Ψ formation is catalyzed by two classes of enzymes, each with distinct modes of substrate recognition. The first class, Box H/ACA snoRNAs, found in archaea and eukaryotes, uses small noncoding RNAs as guides to target the catalytic protein Cbf5/dyskerin to its substrates by base pairing between the guide RNA and the target RNA (Ganot, Bortolin, & Kiss, 1997). This class primarily targets modifications within rRNA. The second class of pseudouridine synthase (Pus) proteins modifies tRNAs as well as additional ncRNAs. The Pus proteins, which are conserved in all domains of life, do not require guide RNAs to bind their targets. Instead, Pus proteins directly recognize various structural and/or sequence elements in their RNA substrates (Arluison, Buckle, & Grosjean, 1999; Urban, Behm-Ansmant, Branlant, & Motorin, 2009). Both types of pseudouridine synthases form Ψ, which is the C5-glycoside isomer of uridine, by breakage of the N1-glycosyl bond followed by 180° base rotation and formation of a C5-glycosyl bond (Fig. 1A). This modification leaves the Watson–Crick edge of uridine unchanged, but frees the hydrogen at position N1 to act as a hydrogen bond donor, which is implicated in the stabilizing effects of Ψ on RNA structure (Charette & Gray, 2000; Hudson, Bloomingdale, & Znosko, 2013).

Dynamic pseudouridylation may play a regulatory role in response to cellular stressors given that the pseudouridine landscape changes substantially in different growth states (Carlile et al., 2014; Courtes et al., 2014; Lovejoy et al., 2014; Schwartz et al., 2014; Wu, Xiao, Yang, & Yu, 2011). Although the functions of endogenous mRNA pseudouridylation events are not yet known, the established effects of Ψ on RNA structure suggest many possibilities for posttranscriptional regulation through the impact of mRNA structure on translation initiation efficiency, ribosome pausing, RNA

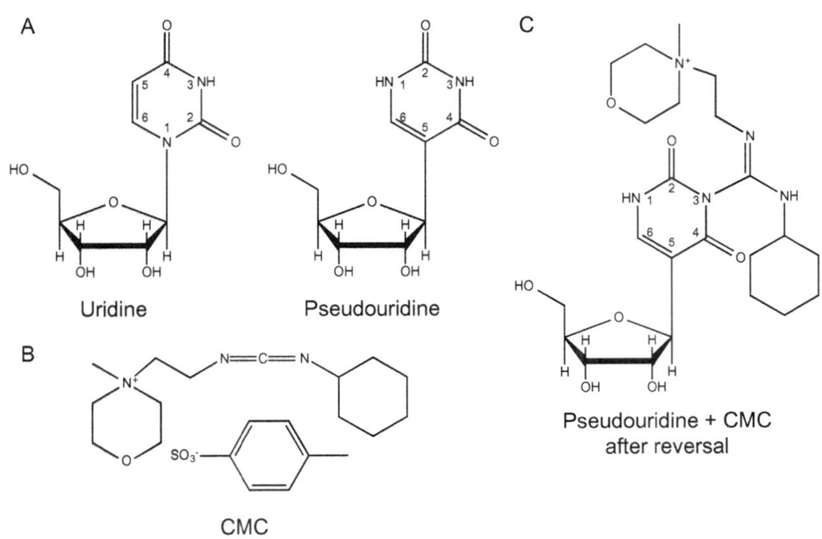

Figure 1 Structures of (A) uridine and pseudouridine, (B) CMC, and (C) Ψ-CMC adduct after CMC treatment and reversal.

localization, and regulation by RNA interference (Jambhekar & DeRisi, 2007; Kudla, Murray, Tollervey, & Plotkin, 2009; Shah, Ding, Niemczyk, Kudla, & Plotkin, 2013; Somogyi, Jenner, Brierley, & Inglis, 1993; Tan et al., 2012). Furthermore, mRNA pseudouridylation may provide a mechanism for dynamically altering the genetic code. Artificially targeted pseudouridylation of stop codons *in vivo* leads to noncanonical base pairing of Ψ in the decoding center of the ribosome and efficient nonsense suppression (Karijolich & Yu, 2011). Structural studies of this surprising phenomenon suggested the possibility of more widespread effects of pseudouridine on decoding (Fernández et al., 2013).

Though regulated pseudouridylation of mRNA has the potential to profoundly affect both protein production and protein function, most of the pseudouridylated sites in mRNAs remain to be discovered. The Pseudo-seq method described here will enable investigations of regulated pseudouridylation events in a variety of contexts, including developmental changes and disease states linked to defects in pseudouridine synthase activity (Anderson, Brewer, Singh, & Boothroyd, 2009; Bykhovskaya, Casas, Mengesha, Inbal, & Fischel-Ghodsian, 2004; Heiss et al., 1998; Mei et al., 2011).

1.2 Detection of Pseudouridine

A variety of techniques have been used to identify and map the locations of Ψ residues within RNAs. Pseudouridine was initially found as an unknown "fifth nucleotide" in preparations of yeast RNA using paper chromatography (Cohn, 1960; Davis & Allen, 1957). Sequencing of tRNAs using RNase digestion, column chromatography, and paper electrophoresis and chromatography led to the first identification of specific sites of pseudouridylation (Holley, Everett, Madison, & Zamir, 1965). Similar techniques were subsequently used to map Ψ positions in rRNAs (Choi & Busch, 1978; Gupta & Randerath, 1979; Tanaka, Dyer, & Brownlee, 1980), but these approaches were limited to very abundant RNAs that could be purified to homogeneity.

The ability to map the positions of Ψ residues with single-nucleotide resolution was greatly improved with the method introduced by Bakin and Ofengand (1993). In this assay, isolated RNA is treated with the carbodiimide *N*-cyclohexyl-*N*′-(2-morpholinoethyl)carbodiimide metho-*p*-toluenesulfonate (CMC) *in vitro* (Fig. 1B) to form covalent adducts at the N1 position of guanidine, N3 position of uridine, and N1 and N3 positions of pseudouridine. Because only the CMC-N3 adduct of pseudouridine is resistant to alkaline cleavage, Ψ residues can be specifically labeled by incubating CMC-derivatized RNA in sodium carbonate buffer, pH 10.4 (Fig. 1C). The bulky Ψ-CMC adduct acts as a strong stop to reverse transcriptase (RT) one nucleotide 3′ of the Ψ, allowing Ψ locations to be mapped using gene-specific primer extension assays (Bakin & Ofengand, 1993). More recently, techniques have been developed to allow site-specific identification and quantification of Ψ using thin-layer chromatography (Zhao & Yu, 2004), as well as mass spectrometry-based techniques that allow detection of normally mass silent Ψ (Durairaj & Limbach, 2008; Popova & Williamson, 2014). While powerful in some contexts, these assays are low throughput and are best suited to identification of Ψs in highly abundant RNAs.

A growing interest in the biological functions of RNA modifications has created a need for new methods for their transcriptome-wide identification. Recent studies have successfully used next-generation sequencing techniques to map various RNA modifications genome-wide, including m^6A (Dominissini et al., 2013; Meyer et al., 2012), m^5C (Squires et al., 2012), inosine (Li et al., 2009), and pseudouridine (Carlile et al., 2014; Lovejoy et al., 2014; Schwartz et al., 2014). In this chapter, we describe Pseudo-seq, a technique that allows efficient, single-nucleotide mapping of Ψs genome-wide (Carlile et al., 2014). Pseudo-seq combines high-throughput sequencing with the ability to specifically derivatize Ψ residues with CMC. Deep sequencing of abortive RT products identifies sites where RT stops, and CMC-dependent RT stops represent sites of pseudouridylation. Here, we provide protocols for isolation of total or poly(A)+ RNA, Pseudo-seq library preparation, and sequencing data analysis. While these procedures have been optimized for performing Pseudo-seq on *Saccharomyces cerevisiae*, they can easily be adapted to other organisms, provided a sufficient quantity of mRNA can be obtained.

2. SAMPLE PREPARATION AND RNA ISOLATION

The first step in Pseudo-seq library preparation is the isolation of high-quality total RNA from a cell type of interest. Total RNA is sufficient for the detection of Ψs in highly abundant RNA species, such as tRNAs, rRNAs, or other very highly expressed ncRNAs. However, Ψ detection in less abundant RNAs, including mRNAs, requires their enrichment from the pool of total RNA. The specific protocols used for isolation of RNA will vary widely depending on the biological system being used and can be found elsewhere. Here, we provide a protocol for the isolation of total RNA from exponentially growing cultures of *S. cerevisiae* using hot acid phenol (Collart & Oliviero, 2001). However, any method of RNA isolation that yields high-quality total RNA will be compatible with Pseudo-seq library preparation. RNA should be prepared from multiple independent biological replicates, as these replicates are essential for high-confidence pseudouridine identification, especially in transcripts of lower abundance. A discussion of the number of biological replicates required can be found in Section 5.2. Additionally, we provide a cost-effective protocol for the enrichment of polyadenylated RNAs using oligo (dT) cellulose beads (Sambrook & Russell, 2001). Other strategies for mRNA enrichment may be substituted as desired (e.g., rRNA depletion may be used for the purification of mRNA from prokaryotes or from tissue samples with significant RNA degradation).

2.1 RNA Isolation from *S. cerevisiae*

Inoculate a 10-ml starter culture in YPAD and grow overnight at 30 °C. The OD_{600} should reach approximately 7.0. Use this starter to inoculate 750 ml of prewarmed YPAD to an OD_{600} of 0.05. Grow this culture at 30 °C in a baffled flask with shaking at 200 rpm to a final OD_{600} of 1.0. Harvest the cells by centrifugation at 16,000 × *g* for 5 min at 4 °C. Resuspend the cell pellet in 25 ml ice-cold water and transfer to a 50-ml conical tube. Pellet cells by centrifugation at 3400 × *g* for 5 min at 4 °C and pour off supernatant. The pellets should be approximately 5 ml and can be either used directly for RNA isolation or snap-frozen in liquid N_2 and stored at −80 °C. The growth conditions described here can be varied to examine pseudouridylation under different conditions, provided a sufficient quantity of cells is harvested.

Isolation of total RNA from yeast cells using hot acid phenol extraction is efficient and yields RNA that is relatively free of DNA contamination. Add 5 ml acid phenol (Sigma P4557, without alkaline buffer) and 5 ml AES Buffer to the cell pellet. Incubate for 30 min in a 65 °C water bath, vortexing for 10 s every minute. Place the tube on ice for 10 min. Add 5 ml chloroform (Sigma C2432) and centrifuge at 3400 × *g* for 5 min at room temperature. Transfer the upper, aqueous phase to a new 15-ml conical tube, add 5 ml of acid phenol:chloroform:isoamyl alcohol (Ambion AM9732; without alkaline buffer), and vortex briefly. Centrifuge and transfer aqueous phase to a new 15-ml conical tube as above. Repeat this extraction step until the interface between the organic and aqueous phases is free of protein (two to three times total). Perform a final extraction by transferring the aqueous phase to a new 15-ml conical tube, adding 5 ml chloroform, and spinning as above. Transfer the aqueous phase to an Oakridge Tube (Thermo 3114-0030) and add 1/9th volume 3 *M* NaOAc, pH 5.3, and 1 volume isopropanol. Spin at 14,000 × *g* for 30 min at 4 °C. Wash the RNA pellet twice with 10 ml ice-cold 70% ethanol, spinning at 14,000 × *g* for 10 min at 4 °C. Air dry the pellet for approximately 3 min. If proceeding directly to RNA fragmentation (Section 3.1), resuspend RNA in 1–2 ml H_2O. If performing poly(A) selection (Section 2.2), resuspend in 1–2 ml TES Buffer. A yield of at least 10 μg of RNA per milliliter of cells at an OD_{600} of 1.0 is expected, and a final RNA concentration greater than 1 μg/μl is desired.

2.2 Poly(A) Selection

This protocol uses oligo (dT) cellulose beads (NEB S1408). However, other commercially available beads for poly(A) selection are suitable for use with Pseudo-seq. It should also be possible to use other available methods for mRNA enrichment, such as those based on subtractive hybridization of rRNA. Such techniques should be used if Ψ detection in lower abundance, nonpolyadenylated transcripts is desired. Yeast mRNAs contain relatively short poly(A) tails compared to other eukaryotic organisms (Subtelny, Eichhorn, Chen, Sive, & Bartel, 2014). We have found that poly(A) selection using oligo (dT) cellulose beads coupled with relatively long incubations yields greater recovery of yeast mRNA than rapid, magnetic bead-based capture methods (Dynabeads), which may be used for organisms with longer poly(A) tails.

A sufficient quantity (1.5 ml of 50% slurry per sample) of oligo (dT) cellulose beads for all samples should be washed in batch in 15 ml conical tubes. Pellet the beads at 3000 × *g* for approximately 30 s at room temperature and remove the supernatant, taking care to minimize disturbance of the beads. Wash three times in two slurry volumes of water, followed by two washes with two slurry volumes of TES + NaCl, pelleting as above between washes. Prior to pelleting the final wash, distribute equal volumes (~750 μl bead bed volume) to individual 15 ml conical tubes for each sample and pellet as above.

Bring 7.5–10 mg of total RNA up to a volume of 4.5 ml in TES (without NaCl) in a 15-ml conical tube. Denature at 65 °C for 15 min and then place on ice for 2 min. To bind poly(A) RNA to the beads, add 563 μl 5 *M* NaCl to the denatured RNA and transfer to the oligo (dT) cellulose pellet.

Vortex to mix and incubate at room temperature for 15 min with rotation. To perform a second round of binding, which increases mRNA yield, pellet the beads as above, transfer the supernatant to a 15-ml conical tube, denature at 65 °C for 10 min, and place on ice for 2 min. Then, add the supernatant back to the beads, vortex to mix, and incubate at room temperature for 15 min with rotation. Pellet the beads as above and discard the supernatant. Wash the beads three times with 5 ml TES + NaCl, incubating each wash at room temperature for 2 min with rotation. Wash once with 2 ml ice-cold water, vortex briefly, pellet as above, and discard the supernatant. This wash should be performed quickly to avoid losing RNA bound to the beads. To elute poly(A) RNA, add 2 ml 55 °C water to the beads and incubate at 55 °C for 5 min. Pellet as above and transfer supernatant to a 15-ml conical tube. Repeat elution as before and pool eluates. Wash the beads once in 5 ml water and once in 5 ml TES + NaCl for reuse in the second round of poly(A) selection.

A second round of poly(A) selection can be performed by rebinding the eluates to oligo (dT) beads; this will decrease rRNA contamination of the purified mRNA. Bring the pooled eluates up to 5 ml total volume in TES (without NaCl) by adding 50 μl 1 *M* Tris, pH 7.6; 10 μl 0.5 *M* EDTA, pH 8.0; 25 μl 20% SDS; and water to 5 ml total. Denature RNA at 65 °C for 10 min and place on ice for 2 min. To bind poly(A) RNA, add 626 μl 5 *M* NaCl to the denatured RNA, transfer to the washed beads, and incubate at room temperature for 15 min with rotation. To perform a second round of denaturation, pellet the beads as above, transfer the supernatant to a 15-ml conical tube, denature at 65 °C for 5 min, and place on ice for 2 min. To rebind, add the supernatant back to the beads, vortex to mix, and incubate at room temperature for 15 min with rotation. Perform the washes and elutions as above, except elute twice in 1.8 ml 55 °C water, and pool. To remove residual oligo (dT) cellulose beads, pass the eluates through a 0.45-μm cellulose acetate filter (VWR 28145-481) with a syringe and transfer in 900 μl aliquots to 2 ml microcentrifuge tubes. To each aliquot, add 1/9th volume 3 *M* NaOAc, pH 5.3; 2 μl GlycoBlue (Invitrogen AM9516); and one volume isopropanol and precipitate at −20 °C for at least 30 min. Spin in a microcentrifuge at max speed at 4 °C for 30 min. Wash the pellet in 750 μl ice-cold 70% ethanol, spin at max speed at 4 °C for 10 min, and air dry for 2 min. Resuspend each pellet in 6 μl of water and pool into a single PCR tube. The expected yield is approximately 2 μg poly(A)+ RNA per sample.

3. PSEUDO-SEQ LIBRARY PREPARATION

Pseudo-seq is derived from the primer extension-based method for Ψ detection described earlier (Bakin & Ofengand, 1993), which we have adapted to the Illumina sequencing platform. This is accomplished by incorporating CMC modification into a sequencing library preparation protocol, and deep sequencing of abortive rather than full-length RT products from both CMC-treated (+CMC) and -untreated (−CMC) RNA samples. Subsequent computational analyses (Section 4) are used to identify reproducible CMC-dependent RT stops, which correspond to Ψ residues (Fig. 2).

RNA is first randomly fragmented to ensure uniform coverage of the transcriptome. After RNA fragmentation, Ψ residues must be specifically derivatized with CMC to allow for their identification. However, since Ψ, U, and G residues are CMC reactive, specific reversal of the U-CMC and G-CMC adducts is needed. This is accomplished by reversal of CMC modification under alkaline conditions, which exploits the resistance of the CMC-N3-Ψ linkage to alkaline hydrolysis. Following CMC modification, a narrow range of RNA fragment sizes is then selected, which is essential for subsequent separation of truncated cDNAs from full-length cDNAs following the RT step. After size selection, a 3′ adapter is ligated onto the RNA fragments, which provides a binding site for a single RT primer, thereby avoiding the requirement for gene-specific primers used

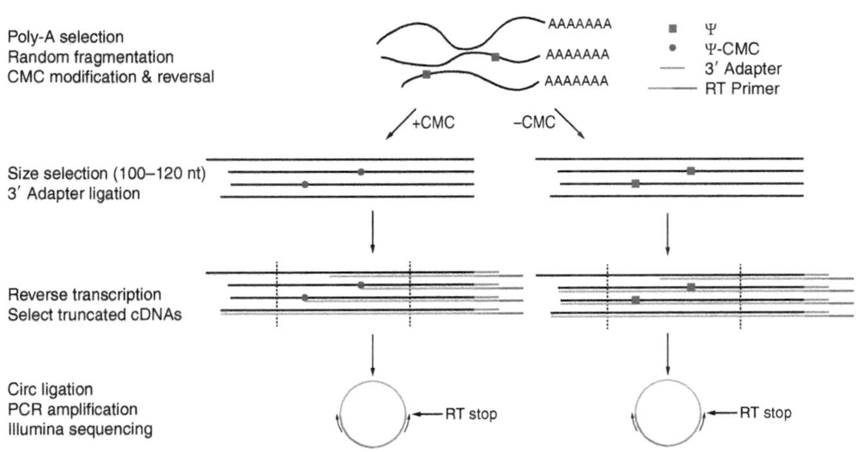

Figure 2 A schematic of Pseudo-seq library preparation. CMC-dependent RT stops correspond to sites of pseudouridylation. See Section 3 for details. (See the color plate.)

in traditional primer extension assays. After cDNA synthesis, truncated RT products are selected by gel purification, enriching for cDNAs whose 3′ ends correspond to sites at which RT stops. Stops due to the presence of Ψ-CMC adducts will be present in the +CMC libraries, while natural stops, such as those caused by RNA secondary structure, will be present in both −CMC and +CMC libraries. Intramolecular ligation with a ssDNA ligase circularizes the cDNA, providing a 5′ binding site for primers for subsequent PCR amplification, as well as for primers for Illumina sequencing. This chapter assumes that quality control steps to assess library concentration and library size distribution will be carried out by the facility performing the sequencing.

3.1 RNA Fragmentation

Randomly fragmented RNA is needed to ensure even coverage of potential pseudouridine sites. RNA can be fragmented in a relatively sequence-independent fashion using divalent zinc cations. However, the efficiency of RNA fragmentation differs depending upon the source and composition of the RNA sample. Thus, optimization of fragmentation time and temperature may be necessary to obtain a sufficient quantity of fragments in the desired size range. For Pseudo-seq, fragments in the range of 60–150 nt range are acceptable, though a narrow size distribution (e.g., 60–70 nt) should be used for a given experiment. The following fragmentation protocols yield sufficient fragments for both +CMC and − CMC libraries.

Fragment all of the pooled poly(A) RNA from Section 2.2. This should be approximately 2 μg of RNA, but up to 25 μg can be efficiently fragmented in this volume. To the poly(A) RNA (in 24 μl), add water to a final volume of 54 μl and place on ice. Add 6 μl of 100 m*M* $ZnCl_2$ (10 m*M* final) and fragment in a thermocycler for 55 s at 94 °C. Quench the reaction by quickly placing on ice, and adding 60 μl of 40 m*M* EDTA (20 m*M* final). Add 1/9th volume 3 *M* NaOAc and one volume isopropanol, and precipitate at −20 °C for at least 30 min. GlycoBlue does not need to be added to the precipitation since it carries over from the poly(A) selection. Spin and wash as described in Section 2.2. Resuspend fragmented RNA in 30 μl of water. Yeast total RNA can be fragmented in 10 m*M* $ZnCl_2$ for 5 min at 94 °C.

3.2 CMC Modification and Reversal

RNA should be modified with freshly made 0.5 *M* CMC in BEU Buffer (212 mg/ml). CMC-treated and mock-treated (−CMC) libraries should be prepared in parallel. To account for the reduced recovery of

CMC-modified RNA by precipitation, transfer 18 and 12 μl of fragmented RNA to microcentrifuge tubes for the + CMC and −CMC libraries, respectively, and bring the total volume of each to 20 μl with water. Add 2.9 μl of 40 m*M* EDTA to each sample (5 m*M* final), denature at 80 °C for 3 min, and place on ice. Add 100 μl of 0.5 *M* CMC in BEU Buffer (0.4 *M* CMC final) to the + CMC sample, and add 100 μl of BEU Buffer to the −CMC sample. Incubate at 40 °C for 45 min at 1000 rpm in a Thermomixer (Eppendorf). To precipitate, add 2 μl GlycoBlue; 50 μl 3 *M* NaOAc, pH 5.3; and 1 ml ethanol and chill at −20 °C for at least 30 min. Spin in a microcentrifuge at max speed at 4 °C for 30 min, and wash twice in 500 μl ice-cold 70% ethanol, spinning at maximum speed at 4 °C for 10 min after each wash. Air dry the RNA pellets for 2 min after washes. These precipitation conditions give a high yield of CMC-modified RNA compared to isopropanol precipitation. The CMC-treated RNA pellets may be more diffuse than the mock-treated pellets.

To reverse the CMC modifications on U and G residues, resuspend the RNA pellet in 30 μl sodium carbonate pH 10.4 Buffer and incubate at 50 °C for 2 h at 1000 rpm in a Thermomixer. Both CMC-modified and mock-treated RNA should be treated in parallel. To precipitate, add 2 μl GlycoBlue, 1/9th volume 3 *M* NaOAc, pH 5.3, and 2.5 volumes ethanol and chill at −20 °C for at least 30 min. The precipitation should be spun and washed as above. Resuspend the RNA pellets in 8 μl of 10 m*M* Tris, pH 8.0.

Careful execution of the CMC modification and reversal steps is important to ensure specific derivatization of Ψ residues by CMC. Thus, fresh CMC and properly stored sodium carbonate pH 10.4 Buffer should be used.

Pseudouridine identification by Pseudo-seq is robust to small changes in modification conditions. We have used CMC concentrations of both 0.2 and 0.4 *M* and have observed similar Ψ signal at both concentrations. However, increasing the extent of CMC modification with higher concentrations may interfere somewhat with the detection of 5′ Ψs in closely spaced groups of modifications due to a shadowing effect of the 3′ Ψs. Thus, for monitoring pseudouridylation in rRNA and tRNA, which contain many closely spaced Ψs, lower concentrations of CMC are recommended.

3.3 3′ End Healing

RNA fragmentation with divalent zinc cations leaves a 2′,3′ cyclic phosphate, which must be removed to make the RNA fragments suitable substrates for subsequent 3′ adapter ligation. These cyclic phosphates can be converted to 3′ phosphates by T4 polynucleotide kinase (PNK), which can then be removed by the action of Calf Intestinal Alkaline Phosphatase (CIP). To 8 μl of RNA, add 0.5 μl of RNasin Plus (Promega N2615), 1.25 μl 10 × PNK Buffer, 1.25 μl T4 PNK (NEB M0201), and 1 μl CIP (NEB M0290) and incubate at 37 °C for 1–2 h. Add 12.5 μl of 2 × RNA Loading Dye to prepare for size selection (Section 3.4).

3.4 RNA Size Selection

After CMC treatment and 3′ end healing, the desired range of RNA fragment sizes is selected. Selection of a narrow range of fragment sizes, spanning 10–20 nt, at this step allows reliable separation of truncated from full-length cDNAs after RT. We have performed Pseudo-seq on a variety of RNA fragment sizes ranging from 60–70 to 120–140 nt and have found that our protocol is robust to changes in RNA fragment size, provided the size range taken is narrow. RNA fragments are size selected by excising and eluting the desired range of RNA fragment sizes from a denaturing PAGE gel.

Prepare 8% TBE/Urea/Polyacrylamide mini-gels (8 × 10 cm), and prerun for 20 min at 200 V. While the gel is prerunning, prepare RNA fragments and 10 bp Ladder (Life Technologies 10821-015) for loading. The 10 bp Ladder should be loaded in both gel lanes flanking RNA fragments to facilitate size selection. For each lane, prepare 20 μl of 10 bp Ladder (0.5 μl 10 bp Ladder, 9.5 μl water, 10 μl 2 × RNA Loading Dye). Denature the RNA fragments and 10 bp Ladder at 95 °C for 2 min, and then place on ice until loading. Load the gel and run for 60 min at 200 V for the fragment sizes indicated below (the bromophenol blue dye front will run off of the gel). Stain the gel for 5 min with SYBR Gold (Invitrogen S-11494) diluted 1:10,000 in 0.5 × TBE, and then visualize by UV transillumination. Excise several ranges of RNA fragment sizes (80–100, 100–120, and 120–140) guided by the 10 bp Ladder (Fig. 3). Proceed with one range of fragment sizes and keep the other size ranges for backup. Backup RNA fragments can be stored as gel slices at −80 °C or can be eluted from the gel slices, and the precipitating eluates can be stored in isopropanol at −20 °C. Elute the RNA fragments from the gel slices as described in Section 6.3.2 and resuspend in 5.5 μl water.

3.5 3′ Adapter Ligation

Ligation of an adenylated DNA adapter to the 3′ ends of RNA fragments provides a uniform primer binding site for cDNA synthesis. Adenylated

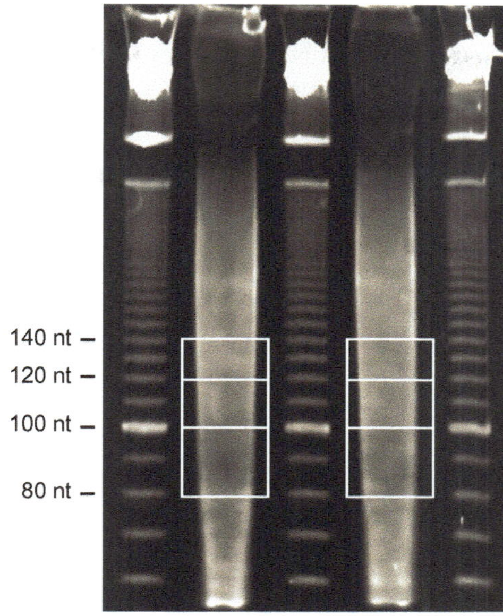

Figure 3 RNA fragment gel purification. Two samples are indicated. Regions representing sizes of 80–100, 100–120, and 120–140 nt were cut from the gel as indicated by bounding boxes.

adapters are available commercially, but are quite expensive. A protocol for adapter adenylation is provided in Section 6.3.3. To the RNA fragments, add 0.5 μl adenylated adapter (100 μ*M*), 1.2 μl 10 × T4 RNA Ligase Buffer, 1 μl RNasin, 3 μl PEG 8000, and 1 μl T4 RNA Ligase (NEB M0204). The absence of ATP in this reaction ensures that only the adenylated adapters are ligated to the 3′ ends of RNA fragments. Incubate at 22 °C for at least 2.5 h. To precipitate, add 30 μl 3 *M* NaOAc, pH 5.3, 260 μl water, 2 μl GlycoBlue, and 300 μl isopropanol. Precipitate at −20 °C for at least 30 min and spin as described in Section 2.2. Resuspend the pellet in 7 μl water. The efficiency of ligation can be checked by running 0.8 μl of the reaction out on an 8% TBE/Urea/Polyacrylamide gel and should be in the range of 70–90%.

3.6 Reverse Transcription and Size Selection

Size selection of truncated cDNAs allows for the sites of RT stops to be determined by sequencing. We have performed Pseudo-seq using both AMV RT (Promega M5108) and ssIII RT (Life Technologies 18080093) enzymes with similar results.

Prepare an annealing mix for each sample by transferring 6.2 μl of each of the ligation reactions to a PCR tube, and adding 1 μl RT Primer, and 0.8 μl 10× RT Buffer w/o Mg^{2+}. In parallel, prepare a no RNA reaction using 6.2 μl of H_2O. To anneal the RT primer to the ligated RNA fragments, incubate in a thermocycler at 65 °C for 4 min, then 55 °C for 2 min, 45 °C for 2 min, and 42 °C for 2 min. Briefly centrifuge the tubes to collect any condensation and then place on ice. Prepare an extension master mix by mixing 0.6 μl 10× RT Buffer w/o Mg^{2+}, 1.4 μl 10 m*M* dNTPs, 0.7 μl 100 m*M* $MgCl_2$, 1.3 μl H_2O, 1.0 μl RNasin Plus, and 1.0 μl AMV RT per sample. Add 6 μl of this master mix to each of the annealing mixes. Incubate in a thermocycler at 42 °C for 1 h. Remove the RNA by adding 1.5 μl 1 N NaOH, and incubating in a thermocycler at 98 °C for 15 min. Neutralize the pH by adding 1.5 μl 1 N HCl. Prepare the samples for size selection by adding 17 μl of 2× RNA Loading Dye.

Prepare 8% TBE/Urea/Polyacrylamide mini-gels, and prerun as described in Section 3.4. Prepare 10 bp ladder as above (Section 3.4). Denature the cDNAs at 95 °C for 2 min, and then place on ice. Load the gel, with each cDNA sample split across two lanes, and run at 200 V for 65 min. The RT primer runs at 85 nt and should be run as close to the bottom of the gel as possible to maximize separation of truncated from full-length cDNAs. For reference, the xylene cyanol dye front runs at approximately 75 nt. Disassemble, stain, and visualize the cDNA gel as described in Section 3.4. Excise gel slices corresponding to truncated cDNAs. These should have at least 25 nt added to the RT primer, since shorter products are nonspecific and should avoid the 105 nt nonspecific product and the full-length cDNA band (Fig. 4). Extract cDNAs from the gel slices as described in Section 6.3.2. Resuspend the cDNA pellets in 15 μl 10 m*M* Tris, pH 8.0.

3.7 Circularization

Intramolecular ligation of cDNAs can be used in lieu of 5′ adapter ligation to provide a 5′ PCR primer binding site. Prepare a circularization master mix. For each sample, mix 2 μl 10 CircLigase Buffer, 1 μl 1 m*M* ATP, and 1 μl 50 m*M* $MnCl_2$. Add 4 μl of this mix to each sample, then add 1 μl 0.5× CircLigase ssDNA Ligase (Epicentre CL4115K) diluted 1:1 in 1× CircLigase Buffer, and mix well. Incubate in a thermocycler at 60 °C for at least 1 h, and then heat inactivate at 80 °C for 10 min. Reactions can be used immediately for PCR or can be stored at −20 °C.

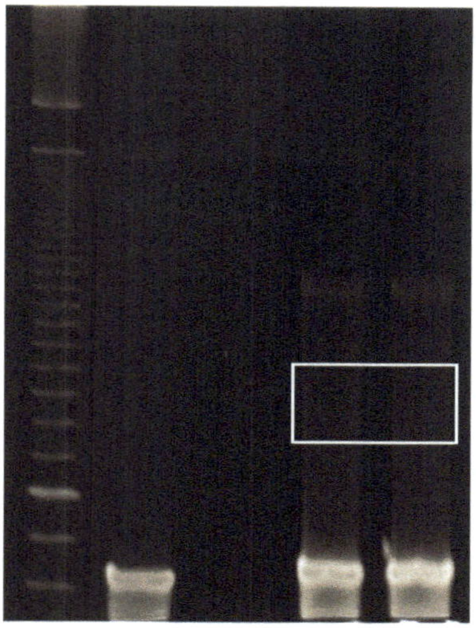

Figure 4 Truncated cDNA gel purification. Excised cDNAs are indicated by bounding boxes. One cDNA sample was loaded in two gel lanes.

3.8 PCR Amplification

Circularized cDNAs are then PCR amplified and gel purified to yield a library suitable for sequencing on the Illumina platform. Several PCR reactions are performed, each using a different number of amplification cycles, to ensure that a reaction with an ideal level of amplification is obtained. For the purposes of this chapter, it is assumed that libraries will be submitted to a facility that will perform quality control and sequencing.

Prepare a PCR Master Mix for each library. Per library add 15 μl HF Buffer, 1.5 μl 10 m*M* dNTPs, 3.78 μl Forward PCR Primer, 3.78 μl Barcoded Reverse PCR Primer, 52.6 μl H_2O, and 0.75 μl Phusion High-Fidelity DNA Polymerase (NEB M0530L); these volumes are sufficient for 4.5 PCR reactions per library. Each library that will be sequenced on the same HiSeq lane should be amplified with a different Barcoded Reverse PCR primer with a unique barcode sequence. To the PCR Master Mix, add 4.5 μl of the circularized cDNA sample, and mix well. Transfer 16.7 μl of the Master Mix to each of four PCR tubes. Perform PCR with the following cycles: (1) 98 °C, 30 s, (2) 98 °C, 10 s, (3) 60 °C, 20 s, (4) 72 °C, 40 s, (5) repeat steps 2–4 17 × (18 cycles total). Remove PCR tubes from the thermocycler and place on ice after the extension phase (step 4) of cycles 12, 14, 16, and 18. Add 3.4 μl of DNA Loading Dye to each PCR reaction. Prepare an aliquot of 10 bp Ladder for each gel by mixing 10 bp Ladder,

15.7 μl H_2O, and 3.3 μl 6 × DNA Loading Dye. Load the reactions on an 8% TBE/Polyacrylamide mini-gel and run at 200 V for 40 min. Disassemble, stain, and visualize the PCR gel as described in Section 3.4. Excise PCR products of the appropriate size, based on the range of RNA fragment sizes used. The 105 bp no insert PCR product should be avoided, and bands should not be cut from lanes with saturated reactions, or those with higher molecular weight species (Fig. 5). Extract PCR reactions from the gel slices as described in Section 6.3.2. Resuspend the PCR products in 10 μl 10 m*M* Tris (pH 8.0). These gel-purified PCR products are suitable for Illumina sequencing and can be submitted to a facility for sequencing. Should insufficient material be recovered, PCR reactions from appropriate cycle numbers can be scaled up to 50 μl.

Avoiding PCR amplification bias during library preparation is important to ensure that composition of library accurately reflects the RNA fragment/cDNA pool, avoiding over or under representation of certain sequences after PCR. This can be done by avoiding saturated PCR products, or those with higher molecular weight bands. The effects of PCR bias

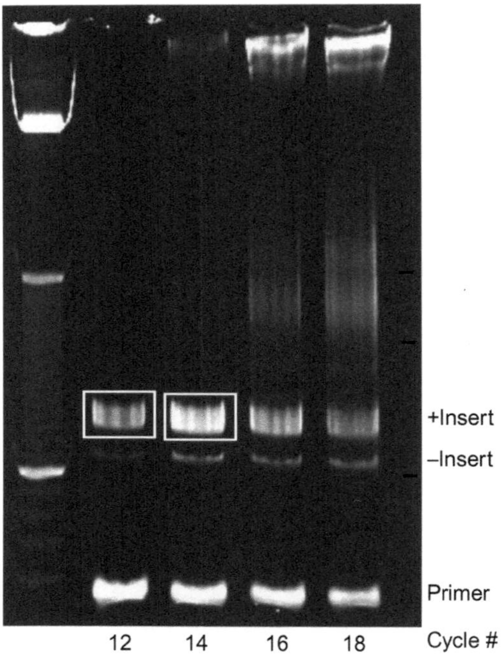

Figure 5 PCR product gel purification. Excised PCR products are indicated by bounding boxes. The lanes for 16 and 18 cycles are too saturated.

can be assessed and minimized by adding at least eight random nucleotides to the RT primer. These random nucleotides serve as barcodes for RNA fragments. Identical reads with identical random barcodes likely arise from PCR amplification bias and can be collapsed into a single read.

4. PSEUDO-SEQ DATA ANALYSIS

Sites of pseudouridylation can be identified through computational analysis of data obtained through Illumina sequencing of Pseudo-seq libraries. The 5′ ends of Pseudo-seq reads, which correspond to the 3′ ends of cDNAs, represent sites of RT stops (Section 3.6). Those RT stops that are reproducibly enriched in + CMC libraries rather than −CMC libraries represent sites of pseudouridylation. This section describes the computational identification of these CMC-dependent RT stops from Pseudo-seq data. The analyses described are carried out using publically available tools designed for the analysis of next-generation sequencing data, and custom scripts that must be prepared by the user. Preparation of these custom scripts requires familiarity with Python or another similar scripting language.

First, the sequencing reads are processed into a form suitable for mapping to the target genome. If barcoded PCR primers were used to multiplex several libraries, allowing multiple libraries to be sequenced on the same flow cell lane, then the reads must be separated by barcode, or index sequence. Barcode parsing allows individual reads to be assigned to the appropriate library. Some fraction of reads will contain some 3′ adaptor sequence that will interfere with mapping, and therefore must be removed from these demultiplexed reads. Reads processed in this manner can then be mapped to the target genome. These steps can easily be combined into an automated pipeline using bash scripts.

Custom scripts are then employed to identify Ψ sites from the 5′ positions (3′ cDNA ends) of these mapped reads. These scripts calculate a Pseudo-seq peak value for each U in the transcriptome, and those sites whose peak values reproducibly exceed a specified threshold in multiple biological replicates are identified as sites of pseudouridylation.

4.1 Library Demultiplexing

The reads generated from an Illumina sequencing run will be output in a FASTQ file. Within this file, each read is represented by four sequential lines, as illustrated below.

```
@HWI-ST1133R:1:1101:2652:2470#AGGTTT/1
AACCGCAGCAGGTCTCCAAGGTGAACAGCCTCTAGTTGAT
+HWI-ST1133R:1:1101:2652:2470#AGGTTT/1
CCCFFFFFHHHHFIJJJJJJHHIJJJJJJJJJJIIJGI
```

The first line, beginning with @, contains the name of the read, which includes an identifier for the sequencing machine and location of the read on the flow cell. Importantly, it also contains the barcode sequence located between the # and / characters, which is the reverse complement of the barcode sequence in the Barcoded Reverse PCR Primer. The second line contains the read itself, and the third and fourth lines contain an alternate name for the read and quality scores, respectively.

To demultiplex FASTQ files, the index for an individual read should be matched to the reverse complements of the barcodes used in PCR amplification (Section 3.8). This matching can be carried out using a custom script that uses simple string comparisons to assign barcodes, allowing one mismatch. This script should write reads to new FASTQ files specific to each barcode used, including a FASTQ file for unmapped barcodes. Alternately, demultiplexing can be performed using Illumina's CASAVA software or with other publically available tools. In some cases, the facility used for sequencing may demultiplex sequencing reads.

These FASTQ files can be quite large and may be compressed to save storage space. The downstream analyses described in Sections 4.2 and 4.3 are compatible with sequence data compressed with the gzip utility. In the command line, a FASTQ file can be compressed with the following command in a UNIX shell to yield a compressed file `sequence_reads.fastq.gz`.

```
gzip sequence_reads.fastq
```

4.2 Trimming of 3′ Adapter Sequences

Some significant fraction of reads will be derived from molecules that are shorter than the read length of the Illumina machine and will contain sequences derived from the 3′ adapter. This adapter sequence can interfere with the mapping of reads to the genome and must be removed from reads prior to mapping. The publically available Cutadapt tool can be used to trim adapter sequence from next-generation sequencing data (Martin, 2011). For a given FASTQ file, adapter sequences can be trimmed with the following command in a UNIX shell:

```
cutadapt -a ADAPTER_SEQ --overlap 3 --minimum-length
18 -o trimmed_reads.fastq.gz input_reads.fastq.gz.
```

This command should be run for each individual FASTQ file of reads. The `-a ADAPTER_SEQ` command specifies the adapter sequence to trim, the `--overlap 3` command requires at least three bases of overlap with the adapter for trimming, the `--minimum-length 18` command discards reads shorter than 18 nt, the `-o trimmed_reads.fastq.gz` command specifies the output file name, and `input_reads.fastq.gz` specifies the input file name.

4.3 Mapping Reads

Reads trimmed of adapter sequences can then be mapped to the target genome using Bowtie2 and to annotated splice junctions using TopHat (Kim et al., 2013; Langmead, Trapnell, Pop, & Salzberg, 2009). Read mapping using these packages requires a bowtie index for the target genome, and a file containing annotated splice junctions in a TopHat readable format. Bowtie indices are available for many commonly used model organisms or can be easily built from a FASTA file of the genomic sequence with the Bowtie2 package, and splice junction annotations for *S. cerevisiae* can be found at SGD (http://www.yeastgenome.org).

To map reads to the target genome and annotated splice junctions, enter the following command in a UNIX shell. TopHat will use BowTie2 to map reads to the genome.

```
tophat2 --no-novel-juncs --no-novel-indels --raw-
juncs splice_juncs bowtie_index
trimmed_reads.fastq.gz
```

This command should be run for each individual FASTQ file of trimmed reads. The `--no-novel-juncs` and `--no-novel-indels` commands specify that TopHat should not map reads to novel splice junctions or indels. The locations of the splice junction file and bowtie index are indicated by `splice_juncs` and `bowtie_index`, respectively. The file `trimmed_reads.fastq.gz` contains the reads trimmed of adapter sequence (Section 4.2).

This command will generate several output files containing mapping information. The relevant file for further analysis is called `accepted_hits.bam`, which contains the mapped reads and information about their mapped positions. This file must be parsed for subsequent analysis. This can be accomplished using the SAMtools package, and a custom script to parse the SAMtools output. This script should determine the number of reads whose 5′ ends (3′ cDNA ends) map to each position in the genome and store this output in a format suitable for subsequent steps (Section 4.4). Additionally, determine the total number of mapped reads and the total number of reads mapped to the rRNA for each library. These values will be used to scale +CMC and −CMC library pairs as described in Section 4.4.

4.4 Computational Identification of Ψ Residues

Ψ identification involves the calculation of a Pseudo-seq peak value for each U in the transcriptome and the identification of sites with reproducibly high peak values. This analysis should be performed with user prepared scripts.

First, the +CMC and −CMC library pairs for each sample should be scaled to the same size. Since Pseudo-seq peak values involve the subtraction of −CMC reads at a given position from +CMC reads, this calculation is sensitive to differences in library size. A larger +CMC library will lead to an increase in calculated peak values, while a larger −CMC library will lead to a decrease in peak values. To scale, the −CMC library multiply the reads at each position in this library to the ratio of reads in the +CMC library to the −CMC library. This ratio should be calculated from the total number of mapped reads (Section 4.3) for Ψ identification in mRNAs. Alternately, the number of rRNA mapping reads can be used for Ψ identification in ncRNAs.

Next, transcripts without sufficient read coverage should be removed from further analysis. Only consider transcripts with a read coverage that exceeds a specified threshold of average reads per nucleotide (*rc*). Empirical determination of a suitable read threshold is described in Section 5.3. With sufficient biological replication, an *rc* value of 0 may be used, allowing analysis of transcripts with low expression levels.

For each U in the transcriptome, calculate the Pseudo-seq peak value for each +CMC and −CMC library pair. Determine the number of reads whose 5′ ends map one nucleotide 3′ of the U (3′ position), which is the expected position of an RT stop induced by a Ψ-CMC adduct, in both +CMC (r^+) and −CMC (r^-) libraries. Additionally, determine the number of reads whose 5′ ends map to a window of size *ws* nucleotides centered at the 3′ position, but exclusive of the reads at that position in both the +CMC (wr^+) and −CMC (wr^-) libraries. Calculate Pseudo-seq peak values according to the equation:

$$\text{peak} = ws \times \frac{r^+ - r^-}{wr^+ + wr^-}$$

Sites of pseudouridylation can be identified in two steps using the calculated peak values. First, for each +CMC and −CMC library pair flag positions with a peak value exceeding a threshold, p. These flagged positions will contain real sites of pseudouridylation amid a large number of false positives. Second, filter out false positives by determining which positions are reproducibly flagged in multiple library pairs. For N biological replicate library pairs, the position should be flagged in at least n replicate pairs. Discussion of biological replication and computational parameters (ws, p, and n) will be discussed in Sections 5.2 and 5.3.

4.5 Genetic Assignment of Ψs to Pseudouridylation Factors

In *S. cerevisiae*, or another genetically tractable organism, Ψs identified by Pseudo-seq can be assigned to known, nonessential pseudouridylation factors by performing Pseudo-seq on strains lacking the factor of interest. Pseudo-seq libraries should be prepared (Section 3) from two biological replicates of a given deletion strain and analyzed as in Sections 4.1–4.4.

For each Ψ identified by Pseudo-seq, the median peak height for all +CMC and −CMC library pairs should be calculated, as well as the median total reads in the window centered at the site (Section 4.4, $wr^{+} + wr^{-}$) for all library pairs. Assign a Ψ to a given factor if the peak heights for both replicate deletion libraries are less than 25% of the median peak height for that Ψ, and if the total window reads in at least one of two replicates is greater than 25% of the median total window reads for that Ψ.

5. EXPERIMENTAL CONSIDERATIONS

There are several factors that should be considered when planning Pseudo-seq experiments and during data analysis. Prior to carrying out experiments, both the sequencing depth needed and the number of biological replicates required should be determined. These factors will affect the transcripts for which Ψ identification is reliable, as well as the computational parameters chosen.

5.1 Read Coverage

The read coverage required for successful identification of Ψ residues will depend upon both the transcriptome size and the abundance of target transcripts. For *S. cerevisiae*, we have found that 12 million reads per +CMC and −CMC library pair (~6 million per library) mapping to coding sequences is sufficient for identification of Ψs in most expressed mRNAs. The number of reads required can be scaled up or down depending on the size of the transcriptome. The read coverage required is also dependent on the abundance of the desired transcripts. For highly abundant transcripts such as tRNAs and rRNAs, the read number can be scaled down substantially.

5.2 Biological Replication

The use of replicate data for Ψ identification is important, since biological replicates allow for efficient filtering of false positives (Section 4.4). Biological replicates are especially important for eliminating false positives in transcripts of low abundance. We have found that use of 14 biological replicates allows robust Ψ identification in mRNAs of a wide range of abundances. While this number of replicates is high, it should be noted that use of strains deleted for known pseudouridylation factors can serve as replicates for control strains. If fewer replicates are used for Ψ identification, more stringent computational parameters can be used to compensate.

5.3 Computational Parameters

The computational parameters used for calling Ψs should be empirically determined from Pseudo-seq data and will depend on both the read coverage attained and the number of biological replicates. These parameters are discussed in Section 4.4 and include the read coverage cutoff for filtering transcripts (*rc*), the window size used for Pseudo-seq peak calculation (*ws*), the peak value threshold (p), and the number of replicate library pairs in which a position must be above p (n).

The use of a larger number of biological replicates will allow more permissive cutoffs to be used, as false positives are more reliably filtered out with a higher number of replicates. Conversely, with fewer biological replicates, more stringent cutoffs should be used. For Ψ calling in an experiment performed in *S. cerevisiae* with 14 biological replicates, the following parameters can be used: $rc=0.0$, $ws=150$, $p=1.0$, and $n=10$.

The appropriate parameters for Ψ calling can be determined empirically. Ψ identification should be performed with several variations of these parameters, and the read distribution surrounding the sites called as pseudouridylation events should be examined. These sites should have a strong enrichment of reads above background one nucleotide 3′ of the identified Ψ. The majority of Ψ sites called with reliable parameters should show this strong, position-specific enrichment of reads. Parameters that yield a large fraction of sites that do not display this enrichment are insufficiently stringent. Known Ψs in rRNAs, tRNAs, or snRNAs will provide examples

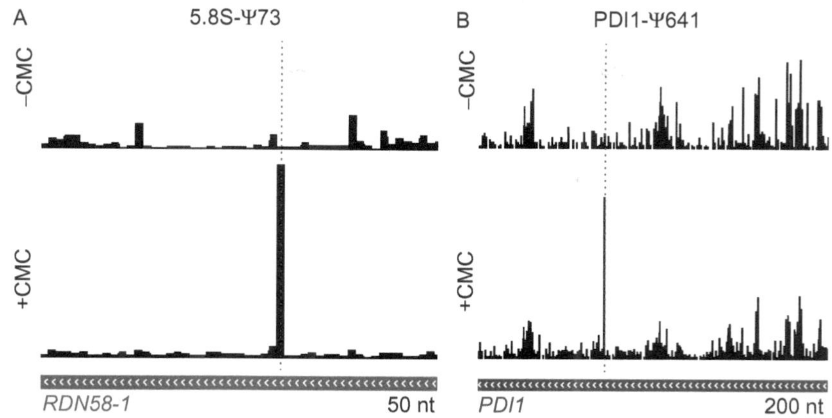

Figure 6 Genome browser views of example Ψs (A) in rRNA and (B) in mRNA. Ψ positions are indicated by dotted red (dark gray in the print version) lines.

of reliable Ψ peaks in abundant transcripts (Fig. 6). Random downsampling of the reads in rRNA followed by Ψ calling from these downsampled reads can also be used to assess the computational parameters chosen.

6. SOLUTIONS, REAGENTS, AND COMMON PROTOCOLS

6.1 Solutions

YPAD: 1% (w/v) Bacto Yeast Extract, 2% (w/v) Bacto Peptone, 2% (w/v) Glucose, 0.004% (w/v) Adenine Sulfate.

AES Buffer: 50 m*M* NaOAc (pH 5.3), 10 m*M* EDTA (pH 8.0), 1% (w/v) SDS.

TES Buffer: 10 m*M* Tris (pH 7.6), 1 m*M* EDTA (pH 8.0), 0.1% (w/v) SDS.

TES + NaCl Buffer: 0.5 *M* NaCl, 10 m*M* Tris (pH 7.6), 1 m*M* EDTA (pH 8.0), 0.1% (w/v) SDS. The SDS may come out of solution over time. If this occurs, redissolve by warming buffer.

BEU Buffer: 50 m*M* Bicine (pH 8.5), 4 m*M* EDTA (pH 8.0), 7 *M* Urea. The final pH will be approximately 9.0.

Sodium Carbonate pH 10.4 Buffer: 50 m*M* Na_2CO_3 (pH 10.4), 2 m*M* EDTA (pH 8.0). Prepare from 1 *M* Na_2CO_3 (pH 10.4) and 0.5 *M* EDTA (pH 8.0). Adjust the pH to 10.4, filter sterilize, and store aliquots at −20 °C.

10× RT Buffer w/o Mg^{2+}: 500 m*M* Tris (pH 8.6), 600 m*M* NaCl, and 100 m*M* DTT. Store at −20 °C.

2× RNA Loading Dye: 95% formamide, 5 m*M* EDTA (pH 8.0), 0.025% (w/v) SDS, 0.025% (w/v) bromophenol blue, 0.025% (w/v) xylene cyanol FF. Store aliquots at −20 °C.

6× DNA Loading Dye: 30% (v/v) Glycerol, 0.025% (w/v) bromophenol blue, 0.025% (w/v) xylene cyanol FF.

RNA Elution Buffer: 300 m*M* NaOAc (pH 5.3), 1 m*M* EDTA (pH 8.0), 100 U/ml RNasin Plus. The RNase inhibitor should be added immediately prior to use.

DNA Elution Buffer: 300 m*M* NaCl, 10 m*M* Tris (pH 8.0).

6.2 Library Oligonucleotide Sequences

3′ Adapter:

/5Phos/TGGAATTCTCGGGTGCCAAGG/3ddC/

RT Primer:

/5Phos/GATCGTCGGACTGTAGAACTCTGAACCTGTCGG TGGTCGCCGTATCATT/iSp18/CACTCA/iSp18/GCCTTGG CACCCGAGAATTCCA

It is important that this oligo be gel purified to ensure that full-length and truncated cDNAs are reliably separated.

Forward PCR Primer:

AATGATACGGCGACCACCGA

Barcoded Reverse PCR Primer:

CAAGCAGAAGACGGCATACGAGATXXXXXXGTGACTG GAGTTCCTTGGCACCCGAGAATTCCA

The XXXXXX sequence indicates a unique barcode sequence.

6.3 Common Protocols

6.3.1 Regeneration of Oligo (dT) Cellulose Beads

Oligo (dT) cellulose beads can be reused multiple times. To regenerate, rotate beads at room temperature for 1 h in greater than two bead volumes of 0.1 N NaOH. Wash beads with water until the eluate reaches a neutral pH, resuspend in TES + NaCl, and store at 4 °C.

6.3.2 Gel Extraction of Nucleic Acids

Place the gel slice in a microcentrifuge tube and add 400 μl of RNA Elution Buffer or DNA Elution Buffer and elute overnight 4 °C or room temperature for RNA and DNA fragments, respectively. Remove the supernatant, and filter through a Spin-X cellulose acetate column (Corning 8162) for 1 min at max speed in a microcentrifuge. Add 2 μl GlycoBlue and 1 volume ispropanol and precipitate at −20 °C for at least 30 min. Spin in a microcentrifuge at maximum speed at 4 °C for 30 min. Wash the pellet in 750 μl ice-cold 70% ethanol, spin at maximum speed at 4 °C for 10 min, and air dry for 2 min. Resuspend the pellet in an appropriate buffer.

6.3.3 Adapter Adenylation

Adenylated oligos with blocked 3′ ends are used as adapters because they reduce sequence bias in RNA fragment capture, and the absence of ATP only allows for the ligation of the adapter to the 3′ ends of RNA fragments (England, Gumport, & Uhlenbeck, 1977; Unrau & Bartel, 1998). Adapter adenylation requires the reagent adenosine 5′-phosphorimidazolide (ImpA), which can be obtained by a straightforward chemical synthesis described elsewhere (Pfeffer, Lagos-Quintana, & Tuschl, 2005). To 420 μl of 50 m*M* ImpA, add $MgCl_2$ to 25 m*M*, 3′ Adapter to 0.2 m*M*, and water to 500 μl. Incubate at 50 °C for 3 h. Run the adenylation reaction out on a 20% denaturing PAGE gel, and gel purify the upper, adenylated species.

ACKNOWLEDGMENTS

We thank members of the Gilbert laboratory for helpful discussions regarding this protocol. This work was supported by grants to W.V.G. from National Institutes of Health (GM094303, GM081399), and the American Cancer Society—Robbie Sue Mudd Kidney Cancer Research Scholar Grant (RSG-13-396-01-RMC). T.M.C. was supported by the American Cancer Society New England Division (Ellison Foundation Postdoctoral Fellowship).

REFERENCES

Anderson, M. Z., Brewer, J., Singh, U., & Boothroyd, J. C. (2009). A pseudouridine synthase homologue is critical to cellular differentiation in *Toxoplasma gondii*. *Eukaryotic Cell*, *8*, 398–409.

Arluison, V., Buckle, M., & Grosjean, H. (1999). Pseudouridine synthetase pus1 of *Saccharomyces cerevisiae*: Kinetic characterisation, tRNA structural requirement and real-time analysis of its complex with tRNA. *Journal of Molecular Biology*, *289*, 491–502.

Bakin, A., & Ofengand, J. (1993). Four newly located pseudouridylate residues in *Escherichia coli* 23S ribosomal RNA are all at the peptidyltransferase center: Analysis by the application of a new sequencing technique. *Biochemistry*, *32*, 9754–9762.

Bykhovskaya, Y., Casas, K., Mengesha, E., Inbal, A., & Fischel-Ghodsian, N. (2004). Missense mutation in pseudouridine synthase 1 (PUS1) causes mitochondrial myopathy and sideroblastic anemia (MLASA). *American Journal of Human Genetics*, *74*, 1303–1308.

Cantara, W. A., Crain, P. F., Rozenski, J., McCloskey, J. A., Harris, K. A., Zhang, X., et al. (2011). The RNA modification database, RNAMDB: 2011 update. *Nucleic Acids Research*, *39*, D195–D201.

Carlile, T. M., Rojas-Duran, M. F., Zinshteyn, B., Shin, H., Bartoli, K. M., & Gilbert, W. V. (2014). Pseudouridine profiling reveals regulated mRNA pseudouridylation in yeast and human cells. *Nature*, *515*, 143–146.

Charette, M., & Gray, M. W. (2000). Pseudouridine in RNA: What, where, how, and why. *IUBMB Life*, *49*, 341–351.

Choi, Y. C., & Busch, H. (1978). Modified nucleotides in T1 RNase oligonucleotides of 18S ribosomal RNA of the Novikoff hepatoma. *Biochemistry*, *17*, 2551–2560.

Cohn, W. E. (1960). Pseudouridine, a carbon-carbon linked ribonucleoside in ribonucleic acids: Isolation, structure, and chemical characteristics. *The Journal of Biological Chemistry*, *235*, 1488–1498.

Collart, M. A., & Oliviero, S. (2001). Preparation of yeast RNA. *Current Protocols in Molecular Biology*, *23*, 13.12.1–13.12.5.

Courtes, F. C., Gu, C., Wong, N. S. C., Dedon, P. C., Yap, M. G. S., & Lee, D.-Y. (2014). 28S rRNA is inducibly pseudouridylated by the mTOR pathway translational control in CHO cell cultures. *Journal of Biotechnology*, *174*, 16–21.

Davis, F. F., & Allen, F. W. (1957). Ribonucleic acids from yeast which contain a fifth nucleotide. *The Journal of Biological Chemistry*, *227*, 907–915.

Dominissini, D., Moshitch-Moshkovitz, S., Schwartz, S., Salmon-Divon, M., Ungar, L., Osenberg, S., et al. (2013). Topology of the human and mouse m6A RNA methylomes revealed by m6A-seq. *Nature*, *485*, 201–206.

Durairaj, A., & Limbach, P. A. (2008). Improving CMC-derivatization of pseudouridine in RNA for mass spectrometric detection. *Analytica Chimica Acta*, *612*, 173–181.

England, T. E., Gumport, R. I., & Uhlenbeck, O. C. (1977). Dinucleoside pyrophosphate are substrates for T4-induced RNA ligase. *Proceedings of the National Academy of Sciences of the United States of America*, *74*, 4839–4842.

Fernández, I. S., Ng, C. L., Kelley, A. C., Wu, G., Yu, Y.-T., & Ramakrishnan, V. (2013). Unusual base pairing during the decoding of a stop codon by the ribosome. *Nature*, *500*, 107–110.

Ganot, P., Bortolin, M. L., & Kiss, T. (1997). Site-specific pseudouridine formation in preribosomal RNA is guided by small nucleolar RNAs. *Cell*, *89*, 565–573.

Ge, J., & Yu, Y.-T. (2013). RNA pseudouridylation: New insights into an old modification. *Trends in Biochemical Sciences*, *38*, 210–218.

Gupta, R. C., & Randerath, K. (1979). Rapid print-readout technique for sequencing of RNA's containing modified nucleotides. *Nucleic Acids Research*, *6*, 3443–3458.

Heiss, N. S., Knight, S. W., Vulliamy, T. J., Klauck, S. M., Wiemann, S., Mason, P. J., et al. (1998). X-linked dyskeratosis congenita is caused by mutations in a highly conserved gene with putative nucleolar functions. *Nature Genetics*, *19*, 32–38.

Holley, R. W., Everett, G. A., Madison, J. T., & Zamir, A. (1965). Nucleotide sequences in the yeast alanine transfer ribonucleic acid. *The Journal of Biological Chemistry*, *240*, 2122–2128.

Hudson, G. A., Bloomingdale, R. J., & Znosko, B. M. (2013). Thermodynamic contribution and nearest-neighbor parameters of pseudouridine-adenosine base pairs in oligoribonucleotides. *RNA*, *19*, 1474–1482.

Jambhekar, A., & DeRisi, J. L. (2007). Cis-acting determinants of asymmetric, cytoplasmic RNA transport. *RNA*, *13*, 625–642.

Karijolich, J., & Yu, Y.-T. (2011). Converting nonsense codons into sense codons by targeted pseudouridylation. *Nature*, *474*, 395–398.

Kim, D., Pertea, G., Trapnell, C., Pimentel, H., Kelley, R., & Salzberg, S. L. (2013). TopHat2: Accurate alignment of transcriptomes in the presence of insertions, deletions and gene fusions. *Genome Biology*, *14*, R36.

Kudla, G., Murray, A. W., Tollervey, D., & Plotkin, J. B. (2009). Coding-sequence determinants of gene expression in *Escherichia coli*. *Science*, *324*, 255–258.

Langmead, B., Trapnell, C., Pop, M., & Salzberg, S. L. (2009). Ultrafast and memory-efficient alignment of short DNA sequences to the human genome. *Genome Biology*, *10*, R25.

Li, J. B., Levanon, E. Y., Yoon, J.-K., Aach, J., Xie, B., Leproust, E., et al. (2009). Genome-wide identification of human RNA editing sites by parallel DNA capturing and sequencing. *Science*, *324*, 1210–1213.

Lovejoy, A. F., Riordan, D. P., & Brown, P. O. (2014). Transcriptome-wide mapping of pseudouridines: Pseudouridine synthases modify specific mRNAs in *S. Cerevisiae*. *PLoS One*, *9*, e110799.

Martin, M. (2011). Cutadapt removes adapter sequences from high-throughput sequencing reads. *EMBnet Journal*, *17*, 10–12.

Mei, Y.-P., Liao, J.-P., Shen, J., Yu, L., Liu, B.-L., Liu, L., et al. (2011). Small nucleolar RNA 42 acts as an oncogene in lung tumorigenesis. *Oncogene*, *31*, 2794–2804.

Meyer, K. D., Saletore, Y., Zumbo, P., Elemento, O., Mason, C. E., & Jaffrey, S. R. (2012). Comprehensive analysis of mRNA methylation reveals enrichment in 3′ UTRs and near stop codons. *Cell*, *149*, 1635–1646.

Pfeffer, S., Lagos-Quintana, M., & Tuschl, T. (2005). Cloning of small RNA molecules. *Current Protocols in Molecular Biology*, *72*, 26.4.1–26.4.18.

Popova, A. M., & Williamson, J. R. (2014). Quantitative analysis of rRNA modifications using stable isotope labeling and mass spectrometry. *Journal of the American Chemical Society*, *136*, 2058–2069.

Sambrook, J., & Russell, D. W. (2001). *Molecular cloning*. Cold Spring Harbour, New York: CSHL Press.

Schwartz, S., Bernstein, D. A., Mumbach, M. R., Jovanovic, M., Herbst, R. H., León-Ricardo, B. X., et al. (2014). Transcriptome-wide mapping reveals widespread dynamic-regulated pseudouridylation of ncRNA and mRNA. *Cell*, *159*, 148–162.

Shah, P., Ding, Y., Niemczyk, M., Kudla, G., & Plotkin, J. B. (2013). Rate-limiting steps in yeast protein translation. *Cell*, *153*, 1589–1601.

Somogyi, P., Jenner, A. J., Brierley, I., & Inglis, S. C. (1993). Ribosomal pausing during translation of an RNA pseudoknot. *Molecular and Cellular Biology*, *13*, 6931–6940.

Squires, J. E., Patel, H. R., Nousch, M., Sibbritt, T., Humphreys, D. T., Parker, B. J., et al. (2012). Widespread occurrence of 5-methylcytosine in human coding and non-coding RNA. *Nucleic Acids Research*, *40*, 5023–5033.

Subtelny, A. O., Eichhorn, S. W., Chen, G. R., Sive, H., & Bartel, D. P. (2014). Poly(A)-tail profiling reveals an embryonic switch in translational control. *Nature*, *508*, 66–71.

Tan, X., Lu, Z. J., Gao, G., Xu, Q., Hu, L., Fellmann, C., et al. (2012). Tiling genomes of pathogenic viruses identifies potent antiviral shRNAs and reveals a role for secondary structure in shRNA efficacy. *Proceedings of the National Academy of Sciences of the United States of America*, *109*, 869–874.

Tanaka, Y., Dyer, T. A., & Brownlee, G. G. (1980). An improved direct RNA sequence method; its application to *Vida faba* 5.8S ribosomal RNA. *Nucleic Acids Research*, *8*, 1259–1272.

Unrau, P. J., & Bartel, D. P. (1998). RNA-catalysed nucleotide synthesis. *Nature*, *395*, 260–263.

Urban, A., Behm-Ansmant, I., Branlant, C., & Motorin, Y. (2009). RNA sequence and two-dimensional structure features required for efficient substrate modification by the *Saccharomyces cerevisiae* RNA:Ψ-synthase Pus7p. *Journal of Biological Chemistry*, *284*, 5845–5858.

Wu, G., Xiao, M., Yang, C., & Yu, Y.-T. (2011). U2 snRNA is inducibly pseudouridylated at novel sites by Pus7p and snR81 RNP. *The EMBO Journal*, *30*, 79–89.

Zhao, X. L., & Yu, Y. T. (2004). Detection and quantitation of RNA base modifications. *RNA*, *10*, 996–1002.

CHAPTER TWELVE

Pseudouridine Chemical Labeling and Profiling

Xiaoyu Li*[,1], Shiqing Ma*[,†,‡,1], Chengqi Yi*[,†,§,2]

*State Key Laboratory of Protein and Plant Gene Research, School of Life Sciences, Peking University, Beijing, China

†Peking-Tsinghua Center for Life Sciences, Peking University, Beijing, China

‡Academy for Advanced Interdisciplinary Studies, Peking University, Beijing, China

§Department of Chemical Biology and Synthetic and Functional Biomolecules Center, College of Chemistry and Molecular Engineering, Peking University, Beijing, China

[2]Corresponding author: e-mail address: chengqi.yi@pku.edu.cn

Contents

[1] These authors contributed equally to this work.

Methods in Enzymology, Volume 560
ISSN 0076-6879
http://dx.doi.org/10.1016/bs.mie.2015.03.010

Abstract

Pseudouridine (Ψ) is the most abundant posttranscriptional RNA modification; yet little is known about its prevalence and function in messenger RNA, mostly due to the challenges in the transcriptome-wide detection of Ψ. Here, we report CeU-Seq—a selective chemical labeling and pull-down method for the comprehensive analysis of transcriptome-wide pseudouridylation; this sequencing method will hopefully pave the way for functional studies of Ψ-mediated biological regulation in the future.

1. THEORY

More than 100 different types of posttranscriptional modifications to RNA molecules have been characterized to date (Machnicka et al., 2013). Among them, pseudouridine (Ψ) is overall the most abundant (Ge & Yu, 2013).

During Ψ formation, a carbon carbon-(C5-C1′) bond is formed between the base and sugar; this creates an extra hydrogen bond donor at the non-Watson-Crick edge (Ge & Yu, 2013). Ψ has been found in tRNA, rRNA, and snRNA. It has been known that rRNA requires Ψ for binding to internal ribosome entry site and translational fidelity. Besides, Ψ in U2 snRNA can fine-tune branch site interactions and affect pre-mRNA splicing. Ψ can also increase base stacking, improve base-pairing, and rigidify the sugar-phosphate backbone (Ge & Yu, 2013).

In eukaryotes, Ψ can be catalyzed via two distinct mechanisms: one is through H/ACA box ribonucleoproteins guided by RNA and the other is through RNA-independent Ψ synthases (Hamma & Ferre-D'Amare, 2006; Kiss, Fayet-Lebaron, & Jady, 2010). Some Ψ synthases (for example, DKC1 and hPUS1) have been shown to be related to human diseases.

In 2011, Yu group reported that targeted pseudouridylation is capable of converting nonsense codons into sense codons (Karijolich & Yu, 2011). Yet evidence and physiological significance of naturally existing mRNA pseudouridylation was unclear. Comparing to ncRNAs, there are significant technical and experimental challenges for the detection of Ψ modifications in mRNA. First, mRNA is of low abundance. Therefore, sensitive detection methods are needed to explore the possibility of Ψ modifications in mRNA. Second, Ψ residues and regular U bases are indistinguishable during reverse transcription. Previous studies have identified a chemical—*N*-cyclohexyl-*N′*-b-(4-methylmorpholinium) ethylcarbodiimide, or CMC—to specifically label Ψ (Bakin & Ofengand, 1998). The CMC–Ψ adduct will cause the reverse transcriptase to terminate one nucleotide 3′ to the CMC–Ψ, while regular U bases can be read-through. Therefore, a profiling method coupling the unique ability of CMC and modern sequencing technology would be ideal to search for Ψ in mRNA. We report here N_3-*C*MC-*e*nriched pse*u*douridine *s*equencing (CeU-Seq) for the comprehensive identification of Ψ sites in the mammalian transcriptome at single-base resolution (Li et al., 2015). CeU-Seq is enabled by a chemically synthesized CMC analogue—N_3-CMC, which pre-enriches Ψ-containing RNA through biotin pull down. We envision such profiling method could assist future functional studies of Ψ-mediated epigenetic regulations.

2. EQUIPMENT

1.7-ml RNase-free tube
10-cm culture dish
−80 °C freezer
−20 °C freezer
50-ml conical polypropylene tubes
Centrifuge
Nanodrop 2000
Agilent 2100 Bioanalyzer
Thermal cycler
Thermo mixer
Qubit
UV light box
Gel electrophoresis equipment

3. MATERIALS

Sodium hydroxide (NaOH)
Lithium chloride (LiCl)
Sodium chloride (NaCl)
Diethylpyrocarbonate (DEPC)
EDTA, disodium
Tris base
Sodium dodecyl sulfate (SDS)
Sodium acetate (NaOAc)

Liquid nitrogen (or dry ice)
Purified water (deionized or Milli-Q filtered)
DEPC-treated water (Sigma)
TRIzol (Invitrogen)
Ethanol
Isopropanol
Chloroform
RNase-free 1.7-ml tubes (Axygen)
Oligo(dT)$_{25}$ dynabeads (Invitrogen)
Glycogen (Invitrogen)
RiboMinus™ Transcriptome Isolation Kit (Human/Mouse) (Invitrogen)
3 *M* NaOAc, pH 5.3
NEBNext® Magnesium RNA Fragmentation Module (NEB)
DBCO-(PEG)$_4$-biotin (Sigma)
Chemically synthesized N_3-CMC
Streptavidin C1 dynabeads (Invitrogen)
Shrimp alkaline phosphatase (rSAP) (NEB)
RNase inhibitor (40 U/μl) (Thermo Scientific)
FastDigest *Bam*HI (Thermo Scientific)
TBE-urea sample buffer (2×) (Invitrogen)
6% TBE-urea gel (Invitrogen)
SYBR green II (Invitrogen)
ZR small-RNA™ PAGE Recovery Kit (Zymo Research)
Superscript III RT (200 U/μl; Invitrogen)
Costar SpinX column (Corning)
CircLigase II (Epicentre)
NEBNext High-Fidelity 2× PCR Master Mix (NEB)

3.1 Solutions and Buffers

Step 2
Binding buffer

Component	Final Concentration	Stock	Amount
Tris–HCl (pH 7.5)	20 m*M*	1 *M*	200 μl
LiCl	1.0 *M*	7.5 *M*	1.33 ml
EDTA	2 m*M*	500 m*M*	40 μl
H_2O			8427 μl

Washing buffer

Component	Final Concentration	Stock	Amount
Tris–HCl (pH 7.5)	10 m*M*	1 *M*	150 μl
LiCl	0.15 *M*	7.5 *M*	300 μl
EDTA	1 m*M*	500 m*M*	30 μl
H_2O			14,520 μl

Step 3
BEU buffer

Component	Final Concentration	Stock	Amount
Urea	7 *M*		4.2 g
Bicine, pH 8.5	50 m*M*	1 *M*	500 μl
EDTA	4 m*M*	500 m*M*	80 μl

Add water to 10 ml.

Sodium carbonate buffer

Component	Final Concentration	Stock	Amount
Na_2CO_3, pH 10.4	50 m*M*	1 *M*	250 μl
EDTA	2 m*M*	500 m*M*	20 μl
H_2O			4730 μl

Step 4
Binding buffer

Component	Final Concentration	Stock	Amount
Tris–HCl (pH 7.5)	10 m*M*	1 *M*	100 μl
NaCl	500 m*M*	5 *M*	1 ml
EDTA	1 m*M*	500 m*M*	20 μl
H_2O			8880 μl

Washing buffer

Component	Final Concentration	Stock	Amount
Tris–HCl (pH 7.5)	10 m*M*	1 *M*	100 μl

NaCl	150 mM	5 M	300 μl
EDTA	1 mM	500 mM	20 μl
H_2O			9580 μl

1 × PNK wash buffer

Component	Final Concentration	Stock	Amount
Tris–HCl (pH 7.5)	20 mM	1 M	200 μl
$MgCl_2$	10 mM	1 M	100 μl
Tween-20	0.2%	50%	40 μl
H_2O			9660 μl

Step 5
10 × TBE

Component	Final Concentration	Stock	Amount
Tris base	89 mM		108 g
Boric acid	89 mM		55 g
EDTA	2 mM	500 mM	4 ml

Add water to 1000 ml.

4. PROTOCOL

4.1 Duration

4–5 days

4.2 Preparation

Design and order all oligonucleotides used for RNA ligation, reverse transcription, annealing, and PCR.

RNA adaptor

5′-phosphate-UGAGAUCGGAAGAGCGGUUCAG-3′-Puromycin
RT primer is
5′-phosphate-NNXXXXNNNNAGATCGGAAGAGCGTCGTGG ATCCTGAACCGCTC-3′ (XXXX represents for barcode)
Anneal oligo: GTTCAGGATCCACGACGCTCTTCaaaa

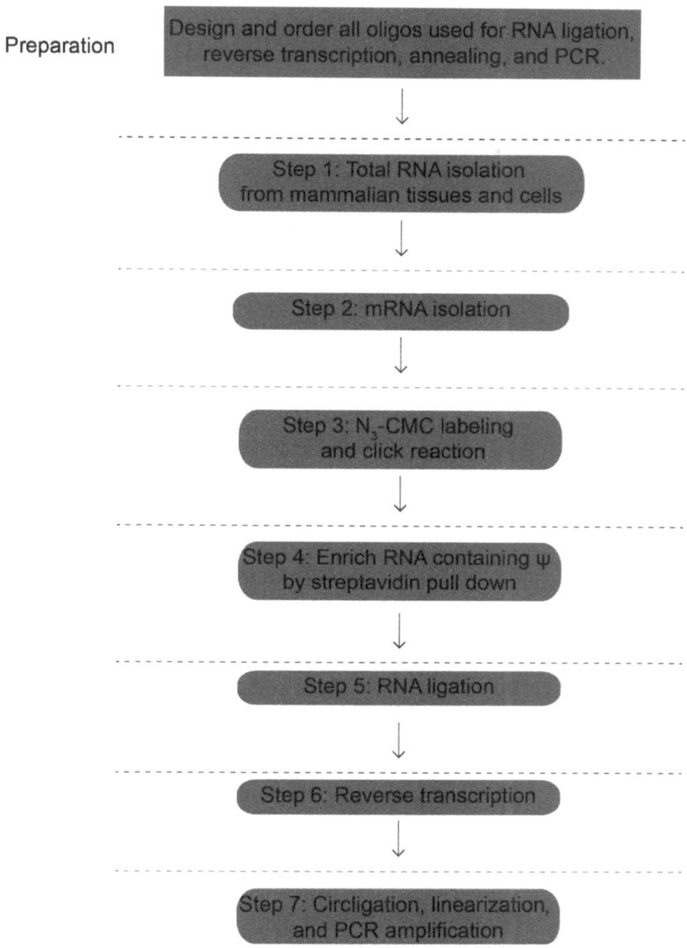

Figure 1 Flowchart of the complete protocol, including preparation.

PCR primers

P5 Solexa:
AATGATACGGCGACCACCGAGATCTACACTCTTTCCC
TACACGACGCTCTTCCGATCT

P3 Solexa:
CAAGCAGAAGACGGCATACGAGATCGGTCTCGGCATT
CCTGCTGAACCGCTCTTCCGATCT

See Fig. 1 for the flowchart of the complete protocol.

5. STEP 1: TOTAL RNA ISOLATION FROM MAMMALIAN TISSUES AND CELLS

5.1 Overview

Total RNA was isolated from mouse tissues or cultured cells. 1 mg total RNA is sufficient for the CeU-Seq library construction.

5.2 Duration

2 h

1.1 A. Harvest 200–500 mg tissues. Wash in cold PBS and snap freeze in liquid nitrogen. Samples should be stored at −80 °C until use. Homogenize tissues in liquid nitrogen. When homogenization is complete, transfer the tissue sample into 50-ml conical polypropylene tubes, and add 1 ml TRIzol per 50–100 mg tissues. Shake tube vigorously by hand for 30 s to ensure that the tissue sample is well homogenized in TRIzol. Incubate the homogenized sample for 5 min at room temperature to permit complete dissociation of the nucleoprotein complex.

1.1 B. Harvest 5–10 10-cm culture dishes. Remove growth media from culture dish, add 5 ml cold PBS per dish, and remove the lid. Add 3–4 ml TRIzol directly to each dish and incubate for 5 min at room temperature and then pipet cells up and down several times to lysate. Transfer the homogenized sample to RNase-free 1.7-ml tubes.

1.2 Add 0.2 ml of chloroform per 1 ml of TRIzol used for homogenization. Shake tubes vigorously by hand for 15 s and incubate for 2–3 min at room temperature.

1.3 Centrifuge the sample at 12,000 × *g* for 15 min at 4 °C. Carefully transfer the supernatant to fresh, RNase-free 1.7-ml tubes.

1.4 Add 0.5 ml of 100% isopropanol to the aqueous phase per 1 ml of TRIzol and incubate at room temperature for 10 min. Centrifuge at 12,000 × *g* for 10 min at 4 °C. Remove the supernatant from the tube and wash the pellet with 1 ml of 75% ethanol per 1 ml of TRIzol. Centrifuge the tube at 7500 × *g* for 5 min at 4 °C. Discard the wash solution.

1.5 Air dry the RNA pellet for 10 min and dissolve the RNA pellet in each tube in 50 μl of RNase-free water.

1.6 Measure the RNA concentration by Nanodrop and analyze the RNA quality using agarose gel electrophoresis or RNA 6000 Pico Chips on an Agilent 2100 Bioanalyzer.

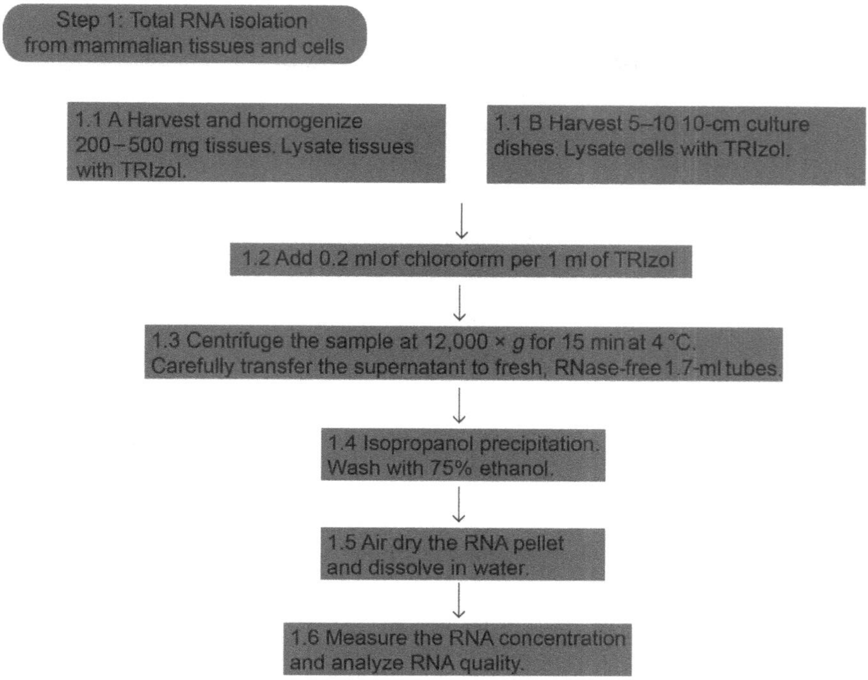

Figure 2 Flowchart of Step 1.

See Fig. 2 for the flowchart of Step 1.

6. STEP 2: mRNA ISOLATION

6.1 Overview

Isolate mRNA from total RNA by two successive rounds of polyA$^+$ selection and one round of rRNA depletion. The final yield of mRNA isolations is ~1%.

6.2 Duration

1 day

2.1 Adjust the volume of 500 μg total RNA to 500 μl with distilled RNase-free water or 10 mM Tris-HCl, pH 7.5. Heat to 65 °C for 2 min and chill on ice.

2.2 Transfer 1 ml of well-resuspended oligo(dT)$_{25}$ dynabeads to a fresh, RNase-free 1.7-ml tube. Place the tube on the magnet for 30 s. Remove the supernatant and wash the beads by adding 1 ml binding buffer and pipetting up and down ~10 times. Put the tube back on the magnet and then remove the supernatant. Resuspend the dynabeads with 500 μl binding buffer.

2.3 Add the total RNA to the suspended oligo(dT)$_{25}$ dynabeads in a 1:1 ratio. Mix thoroughly by pipetting up and down ~15 times. Rotate up to down for 5–10 min at room temperature to allow mRNA to anneal to the oligo(dT)$_{25}$. Place the tube on the magnet until solution is clear. Remove the supernatant.

2.4 Remove the tube from the magnet and wash the mRNA–bead complex twice by adding 1 ml washing buffer and pipetting up and down ~10 times. Put the tube back on the magnet and then remove the supernatant.

2.5 Add 500 μl of 10 m*M* Tris–HCl, pH 7.5, to resuspend the mRNA–bead complex. Incubate at 70 °C for 2 min and place the tube immediately on the magnet. Transfer the eluted mRNA to a new RNase-free tube and put on ice.

2.6 Add 1 ml of 10 m*M* Tris–HCl to the resuspended beads again, incubate for 3 min at 80 °C with >1000 rpm, place the tube immediately on the magnet, and discard the supernatant. Wash the beads as Step 2.2.

2.7 Heat the eluted RNA in Step 2.5 to 65 °C for 2 min again and chill it on ice. Add the RNA into the suspension of Step 2.6. Repeat Steps 2.3 and 2.4.

2.8 Elute the RNA with 200 μl of 10 m*M* Tris–HCl as Step 2.5.

2.9 Add 20 μl of 3 *M* NaOAc and 4 μl glycogen to the eluted RNA and 0.5 ml of 100% ethanol. Mix well and incubate at −20 °C for at least 1 h. Centrifuge at >12,000 × *g* for 20 min at 4 °C. Remove the supernatant from the tube and wash the pellet with 1 ml of 75% ethanol. Centrifuge at >12,000 × *g* for 5 min at 4 °C. Discard the wash without disturbing the pellet.

2.10 Air dry the RNA pellet for 10 min and dissolve the RNA pellet in each tube in 5 μl RNase-free water. Add 4 μl RiboMinus Probe (100 pmol/μl) and 150 μl hybridization buffer to the RNA and mix well.

2.11 Heat the mixture at 75 °C for 5 min and incubate at 37 °C for 30 min to allow rRNA anneal with the RiboMinus Probe by cooling slowly.

2.12 Transfer 250 μl of well-resuspended RiboMinus magnetic beads to a new, RNase-free 1.7-ml tube. Place the tube on the magnet until the solution is clear. Wash the beads with 250 μl RNase free water twice. Wash the beads with 250 μl hybridization buffer once. Resuspend the beads with 100 μl hybridization buffer. Incubate the beads at 37 °C until use.

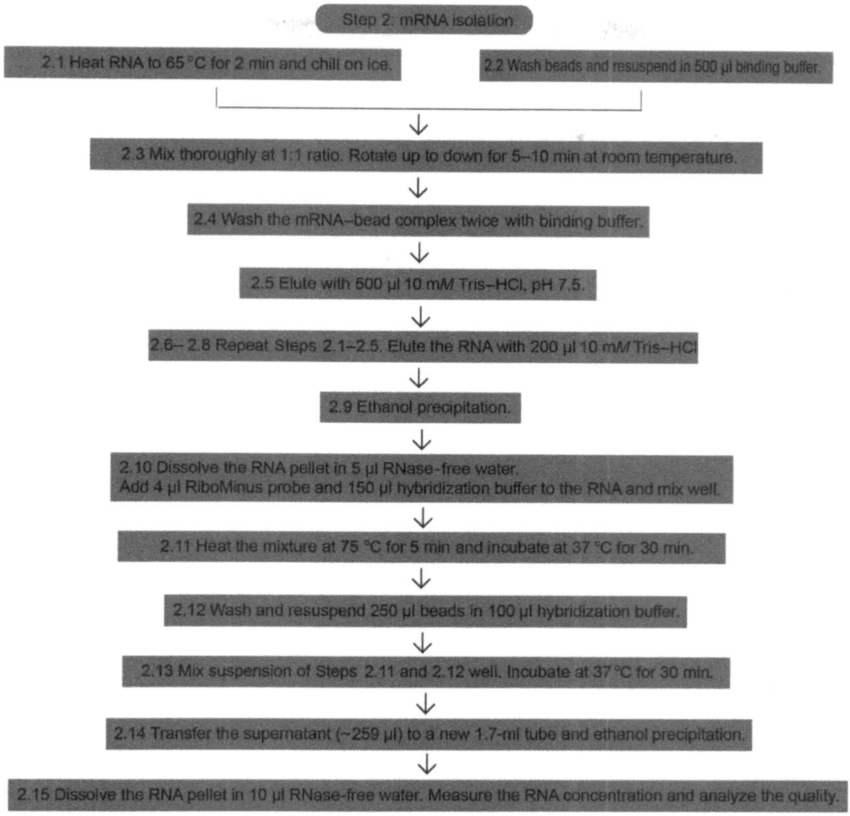

Figure 3 Flowchart of Step 2.

2.13 Transfer ~159 µl of the cooled hybridized sample (from Step 2.11) to the prepared RiboMinus magnetic beads from Step 2.12 and mix well. Incubate the tube at 37 °C for 30 min. During incubation, gently mix the contents from top to bottom to avoid the beads accumulation.

2.14 Place the tube on a magnetic stand until the solution is clear. Transfer the supernatant (~259 µl) to a new, RNase-free 1.7-ml tube. Add 26 µl of 3 *M* NaOAc and 4 µl glycogen to the eluted RNA and 650 µl of 100% ethanol. Mix well and incubate at −20 °C for at least 1 h. Centrifuge at >12,000 × *g* for 20 min at 4 °C. Remove the supernatant from the tube and wash the pellet with 1 ml of 75% ethanol. Centrifuge at >12,000 × *g* for 5 min at 4 °C. Discard the wash without disturbing the pellet. Repeat washing once.

2.15 Air dry the RNA pellet for 10 min and dissolve the RNA pellet in 10 µl RNase-free water. Measure the RNA concentration by

Nanodrop and analyze the RNA quality using RNA 6000 Pico Chips on an Agilent 2100 Bioanalyzer.

See Fig. 3 for the flowchart of Step 2.

7. STEP 3: N_3-CMC LABELING AND CLICK REACTION

7.1 Overview

mRNA was fragmented to 100 ~ 150 nt and ψ in mRNA was reacted with N_3-CMC followed by click reaction.

7.2 Duration

1–2 days

3.1 Add 2 μl RNA fragmentation buffer (10×) to 10 μg mRNA in 18 μl RNase-free water. Incubate at 94 °C for 5 minutes and then chill on ice. Add 2 μl of 10× RNA fragmentation stop solution and mix well. Add 78 μl RNase-free water, 10 μl of 3 *M* NaOAc, and 4 μl glycogen to the fragmented RNA and 250 μl of 100% ethanol. Mix well and incubate at −20 °C for at least 1 h. Centrifuge at >12,000 × *g* for 20 min at 4 °C. Remove the supernatant from the tube and wash the pellet with 1 ml of 75% ethanol. Centrifuge at >12,000 × *g* for 5 min at 4 °C. Discard the wash without disturbing the pellet. Repeat washing once.

3.2 Air dry the RNA pellets for 10 min and dissolve the RNA pellet in 18 μl RNase-free water.

3.3 Add 2 μl of 50 m*M* EDTA to the fragmented mRNA and mix well. Incubate at 80 °C for 5 min and chill on ice to disrupt the mRNA secondary structure.

3.4 Add 10 μl denatured mRNA to 100 μl of 0.2 *M* chemically synthesized N_3-CMC in BEU buffer. And add the other 10 μl denatured mRNA to 100 μl BEU buffer as mock. Incubate at 37 °C for 20 min shaking at 800 rpm. ψ, U, and G residues in RNA can be reacted with N_3-CMC.

3.5 Add 10 μl of 3 *M* NaOAc and 4 μl glycogen, and 250 μl of 100% ethanol to the reaction and mock, respectively. Mix well and incubate at −20 °C for at least 1 h. Centrifuge at >12,000 × *g* for 20 min at 4 °C. Remove the supernatant from the tube and wash the pellet with 1 ml of 75% ethanol. Centrifuge at >12,000 × *g* for 5 min at 4 °C. Discard the wash without disturbing the pellet. Repeat washing once.

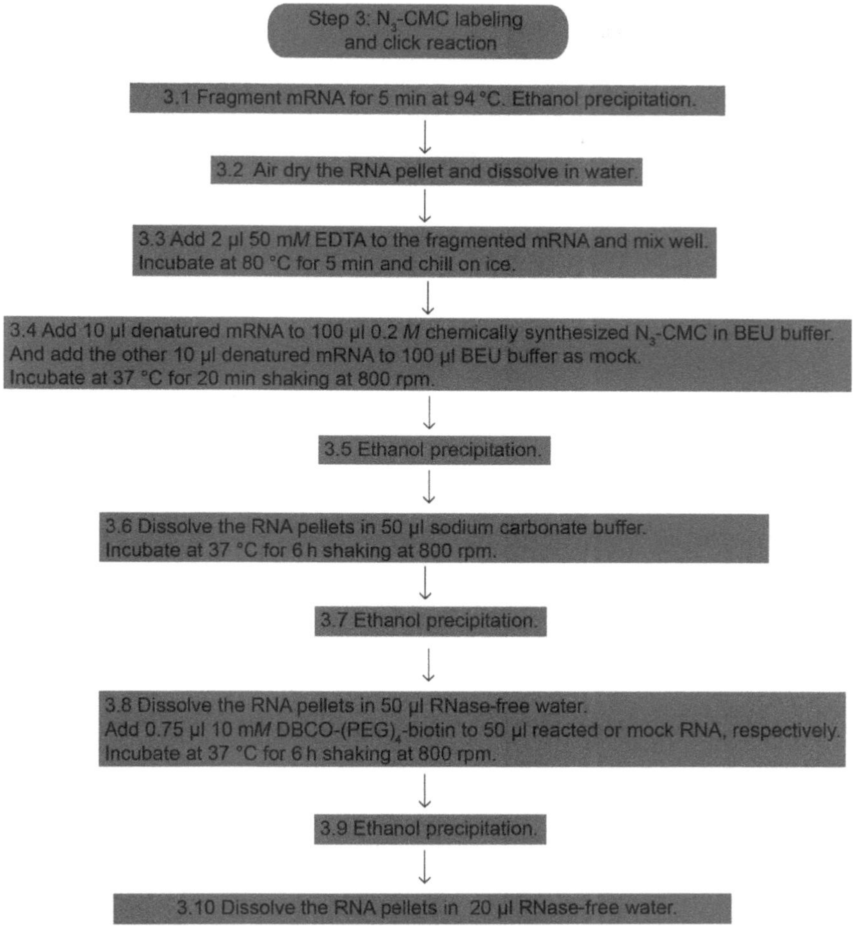

Figure 4 Flowchart of Step 3.

3.6 Air dry the RNA pellet for 10 min and dissolve the RNA pellets in 50 µl sodium carbonate buffer (50 m*M* Na_2CO_3, 2 m*M* EDTA, pH 10.4). Incubate at 37 °C for 6 h shaking at 800 rpm to remove the N_3-CMC from U and G residues.

3.7 Add 5 µl of 3 *M* NaOAc, 2 µl glycogen, and 125 µl of 100% ethanol to the reaction and mock, respectively. Mix well and incubate at −20 °C for at least 1 h. Centrifuge at >12,000 × *g* for 20 min at 4 °C. Remove the supernatant from the tube and wash the pellet with 1 ml of 75% ethanol. Centrifuge at >12,000 × *g* for 5 min at 4 °C. Discard the wash without disturbing the pellet. Repeat washing once.

3.8 Air dry the RNA pellets for 10 min and dissolve the RNA pellet in 50 μl RNase-free water. Add 0.75 μl 10 m*M* DBCO-$(PEG)_4$-biotin to 50 μl reacted or mock RNA, respectively, to a final concentration of 150 μ*M*. Incubate at 37 °C for 2 h shaking at 800 rpm, performing the click reaction, to label RNA containing ψ with CMC–biotin.

3.9 Add 5 μl of 3 *M* NaOAc and 2 μl glycogen, and 125 μl of 100% ethanol to the reaction and mock, respectively. Mix well and incubate at −20 °C for at least 1 h. Centrifuge at >12,000 × *g* for 20 min at 4 °C. Remove the supernatant from the tube and wash the pellet with 1 ml of 75% ethanol. Centrifuge at >12,000 × *g* for 5 min at 4 °C. Discard the wash without disturbing the pellet. Repeat washing once.

3.10 Air dry the RNA pellets for 10 min and dissolve the RNA pellet in 20 μl RNase-free water.

See Fig. 4 for the flowchart of Step 3.

8. STEP 4: ENRICH ψ-CONTAINING RNA FRAGMENTS BY STREPTAVIDIN PULL DOWN

8.1 Overview

ψ in mRNA was reacted with N_3-CMC followed by click reaction, therefore, mRNA containing the adduct of ψ–CMC–biotin could be pulled down by streptavidin dynabeads.

8.2 Duration

1–2 h

4.1 Transfer 50 μl well-resuspended streptavidin C1 dynabeads to a fresh, RNase-free 1.7-ml tube. Place the tube on the magnet until the solution is clear. Remove the supernatant and add 200 μl DEPC-treated 0.1 *M* NaOH, 0.05 *M* NaCl. Incubate at room temperature for 2 min to remove the RNase. Place the tube on the magnet until the solution is clear. Repeat treating once.

4.2 Remove the supernatant and add 200 μl DEPC-treated 0.1 *M* NaCl to resuspend the beads. Place the tube on the magnet until the solution is clear.

4.3 Wash the beads with 200 μl binding buffer twice.

4.4 Resuspend the beads in 200 μl binding buffer. Add the 20 μl reacted RNA into the beads suspension.

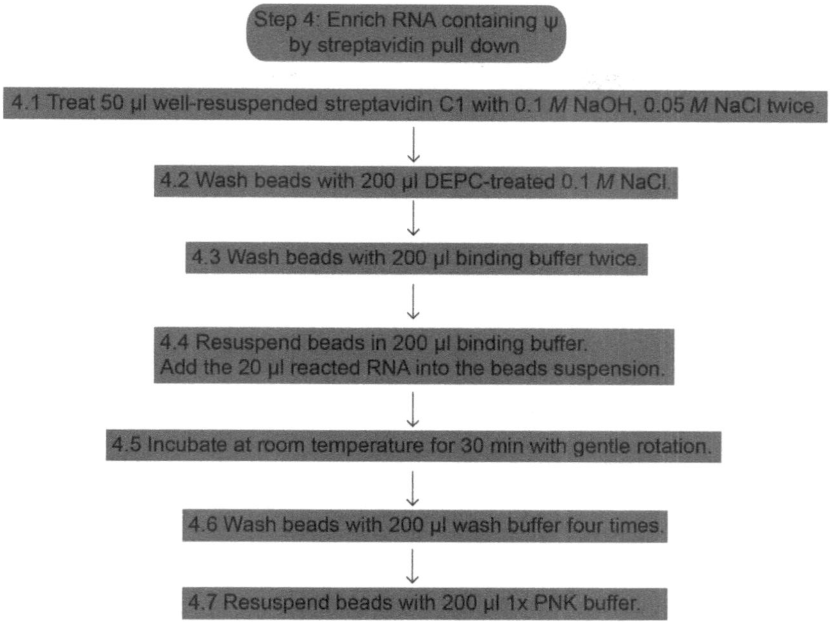

Figure 5 Flowchart of Step 4.

4.5 Incubate at room temperature for 30 min with gentle rotation to allow RNA containing ψ to bind with streptavidin beads.

4.6 Place the tube on the magnet until the solution is clear and remove the supernatant. Add 200 μl wash buffer to the beads, resuspend beads with gentle rotation at room temperature for 5 min to remove the unbounded RNA. Repeat washing three times.

4.7 Place the tube on the magnet until the solution is clear and remove the supernatant. Resuspend beads with 200 μl of 1× PNK buffer.

See Fig. 5 for the flowchart of Step 4.

9. STEP 5: RNA LIGATION

9.1 Overview

For the enriched RNA, the RNA adaptor ligation was performed on beads. For the mock, it was done in solution.

9.2 Duration

1 day

For the enriched RNA, ligation was performed on beads.

5.1 RNA dephosphorylation.

Place the tube on the magnet until the solution is clear and remove the supernatant.

Resuspend the beads in 40 μl of the following mixture:

4.0 μl	CutSmart buffer
2.0 μl	Shrimp alkaline phosphatase (rSAP)
33.0 μl	RNase-free water
1.0 μl	RNase Inhibitor

Incubate for 1 h at 37 °C.

5.2 Place the tube on the magnet until the solution is clear and remove the supernatant. Resuspend beads with 200 μl high-salt wash buffer. Place the tube on the magnet until the solution is clear and remove the supernatant. Resuspend beads with 200 μl of 1× PNK buffer. Repeat wash with 200 μl of 1 × PNK buffer twice to remove the rSAP.

5.3 RNA adaptor ligation on beads.

Place the tube on the magnet until the solution is clear and remove the supernatant.

Resuspend the beads in 40 μl of the following mixture:

21.0 μl	RNase-free water
4.0 μl	10× ligation buffer
4.0 μl	10 m*M* ATP
1.0 μl	T4 RNA ligase 1
1.0 μl	RNase inhibitor
1.0 μl	RNA adaptor (40 μ*M*)
8.0 μl	PEG400

Incubate overnight (~16 h) at 16 °C.

5.4 Place the tube on the magnet until the solution is clear and remove the supernatant. Resuspend beads with 200 μl washing buffer. Repeat wash with 200 μl washing buffer three times to remove the excessive adaptor.

5.5 Resuspend beads with 200 μl of 10 m*M* Tris, 5 m*M* EDTA, 1% SDS. Heat at 95 °C for 5 min and place the tube immediately on the magnet. Transfer the eluted mRNA to a new RNase-free tube and put on ice.

5.6 Add 20 μl of 3 *M* NaOAc and 4 μl glycogen, and 500 μl of 100% ethanol to the reaction and mock, respectively. Mix well and incubate at −20 °C for at least 1 h. Centrifuge at >12,000 × *g* for 20 min at 4 °C. Remove the supernatant from the tube and wash the pellet with 1 ml of 75% ethanol. Centrifuge at >12,000 × *g* for 5 min at 4 °C. Discard the wash without disturbing the pellet. Repeat washing with 1 ml of 75% ethanol once.

5.7 Air dry the RNA pellet for 10 min and dissolve the RNA pellets in 10 μl RNase-free water.

For the mock RNA, ligation was performed in solution.

5.8 RNA dephosphorylation.

Add the following reagents to the 20 μl resuspended mock RNA and mix well.

4.0 μl	CutSmart buffer
2.0 μl	Shrimp alkaline phosphatase (rSAP)
14.0 μl	RNase-free water

Incubate for 30 min at 37 °C. Heat for 5 min at 65 °C to inactive rSAP.

5.9 Add 500 μl TRIzol to the mixture and incubate the sample for 5 min at room temperature to permit complete dissociation of the protein complex. Add 0.1 ml of chloroform and shake tube vigorously by hand for 15 s and incubate for 2–3 min at room temperature. Centrifuge the sample at 12,000 × *g* for 15 min at 4 °C. Carefully transfer the supernatant to fresh, RNase-free 1.7-ml tubes. Add 20 μl of 3 *M* NaOAc, 4 μl glycogen, and 250 μl of 100% isopropanol to the aqueous phase. Mix well and incubate at −20 °C for at least 1 h. Centrifuge at >12,000 × *g* for 20 min at 4 °C. Remove the supernatant from the tube and wash the pellet with 1 ml of 75% ethanol. Centrifuge at >12,000 × *g* for 5 min at 4 °C. Discard the wash without disturbing the pellet. Repeat washing with 1 ml of 75% ethanol once.

5.10 Air dry the RNA pellet for 10 min and dissolve the RNA pellets in 12 μl RNase-free water.

5.11 RNA ligation in solution.

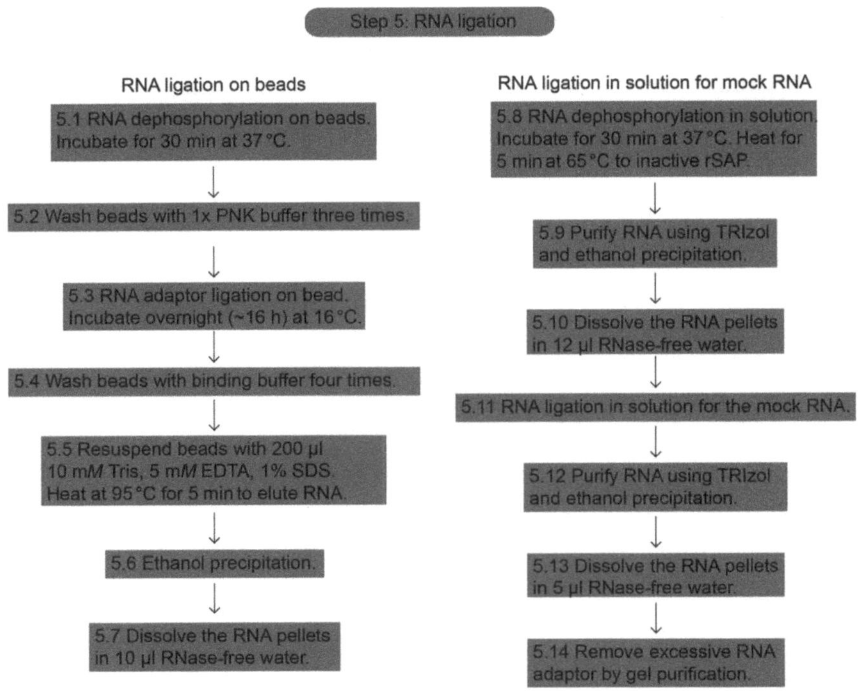

Figure 6 Flowchart of Step 5.

Add the following reagents to the 10 μl resuspended dephosphorylated mock RNA and mix well.

11.0 μl	RNase-free water
4.0 μl	10× ligation buffer
4.0 μl	10 m*M* ATP
1.0 μl	T4 RNA ligase 1
1.0 μl	RNase inhibitor
1.0 μl	RNA adaptor (40 μ*M*)
8.0 μl	PEG400

Incubate overnight at 16 °C.

5.12 Add 500 μl TRIzol to the mixture and incubate the sample for 5 min at room temperature to permit complete dissociation of the protein complex. Add 0.1 ml of chloroform and shake tube vigorously by hand for 15 s and incubate for 2–3 min at room temperature. Centrifuge the sample at 12,000 × *g* for 15 min at 4 °C. Carefully transfer the supernatant to fresh, RNase-free 1.7-ml tubes. Add 20 μl of 3 *M* NaOAc, 4 μl glycogen, and 250 μl of 100% isopropanol to the aqueous phase.

Mix well and incubate at −20 °C for at least 1 h. Centrifuge at >12,000 × *g* for 20 min at 4 °C. Remove the supernatant from the tube and wash the pellet with 1 ml of 75% ethanol. Centrifuge at >12,000 × *g* for 5 min at 4 °C. Discard the wash without disturbing the pellet. Repeat washing with 1 ml of 75% ethanol once.

5.13 Air dry the RNA pellet for 10 min and dissolve the RNA pellets in 5 μl RNase-free water.

5.14 Remove excessive RNA adaptor by gel purification.

A. Add 5 μl TBE-urea sample buffer (2×) to 5 μl resuspended RNA.

B. Pre-run 6% TBE-urea gel for 10 min.

C. Load the 10 μl sample. Run the gel until the lower (dark blue) dye is close to the bottom.

D. Stain it by incubation for 10 min and shake in 10 ml TBE buffer with 2 μl gel-safe dye. Visualize by UV and cut the sizes of 40–200 nt to remove the excessive adaptor.

E. Recover RNA from TBE-urea gel using ZR small-RNA™ PAGE Recovery Kit.

See Fig. 6 for the flowchart of Step 5.

10. STEP 6: REVERSE TRANSCRIPTION

10.1 Overview

Chemically enriched RNA and mock RNA was reverse transcribed to cDNA, and excessive RT primers were removed by gel purification.

10.2 Duration

1 day

6.1 Add 1 μl RT primer (0.5 pmol/μl) and 1 μl of 10 m*M* dNTP mix to the 12 μl resuspended or eluted RNA, respectively. Heat at 70 °C for 5 min and keep at 25 °C.

6.2 Add the following mixture to the denatured RNA.

4.0 μl	5 × RT buffer
1.0 μl	0.1 *M* DTT
1.0 μl	Superscript III RT (200 U/μl)

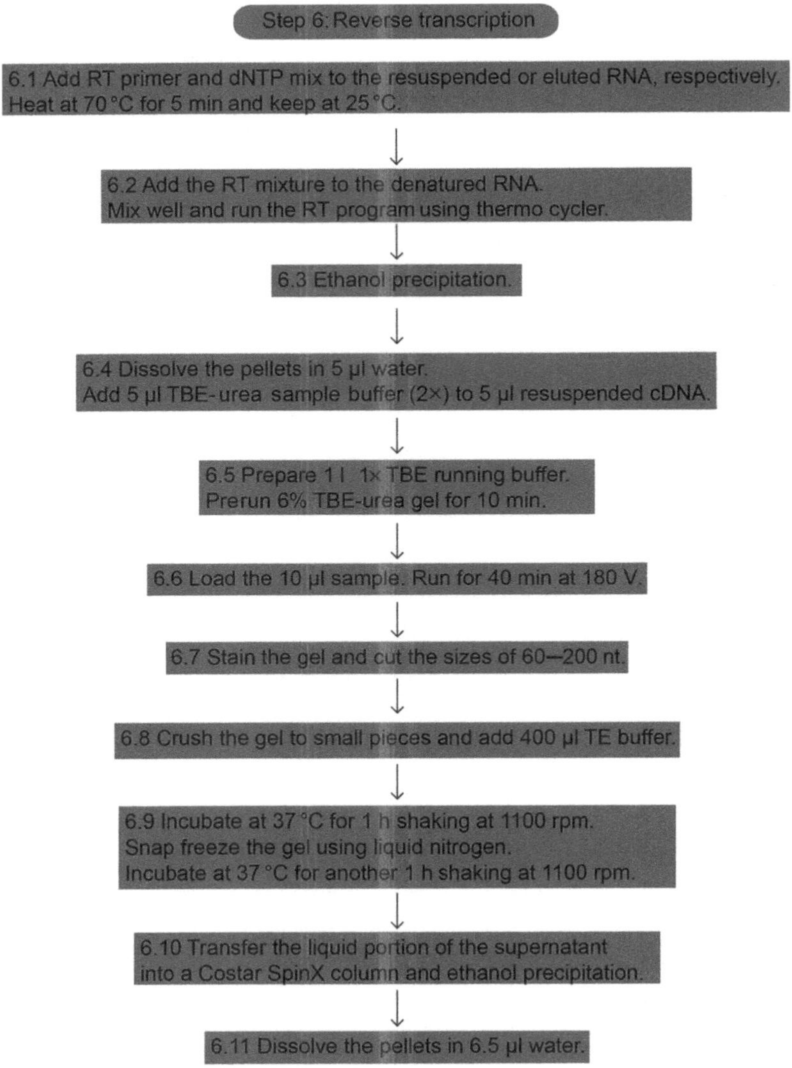

Figure 7 Flowchart of Step 6.

Mix well and run the following program using thermo cycler:

25 °C	5 min
50 °C	60 min
80 °C	5 min
4 °C	Hold

6.3 Add 60 μl water, 8 μl of 3 *M* NaOAc, 4 μl glycogen, and 200 μl of 100% ethanol to the mixture. Mix well and incubate at −20 °C for at least 1 h. Centrifuge at >12,000 × *g* for 20 min at 4 °C. Remove the supernatant from the tube and wash the pellet with 1 ml of 75% ethanol. Centrifuge at >12,000 × *g* for 5 min at 4 °C. Discard the wash without disturbing the pellet. Repeat washing with 1 ml of 75% ethanol once.

6.4 Air dry the cDNA pellet for 10 min and dissolve the pellets in 5 μl water. Add 5 μl TBE-urea sample buffer (2×) to 5 μl resuspended cDNA.

6.5 Prepare 1 l of 1× TBE running buffer. Prerun 6% TBE-urea gel for 10 min.

6.6 Load the 10 μl sample. Run for 40 min at 180 V until the lower (dark blue) dye is close to the bottom.

6.7 Stain the gel. Cut the sizes of 60–200 nt to remove the excessive RT primers.

6.8 Transfer the gel into a new fresh 1.7-ml tube. Crush the gel to small pieces and add 400 μl TE buffer.

6.9 Incubate at 37 °C for 1 h shaking at 1100 rpm. Snap freeze the gel using liquid nitrogen. Incubate at 37 °C for another 1 h shaking at 1100 rpm.

6.10 Transfer the liquid portion of the supernatant using Costar SpinX column. Transfer the flow through to a new fresh 1.7-ml tube. Add 40 μl of 3 *M* NaOAc, 4 μl glycogen, and 1 ml of 100% ethanol to the mixture. Mix well and incubate at −20 °C for at least 1 h. Centrifuge at >12,000 × *g* for 20 min at 4 °C. Remove the supernatant from the tube and wash the pellet with 1 ml of 75% ethanol. Centrifuge at >12,000 × *g* for 5 min at 4 °C. Discard the wash without disturbing the pellet. Repeat washing with 1 ml of 75% ethanol once.

6.11 Air dry the cDNA pellet for 10 min and dissolve the pellets in 6.5 μl water.

See Fig. 7 for the flowchart of Step 6.

11. STEP 7: CIRCLIGATION, LINEARIZATION, AND PCR AMPLIFICATION

11.1 Overview

cDNA was circligated and linearized by *Bam*HI digestion. The linearized cDNA was amplified by PCR.

11.2 Duration

1 day

7.1 Add the following mixture to the resuspended cDNA and mix well:

0.8 μl	10× CircLigase buffer II
0.4 μl	50 m*M* $MnCl_2$
0.3 μl	CircLigase II

Incubate at 60 °C for 1 h.

7.2 Anneal an anneal oligo to the circligated cDNA. Add the following anneal mix:

26 μl	water
3 μl	FastDigest buffer
1 μl	10 μ*M* anneal oligo

Mix well and run the following program using thermo cycler:

95 °C	2 min

Successive cycles of 30 s, starting from 95 °C and decreasing the temperature by 0.5 °C each cycle down to 25 °C.

25 °C	Hold

7.3 Add 2 μl FastDigest *Bam*HI and mix well. Incubate at 37 °C for 30 min to linearize the circligated cDNA.

7.4 Add 40 μl water, 8 μl of 3 *M* NaOAc, 4 μl glycogen, and 200 μl of 100% ethanol to the mixture. Mix well and incubate at −20 °C for at least 1 h. Centrifuge at >12,000 × *g* for 20 min at 4 °C. Remove the supernatant from the tube and wash the pellet with 1 ml of 75% ethanol. Centrifuge at >12,000 × *g* for 5 min at 4 °C. Discard the wash without disturbing the pellet. Repeat washing with 1 ml of 75% ethanol once.

7.5 Air dry the cDNA pellet for 10 min and dissolve the pellets in 23 μl water.

7.6 Add the following PCR mix to the resuspended cDNA:

25.0 μl	NEBNext High-Fidelity 2× PCR Master Mix
1.0 μl	25 μ*M* P3 Solexa Primer
1.0 μl	25 μ*M* P5 Solexa Primer

Mix well and run the following program using thermo cycler:

98 °C	30 s
18 cycles of:	
98 °C	10 s
65 °C	30 s
72 °C	30 s
72 °C	5 min
4 °C	Hold

7.7 Purify the PCR products using AMPure XP beads. Add 90 μl (1.8×) of resuspended AMPure XP beads and mix well by pipetting up and down at least 10 times. Incubate for 5 min at room temperature. Place the tube on the magnet until the solution is clear. Carefully remove the supernatant. Add 200 μl of freshly prepared 80% ethanol to the tube while in the magnetic stand without disturbing the beads. Incubate at room temperature for 30 s and then carefully remove and discard the supernatant. Repeat washing with 200 μl of freshly prepared 80% ethanol once. Air dry beads for 10 min while the tube is on the magnetic stand with lid open. Add 25 μl nuclease-free water to the beads. Mix well by pipetting up and down, and put the tube in the magnetic stand until the solution is clear. Transfer the supernatant to a clean PCR tube.

7.8 Size select the purified cDNA library using gel purification. Mix the purified 25 μl PCR product with 5 μl of 6× loading dye.

7.9 Prepare 1 l of 1× TBE running buffer. Prerun 6% TBE gel for 10 min.

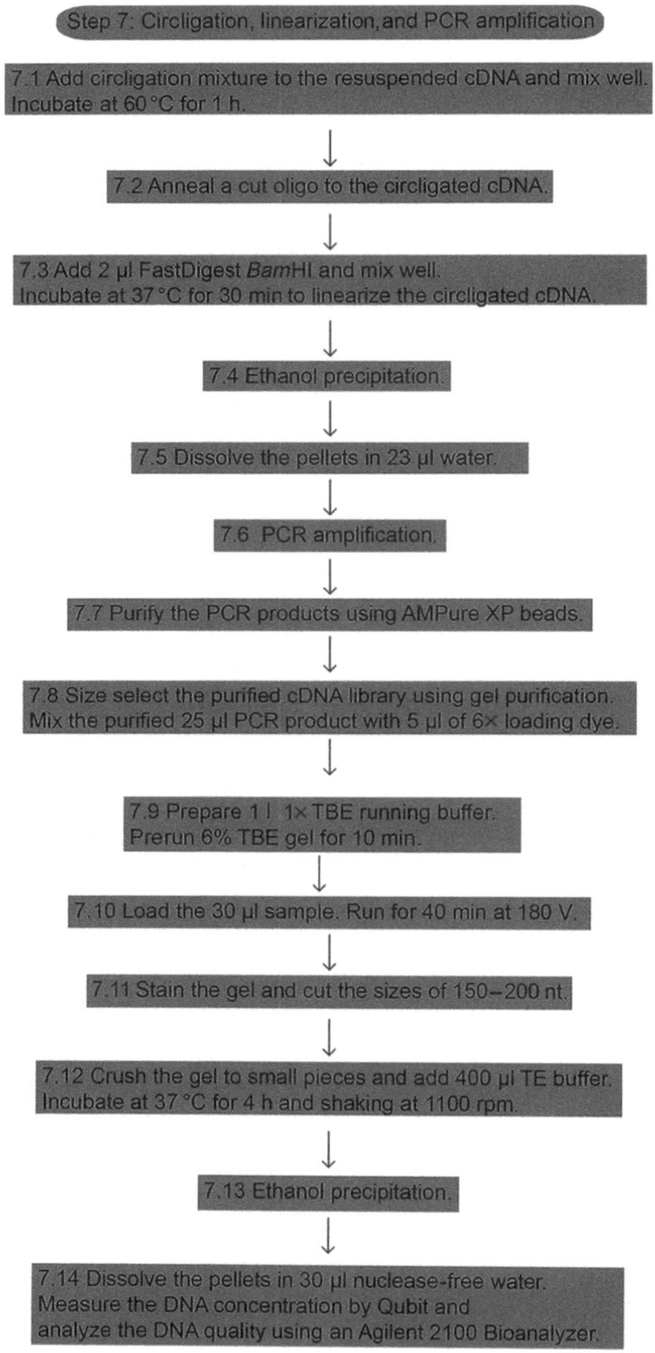

Figure 8 Flowchart of Step 7.

7.10 Load the 30 μl sample. Run for 40 min at 180 V.

7.11 Stain the gel. Cut the sizes of 150–200 nt to remove the primer dimers.

7.12 Transfer the gel into a new fresh 1.7-ml tube. Crush the gel to small pieces and add 400 μl TE buffer. Incubate at 37 °C for 4 h shaking at 1100 rpm.

7.13 Transfer the liquid portion of the supernatant using Costar SpinX column. Transfer the flow through to a new fresh 1.7-ml tube. Add 40 μl of 3 *M* NaOAc, 4 μl glycogen, and 1 ml of 100% ethanol to the mixture. Mix well and incubate at −20 °C for at least 1 h. Centrifuge at >12,000 × *g* for 20 min at 4 °C. Remove the supernatant from the tube and wash the pellet with 1 ml of 75% ethanol. Centrifuge at >12,000 × *g* for 5 min at 4 °C. Discard the wash without disturbing the pellet. Repeat washing with 1 ml of 75% ethanol.

7.14 Air dry the cDNA pellet for 10 min and dissolve the pellets in 30 μl nuclease-free water. Measure the DNA concentration by Qubit and analyze the DNA quality using an Agilent 2100 Bioanalyzer.

7.15 PCR products are sequenced on Illumina Hiseq 2500 with 50 bp single-end reads. Raw sequencing data are demultiplexed to each sample according to their barcode information. Trim galore is used to remove or trim adaptor-contained reads and low-quality reads, requiring the minimum length of trimmed reads to be 20 bases. Remaining reads can be mapped to corresponding transcriptome using BWA (Li & Durbin, 2010). ψ sites are called according to the ratio of stop reads and read-through reads at a U base.

See Fig. 8 for the flowchart of Step 7.

REFERENCES

Bakin, A. V., & Ofengand, J. (1998). Mapping of pseudouridine residues in RNA to nucleotide resolution. *Methods in Molecular Biology*, *77*, 297–309.

Ge, J., & Yu, Y. T. (2013). RNA pseudouridylation: New insights into an old modification. *Trends in Biochemical Sciences*, *38*, 210–218.

Hamma, T., & Ferre-D'Amare, A. R. (2006). Pseudouridine synthases. *Chemistry & Biology*, *13*, 1125–1135.

Karijolich, J., & Yu, Y.-T. (2011). Converting nonsense codons into sense codons by targeted pseudouridylation. *Nature*, *474*, 395–398.

Kiss, T., Fayet-Lebaron, E., & Jady, B. E. (2010). Box H/ACA small ribonucleoproteins. *Molecular Cell*, *37*, 597–606.

Li, H., & Durbin, R. (2010). Fast and accurate long-read alignment with Burrows-Wheeler Transform. *Bioinformatics*, *Epub.*, *26*(5), 589–595.

Li, X., Zhu, P., Ma, S., Song, J., Bai, J., Sun, F., et al. (2015). Chemical pull-down reveals dynamic pseudouridylation of the mammalian transcriptome. *Nature Chemical Biology* (in press).

Machnicka, M. A., Milanowska, K., Osman Oglou, O., Purta, E., Kurkowska, M., Olchowik, A., et al. (2013). MODOMICS: A database of RNA modification pathways—2013 update. *Nucleic Acids Research*, *41*, D262–D267.

CHAPTER THIRTEEN

Experimental Approaches for Target Profiling of RNA Cytosine Methyltransferases

Vahid Khoddami[*,1], **Archana Yerra**[*], **Bradley R. Cairns**[*,2]
[*]HHMI, Department of Oncological Sciences, Huntsman Cancer Institute, University of Utah School of Medicine, Salt Lake City, Utah, USA
[1]Current address: Department of Cell Biology, Howard Hughes Medical Institute, Harvard Medical School, 240 Longwood Avenue, Boston, MA 02115, USA
[2]Corresponding author: e-mail address: brad.cairns@hci.utah.edu

Contents

Abstract

RNA cytosine methyltransferases (m^5C-RMTs) constitute an important class of RNA-modifying enzymes, methylating specific cytosines within particular RNA targets in both coding and noncoding RNAs. Almost all organisms express at least one m^5C-RMT, and vertebrates often express different types or variants of m^5C-RMTs in different cell types. Deletion or mutation of particular m^5C-RMTs is connected to severe pathological manifestations ranging from developmental defects to infertility and mental retardation. Some m^5C-RMTs show spatiotemporal patterns of expression and activity requiring careful experimental design for their analysis in order to capture their context-dependent targets. An essential step for understanding the functions of both the enzymes and the modified cytosines is defining the one-to-one connection between particular m^5C-RMTs and their target cytosines. Recent technological and methodological advances have provided researchers with new tools to comprehensively explore RNA cytosine methylation and methyltransferases. Here, we describe three complementary approaches applicable for both discovery and validation of candidate target sites of specific m^5C-RMTs.

Methods in Enzymology, Volume 560
ISSN 0076-6879
http://dx.doi.org/10.1016/bs.mie.2015.03.008

1. INTRODUCTION

RNA modifications are abundant, varied, and ubiquitous—and can affect the RNA in various ways including structure, folding, and interaction with proteins, DNA, ions, or other RNAs (Machnicka et al., 2013). RNA modifications have been most intensively studied in noncoding RNAs (ncRNAs) involved in translation (e.g., tRNAs and rRNAs), where they have been shown to affect RNA folding and stability—and can potentially expand the coding capacity of mRNAs. RNA cytosine methylation is one of the many (>100) known nucleotide modifications in RNA. It is present in bacteria and archaea and in all classes of eukaryotes from yeasts to human (Motorin, Lyko, & Helm, 2010). Although first discovered in ribosomal RNA and tRNA, where it remains best studied mechanistically, recent studies show that this modification occurs in many other classes of RNA (Edelheit, Schwartz, Mumbach, Wurtzel, & Sorek, 2013; Squires et al., 2012). Presently, we know little about the full repertoire of RNA cytosine methylation in particular cell types, nor do we know whether and how the methylation landscape changes in development, or in response to environmental changes. Furthermore, most vertebrates contain multiple cytosine RNA methyltransferases (m^5C-RMTs), only a portion of which has been linked to particular substrates (Motorin et al., 2010). Defects arising from a loss of m^5C-RMT range from subtle to fatal—best illustrated in studies of individual m^5C-RMTs (Table 1). Studies in mice show that knockdown or knockout of certain m^5C-RMTs is lethal individually (NSUN4) (Metodiev et al., 2014) or in combination (NSUN2 and DNMT2) (Tuorto et al., 2012). In humans, defects or misregulation of m^5C-RMTs are associated with cancers (Fonagy et al., 1993; Frye et al., 2010; Job et al., 2010; Okamoto et al., 2012), infertility (Khosronezhad et al., 2014), and mental retardation (Abbasi-Moheb et al., 2012; Khan et al., 2012; Martinez et al., 2012), prompting thorough analysis of different members of this important class of enzymes. Clearly, the "big question" is "how does cytosine methylation affect and regulate RNA function?" However, in order to answer this question, we must first know the full scope of RNA cytosine methylation in different cell types, and we must link those modifications to particular m^5C-RMTs. These two types of information can then be combined to greatly inform the design and interpretation of functional experiments. Here in this chapter, we explain three different approaches for target profiling and validation of particular m^5C-RMTs. The first

Table 1 Features of Known Human m^5C-RMTs

Human m^5C-RMTs	Targets	Biological Process and Phenotypes	Selected References
NSUN1	rRNA	Cell proliferation Upregulated in cancers Ribosomal maturation and assembly (*Saccharomyces cerevisiae*)	Fonagy et al. (1993), Hong, Wu, Brockenbrough, Wu, and Aris (2001), and Sharma, Yang, Watzinger, Kotter, and Entian (2013)
NSUN2	tRNA, ncRNA, mRNA	Intellectual disability (*Homo sapiens*) tRNA stability and translation Mitotic spindle stability Stem cell differentiation, testis differentiation (*Mus musculus*)	Abbasi-Moheb et al. (2012), Blanco et al. (2011), Frye and Watt (2006), Hussain, Sajini, et al. (2013), Hussain, Tuorto, et al. (2013), Khan et al. (2012), Khoddami and Cairns (2013), Martinez et al. (2012), Squires et al. (2012), and Tuorto et al. (2012)
NSUN3	Unknown	Loss of function in cancer	Job et al. (2010)
NSUN4	Mitochondrial rRNA	Mitochondrial ribosome maturation, assembly Knockout: embryonic lethal (*M. musculus*)	Camara et al. (2011), Metodiev et al. (2014), Spahr, Habermann, Gustafsson, Larsson, and Hallberg (2012), and Yakubovskaya et al. (2012)
NSUN5	rRNA	Deleted in Williams–Beuren's syndrome (*H. sapiens*) Modulating lifespan	Doll and Grzeschik (2001), Schosserer et al. (2015), and Sharma et al. (2013)
NSUN6	Unknown	Unknown	

Continued

Table 1 Features of Known Human m^5C-RMTs—cont'd

Human m^5C-RMTs	Targets	Biological Process and Phenotypes	Selected References
NSUN7	Unknown	Catalytic mutant: reduced sperm motility (*M. musculus*) Misfolding mutant: male infertility (*H. sapiens*)	Harris, Marquez, Suarez, and Schimenti (2007) and Khosronezhad, Colagar, and Jorsarayi (2014)
DNMT2	tRNAs (Asp, Gly, Val)	Developmental defects (*Danio rerio*) Immunity against RNA viruses, proper sister chromatid segregation, and retrotransposon silencing (*Drosophila melanogaster*) Transgenerational inheritance (*M. musculus*) Stress response (*Schizosaccharomyces pombe*)	Becker et al. (2012), Durdevic et al. (2013), Khoddami and Cairns (2013), Kiani et al. (2013), Phalke et al. (2009), Rai et al. (2007), Schaefer et al. (2010), Tuorto et al. (2012), and Yadlapalli and Yamashita (2013)

approach involves in-parallel bisulfite treatment and sequencing of RNA extracted from wild-type and knockout/-down cells or organisms. The second approach represents application of a mechanism-based enrichment technique, Aza-IP, for simultaneous isolation and identification of both target RNAs and target cytosines of particular m^5C-RMTs. Finally, the third approach describes an alternative way to validate target molecules/sites through *in vitro* reconstitution of the methyltransferase reaction.

2. METHODS

2.1 RNA Bisulfite Sequencing

Cytosine and 5-methyl-cytosine (m^5C) nucleotides react differently to an acidic sodium bisulfite solution—a property that has provided a valuable tool for sequencing-based methylation profiling of both DNA and RNA (Motorin et al., 2010). Following bisulfite treatment, all of the unmethylated cytosines (C) are deaminated to uracil (U), appearing as thymine (T) after sequencing, while m^5Cs are refractory to this deamination and remain as C, representing sites of cytosine methylation (Fig. 1A and B). Therefore, for accurate mapping of m^5Cs within the genome or transcriptome, complete conversion of virtually all unmethylated Cs is mandatory, as incomplete conversion results in false positives. Here, one key for complete conversion is efficient denaturation of nucleic acids, as bisulfite ions only attack cytosines in the single- rather than double-stranded form. For genomic DNA, denaturation is efficiently achieved by incubating the DNA in alkaline pH, a treatment that must be avoided when working with RNA as it causes severe degradation and loss of RNAs. To overcome this, protocols have been established to treat the RNA at elevated temperatures in order to enhance denaturation of RNAs, increasing the conversion rates to >95% (Schaefer, Pollex, Hanna, & Lyko, 2009; Squires et al., 2012). Here, we present a novel alternative RNA bisulfite-sequencing protocol utilizing heat denaturation in the presence of formamide (a strong denaturant) for improving both the conversion rate and RNA recovery yield. Additionally, we use random fragmentation of RNA samples prior to bisulfite treatment to further improve the conversion rate, likely through resolving the local intramolecular secondary structures within individual RNA molecules. Our analysis on pure synthetic oligonucleotides and high-throughput sequencing of total RNAs from different sources shows greater RNA recovery and near-complete (>99%) conversion rates. The step-by-step protocol for bisulfite sequencing is described below:

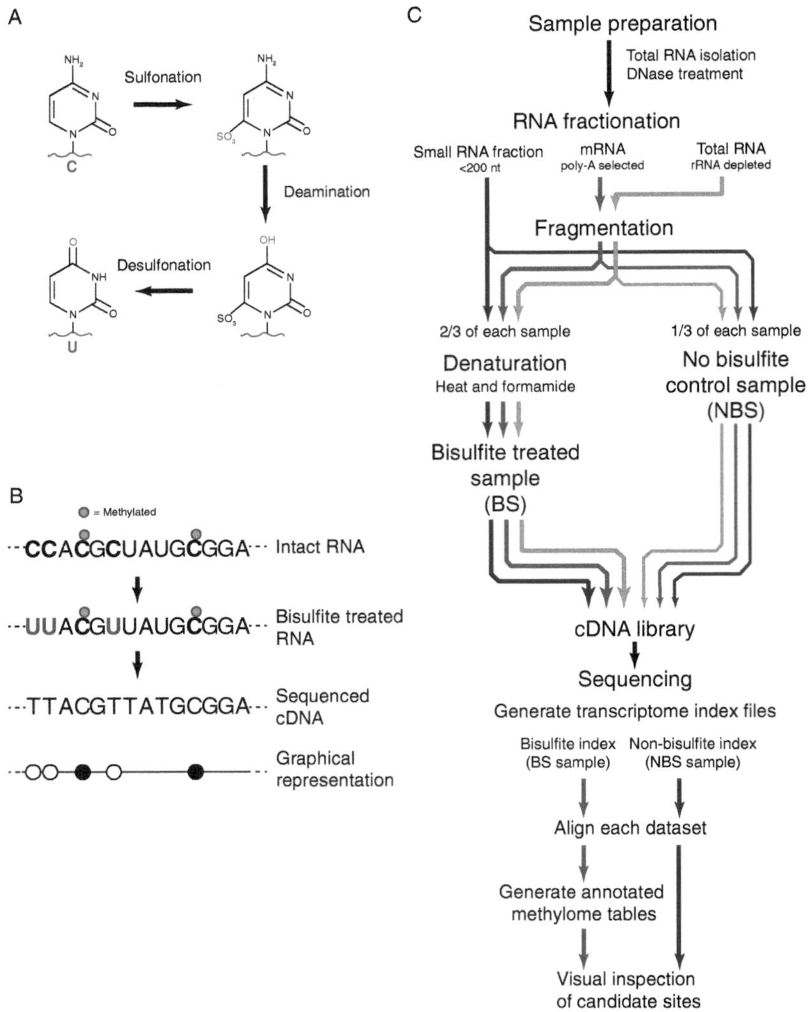

Figure 1 (A) Principle of bisulfite conversion. Unmethylated cytosine is sulfonated by sodium bisulfite and is subsequently deaminated to uridine. Methylated cytosine is refractory, and remains as cytosine. (B) Impact of cDNA synthesis. Upon conversion to cDNA, unmethylated cytosine is read by sequencing as thymine. (C) Bisulfite treatment of RNA. A flowchart of RNA bisulfite sequencing and analysis is depicted. Fractionation of RNA samples is recommended for full coverage of all possible sites in different RNA populations. It is also useful to have a matched nontreated control for each fraction to filter out false positives during the analysis phase. For high-throughput bisulfite sequencing and analysis, it is recommended to align both BS and NBS samples. Once methylome tables are calculated for the BS samples, these can be compared to the NBS samples to filter out false positives arising from incorrect mapping. (See the color plate.)

1. *Sample RNA preparation*. Bisulfite sequencing can be applied to either validate candidate sites or discover novel target sites for a particular RNA cytosine methyltransferase. Here, RNA samples from wild-type or mutant cells/organisms (lacking a particular enzyme) undergo bisulfite treatment and sequencing. Analytical comparison between the RNA methylome of the two groups will indicate the candidate targets of the methyltransferase under study. The RNA sample is typically (but not limited to) total RNA extracted from cells or tissues utilizing any of the commercial RNA purification kits. Regardless, it is an absolute requirement to remove DNA contamination by treating the samples with RNase-free DNase. Before using the DNase-treated RNA, depending on the expected targets, purification or fractionation of RNAs might be required. For example, when studying methylation patterns in abundant RNA species such as tRNAs and rRNAs as well as some other species like snRNAs and snoRNAs, the raw total RNA samples can be used. However, to study the less-abundant RNA species, enrichment of the target RNA populations from total RNA is highly recommended. A primary way for enrichment of the low-copy RNAs is to conduct rRNA depletion using commercial kits (e.g., Ribominus or Ribozero). Another mode of enrichment involves fractionation. For example, to study methylation in the small RNA species, commercial kits can be used to purify the small RNA fractions (typically <200 bp) from the total RNA, and particularly for the analysis of mRNA methylation, we recommend performing poly-A selection to specifically enrich mRNAs. Finally, to obtain a comprehensive high-throughput transcriptome-wide RNA methylome of a particular sample with high resolution and precision, if the initial materials are not limiting, we recommend performing bisulfite sequencing separately on three different groups: (1) rRNA-depleted total RNA, (2) the small RNA fraction, and (3) the poly-A-selected RNA (Fig. 1C). This combination will maximize the level and accuracy of information one can obtain from a particular sample.
2. *RNA fragmentation*. After RNA sample preparation, for all different types of RNA fractions (except for the small RNA sample), we recommend an additional step: random RNA fragmentation for the reasons explained above. Here, up to 5 μg of the DNase-treated RNA in 9 μl is mixed with 1 μl of 10 × RNA fragmentation reagent (Ambion) (total of 10 μl) and heated in a thermal cycler at 70 °C for 15 min, terminated by addition of 1 μl of the 10 × stop solution, and kept on ice. The fragmented RNA is then recovered by ethanol precipitation. Briefly, by addition of

nuclease-free water the volume is adjusted to 175 μl and is mixed with 5 μl of the 15 mg ml^{-1} GlycoBlue, followed by addition of 20 μl of 3 *M* sodium acetate and mixing (total of 200 μl). Next, 2.5 × volume (500 μl) of absolute ethanol is mixed well with the contents and the tube is incubated at −80 °C for >30 min, followed by centrifugation at ~21,000 × *g* for 20 min at 4 °C. The pellets are then washed once with 1 ml of 70% (vol/vol) ice-cold ethanol, air-dried and dissolved in 45 μl of RNase-free water.

3. *Bisulfite treatment.* The bisulfite treatment involves three steps: (1) denaturation, (2) sulfonation, and (3) desulfonation. In the denaturation step, the RNA molecules are unfolded to make all cytosines accessible to bisulfite ions. The conversion rate and precision of a typical bisulfite-sequencing technique relies almost entirely on how well the RNA molecules are denatured. Here, to achieve the maximum conversion rate, we enforce the RNA denaturation by heating the RNA in the presence of formamide. For this, in a 2 ml RNase-free-certified tube, 45 μl of the sample (containing up to 5 μg of fragmented DNase-treated RNA) is mixed well with 240 μl of deionized formamide by pipetting up and down several times (note: the formamide concentration is now ~84%), and incubated in preheated water bath at 95 °C for 5 min followed by at least 2 min incubation on ice. In the sulfonation step, sodium bisulfite is added to the mixture resulting in the attack of the bisulfite ions at the sixth position of the cytosine base, triggering deamination and conversion of Cs (but not m^5Cs) to Us (Fig. 1A). For this, 312 μl of 5 *M* sodium bisulfite (pH 5.0) and 3 μl of 100 m*M* hydroquinone are added to the denatured RNA (total volume 600 μl) and mixed well by pipetting up and down several times (note: the formamide concentration is now ~40%), and the tube's lid is sealed strongly using both Parafilm and microcentrifuge tube lock (to prevent opening of the tube during incubation). The mixture is now incubated at 50 °C, rotating for 16 h in a rotator hybridization oven. In the desulfonation step, the bisulfite ions connected covalently to the cytosine bases should be removed (Fig. 1A). To achieve this, first the excess amount of bisulfite ions is removed from the mixture by passing the 600 μl mixture entirely through an equilibrated NAP-10 column (see the manufacturer's protocol for column equilibration and sample application and elution), then apply 400 μl RNase-free water to the column and wait until the last drop is passed through, then add 1 ml of RNase-free water and collect all of the flow-through in a clean RNase-free-certified tube. The covalent

connection between the bisulfite ions and cytosines is spontaneously reversed in water at a slow rate, which needs to be enhanced at slightly basic pH. For this, we split the 1 ml sample into two tubes (0.5 ml each), and to each tube add 0.5 ml of 2 *M* Tris buffer (pH 9.0), mix by vortexing, and then incubate for 2 h at 37 °C in a water bath.

4. *RNA recovery*. Upon completion of bisulfite treatment, the RNA is recovered and concentrated by ethanol precipitation for subsequent cDNA synthesis and sequencing. For this, to each one of the tubes from the previous step, 20 μl of 15 mg/ml GlycoBlue and 111 μl of 3 *M* sodium acetate (pH 5.5) are added and mixed well by vortexing, then the content of each tube is split into three tubes each receiving ~377 μl of the mixture (total of six tubes for each sample), and each is mixed with 950 μl of absolute ethanol and incubated at −80 °C for >30 min, followed by centrifugation at ~21,000 × *g* for 20 min at 4 ° C. The pellets are then washed once with 1 ml of 70% (vol/vol) ice-cold ethanol, air-dried and dissolved in 5 μl of RNase-free water. The contents of all six tubes (total of 30 μl) are then collected into a single tube and stored at −80 °C until used.

5. *High-throughput sequencing*. For comprehensive transcriptome-wide target profiling of a given cytosine, methyltransferase bisulfite-treated RNA from wild-type and mutant cells/organisms can be subjected to library preparation and high-throughput sequencing (Fig. 1C) according to standard protocols. There are, however, a few recommendations: (1) For library preparation, it is best to apply kits that can efficiently recover the smaller size RNAs. This is particularly important as most RNA molecules are fragmented as described above. (2) It is necessary to perform "directional" library preparation and sequencing. This is especially important for many cases where RNA is transcribed from both sense and antisense strands. (3) In general, for a typical high-throughput sequencing experiment, we highly recommend sequencing both bisulfite-treated (BS) and -untreated (NBS) samples of the same type. For this, one can take ~1/3 of the original RNA sample as NBS and use the rest (~2/3) for bisulfite treatment (BS), and then subject, in parallel, both of the BS and NBS samples to library preparation and sequencing. Sequencing of the NBS sample is particularly important at the time of analysis (see below) to distinguish between the true- and false-positive methylation sites. (4) To obtain a comprehensive high-resolution RNA methylome, it is important to have a high read coverage by: (1) sequencing the samples in several lanes and

(2) sequencing at least 75 bp to enable efficient mapping. Note: for the BS–NBS sample pairs, the NBS samples need not be sequenced at the same level, as they are only serving as the control for specific regions of the genome, so ½ or ¼ of BS sequencing level should suffice.

6. *Analyzing the high-throughput data.* The steps for analysis are as follows:
   a. *Generating transcriptome index file.* The final output of the high-throughput sequencing is provided in FASTQ files, in which information about individual sequenced reads corresponding to individual RNA molecules is stored. For extracting the methylation data from FASTQ files, they first need to be matched with the genomic sequences to index the exact coordinates that each individual sequenced read is mapped to. Here, because protein-coding genes undergo splicing to produce mature mRNAs, by mapping the FASTQ files only to the genome, those sequenced reads spanning the exon–exon junctions are discarded as unmappable, thus their data are lost. The solution is to map the FASTQ files to an extended transcriptome index that includes both the entire genome, and also artificial chromosomes which contain all individual exon–exon junctions in the genome. This will ensure keeping such reads in the final analysis. To make the extended transcriptome, we recommend using the USeq package and following its guidelines (Nix, Courdy, & Boucher, 2008). For analyzing the bisulfite-untreated (NBS) and -treated (BS) datasets, two different transcriptome index files should be made using the "Novoindex" application from the commercial Novocraft package according to the instructions (http://www.novocraft.com). To generate the index file for NBS dataset, the index file should be made in the normal mode and for the BS dataset an option for *in silico* bisulfite conversion of the transcriptome should be used in the command line to include the C–U conversions.
   b. *Sequence alignment.* For alignment, the "Novoalign" application from the Novocraft package is used by setting the parameters to keep the sequenced reads $\geq$18 bp, and allowing mismatches.
   c. *Converting the coordinates of the splice junctions.* After sequencing, the output files will be first processed by USeq package to convert the coordinates of the reads mapped to exon–exon junction short artificial chromosomes into real genomic coordinates.
   d. *Making the annotated methylome tables.* Using custom programs, all of the sites showing at least one occurrence of maintained

(unconverted) cytosine is indexed and annotated along with recording the total number of times that particular site is sequenced (as either of C or T) to calculate: (1) the read coverage, (2) *P* value, and (3) non-conversion (possibly methylation) rate of each particular site. Usually, sites showing *P* value of <0.05, read coverage of $>10\times$, and non-conversion rate of $>20\%$ are considered as candidate methylation sites that need to be visually inspected, and if required experimentally validated. Those sites that are present in the wild-type sample but are completely absent in the samples from knockout experiments, or show decreased levels in the samples from knockdown experiments, will be considered as the candidate target sites of the methyltransferase under study.

e. *Visual inspection of the candidate methylated sites.* To visualize the alignment files, the same files used to make the methylome tables will be processed by USeq package to make index files (BAM/BAI files) that can be uploaded into genome browsers such as IGV (Integrative Genomics Viewer; Robinson et al., 2011) in order to inspect the candidate methylated sites at base-pair resolution. Visual inspection of candidate sites will be helpful especially to remove the possible false-positive calls from the table. There are two major types of possible false positives:

   i. *False positives that appear due to incomplete conversion.* Sites that are surrounded by several unconverted cytosines in similar sequenced reads (clustered-unconverted Cs) might indicate incomplete conversion due to strong secondary structures that remain undenatured even after heating in the presence of the strong denaturant formamide (Fig. 2A). This type of artifact is quite common and such clustered sites are better discarded, unless a paired wild-type–mutant analysis demonstrates that one or more of the sites within such clusters are true-methylated sites.

   ii. *False positives that appear due to mapping error.* It is important to note that a bisulfite-converted sequence has lower base complexity due to conversion of all Cs (except m^5Cs) to Us (Ts after sequencing). This lower complexity increases the mapping error rate especially for very short RNA reads. Such cases are frequently observed in this type of transcriptome analysis. Here, one efficient way to filter such artifacts is to sequence both of the BS and NBS samples as explained above. In our experience,

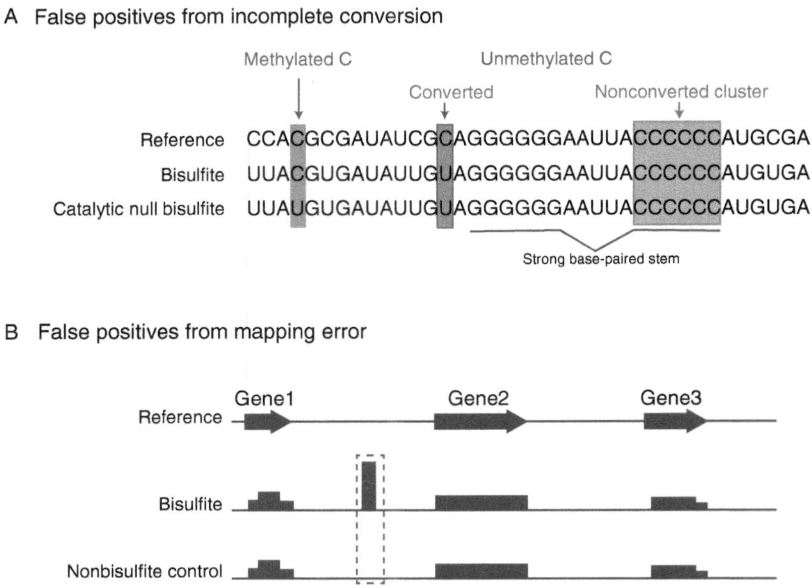

Figure 2 Examples of false positives observed during visual inspection of candidate methylation sites. (A) Nonconverted clusters of cytosines can occur due to incomplete denaturation of strong secondary structures. (B) Mapping errors can occur due to the lower complexity of bisulfite-converted RNA, resulting in the alignment of reads to an incorrect genomic locus. Any sites seen as methylated in these regions are likely false positives, and can be filtered out if a nonbisulfite-treated control is processed in parallel.

in most cases, if an artifact is generated due to mapping error, only the BS but not the NBS dataset will show coverage for that particular site (Fig. 2B). We found the BS–NBS paired comparison a very efficient way to filter a significant number of such artifacts in our analysis.

7. *Target validation by Sanger sequencing.* For small-scale analysis and particularly for validation of the candidate target cytosines of a given methyltransferase enzyme, bisulfite-treated RNA samples from wild-type and mutant sources can be subjected to cDNA synthesis, followed by PCR amplification, cloning, and conventional Sanger sequencing. Almost any available kits can be used for the first-strand cDNA synthesis using either gene-specific or random hexamer (note: oligo-dT is not recommended as the mRNAs are fragmented). For PCR amplification, two important parameters should be considered: first, as the bisulfite-treated RNAs are usually short (even when the RNA is not fragmented prior to treatment, as bisulfite treatment itself causes fragmentation of RNAs), the PCR primers should be designed for amplification of short

amplicons, e.g., 100–150 bp. Second, in most (if not all) cases, the first primer pairs fail to amplify the desired amplicon due to lower nucleotide complexity of amplicons (almost all Cs are converted to Us) causing difficulties for amplification of the target. To overcome this, we recommend designing two primer pairs for a nested PCR. Using a tiny amount (2–5 μl) of the output of the first PCR as the sample for the second PCR will almost always guarantee successful amplification of the target region. After obtaining the desired amplicon, to assess the methylation status of the original individual RNA molecules, the PCR product should be subjected to cloning (using the cloning system of choice), followed by plasmid purification and sequencing of individual plasmids using universal primers on the cloning vector. After sequencing, the sequences of each individual clone from wild-type or mutant samples will be matched with the virtually bisulfite-converted sequences of the target sequence. Any remaining Cs in each molecule will be counted and can be presented as closed circles in a cartoon format (Fig. 1B). Comparison of the wild type and mutant will show the target site(s) of the methyltransferase under study.

2.2 Aza-IP

As explained above, target profiling of m^5C-RMTs with RNA bisulfite sequencing requires deep sequencing of RNAs purified from a wild-type sample paired with those from cells/organisms lacking a particular m^5C-RMT. This approach, though useful in many cases, suffers from at least four issues: (1) generating knockout cells/organisms by inactivating/deleting the methyltransferase gene, or efficient depletion of its transcripts (via RNAi technology), is a requirement which can be challenging, (2) high-throughput sequencing of two sample series deep enough for the detection of low-copy RNAs and those with low modification penetrance require multiple lanes of sequencing per sample which can be very expensive, especially for organisms with large genomes like humans, (3) the bisulfite-sequencing method suffers from false positives as a result of incomplete conversion of cytosines in regions with strong secondary structures, which are resistant to even harsh denaturing conditions/buffers, and (4) the output of this approach by itself is insufficient to resolve whether the identified candidate cytosines are the direct or indirect targets of the deleted/depleted m^5C-RMT under study, requiring further experiment to address this. To overcome all these issues at once, we established the Aza-IP technique (Khoddami & Cairns, 2013, 2014), which combines four different features in one technique. Aza-IP: (1) does not rely on wild-type–mutant paired samples and can be done using just the wild-type cells/organisms, (2) is designed to isolate the m^5C-RMT targets by enrichment, overcoming the target RNA copy number and methylation penetrance issues, (3) determines the exact target sites of the m^5C-RMTs within the same experiment by finding the C > G transversion sites induced by the technique at the precise target sites of the enzymes (see below), and (4) isolates only the direct targets (substrates) of the m^5C-RMT under study.

Aza-IP is a mechanism-based high-throughput enrichment technique that links the enzyme to the spectrum of RNAs and specific target sites within those RNAs (Khoddami & Cairns, 2013). This technique relies on the catalytic mechanism by which m^5C-RMTs methylate cytosines. All m^5C-RMTs during catalysis form a covalent bond with the target cytosine positioned in their active sites. Upon completion of the methylation reaction, the covalent bond is broken leaving methylated target RNAs and free enzymes capable of methylating their other substrates. Certain inhibitors, such as nucleotide analog 5-azacytidine (5-aza-C), function as suicide inhibitors of m^5C-RMTs by locking the enzyme into an enzyme–substrate covalent adduct during the catalytic process, preventing release (Fig. 3A). Aza-IP exploits this phenomenon in cells treated with 5-aza-C to covalently link particular m^5C-RMTs specifically to their direct target RNAs during catalysis. The stabilized m^5C-RMT–RNA adducts are then subjected to immunoprecipitation followed by RNA fragmentation to release and sequence the enriched RNAs. Interestingly, enrichment of m^5C-RMT targets by Aza-IP leaves a C > G transversion signature after sequencing precisely at the target site of the enzyme within the enriched RNAs, facilitating their isolation in the same experiment. The transversion signature is generated during Aza-IP procedure through ring opening of the 5-aza-C base incorporated at the precise target site. The ring-opened 5-aza-C tends to base-pair more frequently with cytosine at the time of cDNA synthesis, thus appears as guanosine in the cDNA sequence (Fig. 3B). In principle, to identify the targets of almost any type of m^5C-RMTs or their particular splice variants, Aza-IP can be applied in almost any cultured cells, or single-cell organisms such as bacteria or yeast, and also in multicellular model organisms such as mouse, if effective concentration of 5-aza-C can be provided for the cells *in vivo*. Here, we briefly introduce the Aza-IP procedure as we have recently covered comprehensively all conceptual and technical aspects of the technique accompanied with a detailed step-by-step protocol, which can be used as a reference for interested researchers (Khoddami & Cairns, 2014).

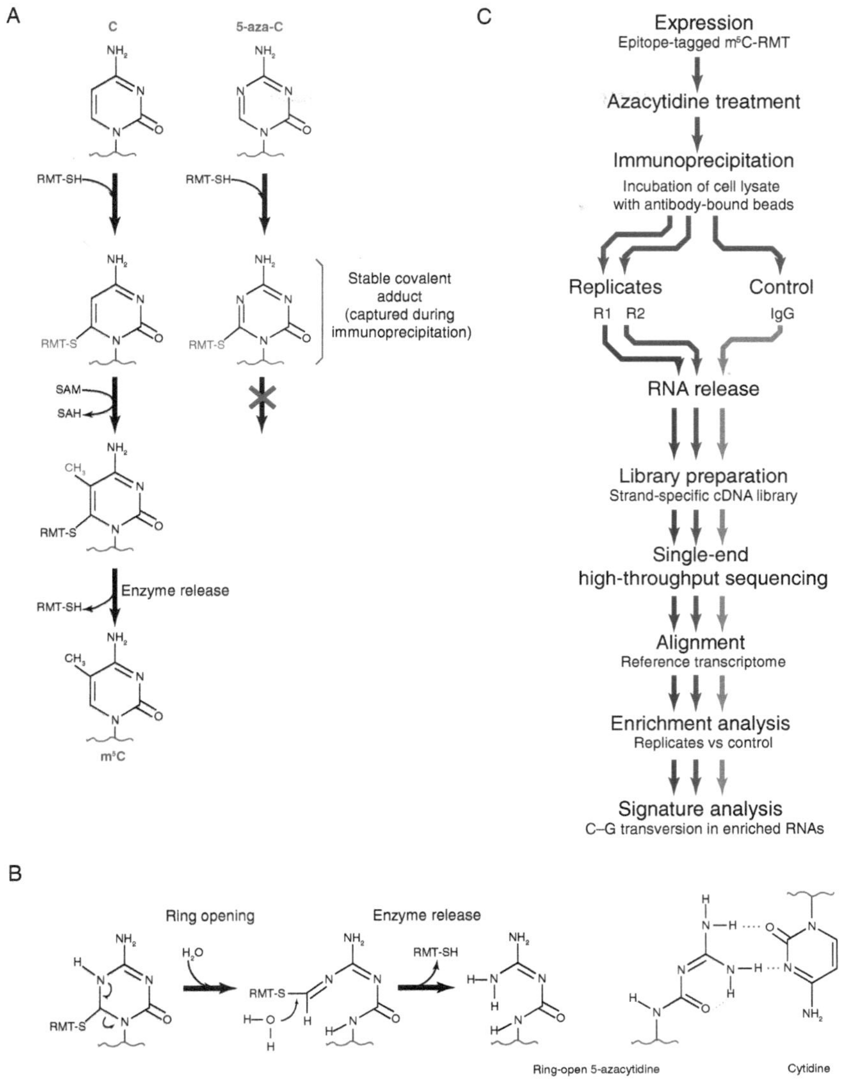

Figure 3 (A) Catalytic mechanism of m^{5}C-RMTs. During catalysis, the m^{5}C-RMT forms a covalent bond with the target cytosine, and is released upon catalyzing the methylation reaction. The nucleotide analog 5-aza-C inhibits the final enzyme release, leaving the enzyme covalently bound to its substrate and unable to participate in further catalysis. (B) C > G transversion. When the enzyme is subsequently released from the RNA, it often leaves the 5-aza-C in a "ring-open" conformation, which can then base-pair with another cytosine, leading to a C > G transversion upon sequencing. (C) Aza-IP. For Aza-IP, it is recommended to set up two replicate samples (R1 and R2) and a matched negative control pulldown using a nonspecific antibody (e.g., IgG), and the same lysate as the pulldown. During analysis, this negative control sample can be used both to calculate true enrichment rates of transcripts, and to filter SNPs for identifying the true C > G transversions. (See the color plate.)

Aza-IP requires a specific, strong antibody to the m^5C-RMT under study, which retains its activity under immunoprecipitation (IP) conditions. If such an antibody is available, Aza-IP can be readily applied on the cell extracts to study the endogenous enzyme. However, in the absence of suitable antibodies to endogenous protein, expressing an epitope-tagged enzyme and performing the Aza-IP with anti-tag antibody would be the best alternative. Here, we explain the critical steps of Aza-IP procedure and analysis for exogenously expressed V5-tagged enzymes expressed in HeLa cells (Fig. 3C).

2.2.1 Expression of Epitope-Tagged m^5C-RMT

First, the full-length m^5C-RMT coding sequence is amplified from cDNA and fused to the tag of interest (here $1 \times$ V5 tag) via standard molecular cloning procedures and after validation by sequencing is subcloned into an appropriate expression vector. There are multiple expression systems that can be used which are designed for transient (e.g., plasmids) or stable (e.g., lentiviral systems) expression, driving the expression of transgene under constitutive (e.g., cytomegalovirus (CMV)) or regulated (e.g., Tet-inducible) promoters, which may or may not have selection (e.g., antibiotic resistance) or reporter (e.g., green fluorescent protein (GFP)) markers. For the best results, we recommend making stable cell lines using Tet-ON lentiviral expression systems to drive the expression of the transgene under the control of Tet-ON system, and selecting the stable cell lines with antibiotic markers (preferentially puromycin, for its faster and cleaner selection features). There are at least three major advantages when using this in comparison to other systems: (1) the amount of viral particles required for making the stable cell line is minimized, as it can be done at a very small scale and after selection the cells can be expanded (virus-free) to the numbers required for the experiment, (2) the stable cell lines, once made, can be kept as frozen stocks for future applications, and (3) most importantly, the transgene overexpression can be induced (by addition of Doxycycline) shortly before performing the Aza-IP experiment, thus minimizing any side effects the enzyme overexpression may have on the cells. After making the stable cell line with the Tet-inducible system, it is necessary to evaluate the expression of the V5-tagged enzyme by western blot.

2.2.2 Azacytidine Treatment

Aza-IP relies on formation of m^5C-RMT–RNA stable adducts *in vivo*. This is achieved by treating the cells for 12–24 h with 5–10 μM 5-aza-C to allow transcription-dependent incorporation of 5-aza-C into nascent RNA transcripts. For this, cultured cells in growth medium are supplemented with freshly prepared 5-aza-C (as it is unstable in solution). 5-aza-C is taken up by the cells, and metabolized enzymatically to its triphosphate form, when it is randomly incorporated into newly synthesized RNA transcripts in place of cytosine. When the m^{5}C-RMT encounters a target RNA in which the precise target site of the enzyme contains 5-aza-C, the covalent adduct is formed which can be pulled down in the IP step. The 5-aza-C-treated cells are used either immediately in the IP step or stored at −80 °C.

2.2.3 Immunoprecipitation

The immunoprecipitation step of Aza-IP is similar to typical IP protocols; however, additional care is required to maintain integrity of RNA throughout the process by using RNase-free reagents and equipment, and an RNase inhibitor in buffers alongside protease inhibitors to keep the enzymes intact all through the procedure. Also to obtain the most accurate and reliable results, we recommend preparing at least two replicate samples of epitope antibody (anti-V5) beads, and one control sample of nonspecific antibody beads (IgG; matched to the epitope antibody of choice; Fig. 3C). For this, the anti-V5 antibody enough for two biological replicates (Rep1 and Rep2) is mixed with magnetic beads conjugated to pan-mouse anti-IgG and washed 3 × with, and resuspended in, diluted modified radioimmunoprecipitation assay (RIPA) buffer [50 m*M* Tris–HCl, pH 7.5; 1% (vol/vol) NP-40; 0.1% (wt/vol) Na-DOC; 0.025% (wt/vol) SDS; 1 m*M* EDTA, pH 8.0; 150 m*M* NaCl] supplemented with 5 mg/mL bovine serum albumin (BSA) 1 × protease and RNase inhibitors. In parallel, the control IgG antibodies, enough for one control experiment (IgG), are washed and resuspended with the same protocol and buffer. Upon completion of 5-aza-C treatment, cells are lysed by sonication in modified RIPA buffer [50 m*M* Tris, pH 7.5; 1% (vol/vol) NP-40; 0.2% (wt/vol) sodium deoxycholate; 0.05% (wt/vol) SDS; 1 m*M* EDTA, pH 8.0; 300 m*M* NaCl] supplemented with protease and RNase inhibitors. After removing the debris by centrifugation, the cell lysate is diluted by addition of 1 × dilution buffer [50 m*M* Tris–HCl, pH 7.5; 1% (vol/vol) NP-40; 1 m*M* EDTA, pH 8.0]. Next the lysate is precleared for 2 h at room temperature with beads (no antibody at this step) in diluted modified RIPA buffer supplemented with 1 × protease and RNase inhibitors. The precleared lysate is then split into three parts, labeled as Rep1, Rep2, and IgG, and mixed with corresponding antibody-bound beads prepared earlier (above) in 1 ml microcentrifuge tubes. After 4 h incubation at room temperature, the beads of each group are separately washed, in parallel, three times with 850 μl of diluted modified RIPA buffer containing protease and RNase inhibitors. Next, 500 μl of the same buffer is added to each tube and the contents are collected and transferred into correspondingly labeled clean 0.65-ml microcentrifuge tubes. The buffer is then removed by using magnetic stands to collect the beads at the bottom of the tubes.

2.2.4 RNA Release, Library Preparation, and Sequencing

To release the RNA from the enzyme, we subject the beads containing the m^5C-RMT–RNA adducts immediately to RNA fragmentation, by addition of Ambion RNA fragmentation reagent, and incubation for 5 min at 94 °C in a thermocycler, followed by stopping the reaction immediately by addition of the stop solution (Ambion). The supernatant (containing fragmented RNA) is then collected and subjected to ethanol precipitation followed by RNA extraction using the Trizol reagent. RNA samples from each of Rep1, Rep2, and IgG samples are then subjected separately to library preparation and high-throughput sequencing according to standard protocols. However, we highly recommend using the directional library preparation kits to facilitate identification of the m^5C-RMT target sites via signature analysis (below). Also, sequencing length of 50 bp or more recommended to increase the coverage and single-end sequencing is appropriate.

2.2.5 Sequencing Analysis

The purpose of sequencing analysis is to identify the RNA targets of m^5C-RMT under study and also to isolate the precise target cytosines within them through three consecutive steps (Fig. 3C):

1. *Alignment*. Sequence alignment can be done using any next-generation sequencing analysis package that can handle mismatches. We recommend using Novoalign application from Novocraft package. First make extended transcriptome index file containing the splicing events using USeq package and Novoindex application (see above, section 2.1), and then align the FASTQ files using Novoalign application. It is also important to set the alignment parameters to trim the library preparation adaptors, allow for 1–3 mismatches per read (to enable observing the C > G transversions), and also allow mapped reads as short as 18 bp in the alignment output (because the RNA is fragmented in the procedure).
2. *Enrichment analysis*. After the test (Rep1 and Rep2) and control (IgG) datasets are aligned, enrichment analysis is performed (we recommend the USeq package) to find transcripts that are enriched in both Rep1 and Rep2 samples in comparison to the nonspecific IgG control sample.

3. *Signature analysis.* The list of enriched transcripts obtained in the previous step is subjected to signature analysis. In the signature analysis, we are looking for significant C > G transversions that are caused by Aza-IP (not by SNPs) as indicators of the methylated sites. For this analysis, genomic variant detection packages can be used. We recommend using VarScan (Koboldt et al., 2009) to identify the transversion sites and calculate the variation rates and *P* values for individual sites.

Finally, it is important to validate a subset of identified candidate target sites by doing either bisulfite sequencing (section 2.1) or, if possible, methyltransferase assay (section 2.3).

2.3 RNA Methyltransferase Assay

An alternative way for validating candidate target sites of a particular methyltransferase is *in vitro* reconstitution of the methyltransferase reaction using purified enzyme and target RNAs. Here, we provide an example of RNA methyltransferase assay for target validation of the human DNMT2 (hDNMT2) enzyme. There are four components for a typical RNA methyltransferase assay:

1. *The enzyme.* A successful assay requires a catalytically active enzyme that can be obtained either through commercial suppliers (if available) or by expression and purification. Functional His-tagged-hDNMT2 can be expressed in and purified from *E. coli* as described previously (Dong et al., 2001).
2. *The substrate.* The target (substrate) RNA can be made either synthetically or by *in vitro* transcription using commercial T7-RNA polymerase kits.
3. *The cofactor.* The cofactor required for this assay is the methyl donor *S*-adenosylmethionine (SAM) which depending on the analysis method can be either "cold" (as it is) or "hot" (i.e., tritium-labeled SAM (*S*-[methyl-^{3}H]-adenosyl-L-methionine (^{3}H-SAM)). The cold SAM is used when the assay is to be analyzed by bisulfite sequencing (either high-throughput or Sanger sequencing) and the hot ^{3}H-SAM is used when the methylation of the target RNAs is to be measured by scintillation assay.
4. *The buffer.* The choice of buffer might depend on the enzyme to ensure that the enzyme is catalytically active and the target molecule is both properly folded and also protected from degradation. For the RNA methyltransferase assay using hDNMT2, we use our own buffer containing (final concentration in the reaction) 100 m*M* Tris–HCl, pH

7.5, 5% glycerol, 5 m*M* $MgCl_2$, 100 m*M* NaCl, 1 m*M* DTT supplemented with RNase inhibitor. We note here that our recipe has a major difference with a previously published buffer system by others (Goll et al., 2006); we replace EDTA by $MgCl_2$ (5 m*M*), which we have demonstrated increases the *in vitro* methyltransferase efficiency of DNMT2 by >30-fold, providing a more sensitive assay for this enzyme, and likely other RNA methyltransferases (Khoddami & Cairns, 2013).

Depending on the type of analysis, the methyltransferase assay can be performed and analyzed in either of the following formats (Fig. 4A and B):

1. *Analysis by scintillation counting*. For this, in an RNase-free-certified tube, mix 400 ng of the purified hDNMT2, 5 μg of the RNA substrate of interest (e.g., synthetically made or purified *in vitro*-transcribed $tRNA^{Asp}$), and 1 μCi of hot ^{3}H-SAM (PerkinElmer) with 8 μl of 5 ×

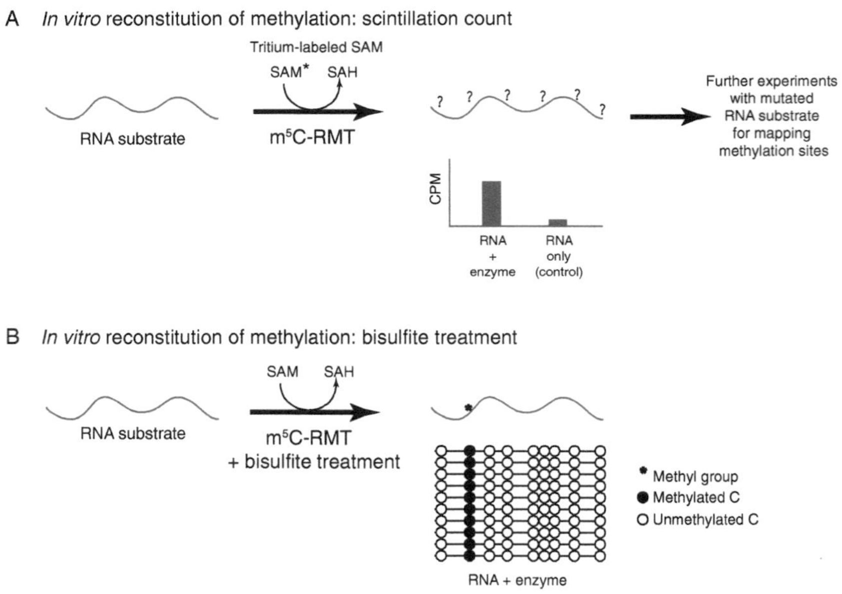

Figure 4 *In vitro* reconstitution of m^5C methylation. (A) Analysis by scintillation counting. Tritium-labeled SAM is used as a cofactor for the *in vitro*-reconstituted methylation reaction, resulting in the RNA substrate being labeled upon methylation. While the presence or absence of methylation on the RNA can be measured by scintillation counts against a negative control, further experimentation—such as the use of RNA substrates with mutated cytosines—is required to determine the exact target sites of the enzymes. (B) Analysis by bisulfite treatment. By subjecting the RNA output from the *in vitro* methylation reaction to bisulfite treatment and sequencing, the exact target sites will be readily determined from a single experiment.

methyltransferase buffer (500 mM Tris–HCl, pH 7.5, 25% glycerol, 25 mM $MgCl_2$, 500 mM NaCl, 5 mM DTT), plus 5 U RNaseIN (Promega). Bring volume to 40 μl by addition of RNase-free water, and incubate at 37 °C for 5 h. To stop the reaction, the unincorporated ^{3}H-SAM is removed by passing the mixture through Micro Bio-Spin Columns (BIO-RAD), followed by concentrating the RNA by ethanol precipitation. The enzymatically catalyzed incorporation of the tritium-labeled methyl group into the substrate RNA is then measured by scintillation counting, and the result is reported and plotted as counts per minute. It is important to note that analyzing the methyltransferase activity of a particular enzyme via scintillation counting using the wild-type substrate RNA can only tell us whether that substrate RNA used in the assay is a target for the enzyme, but it does not show the exact cytosine(s) targeted by the enzyme. To address this, one can run methyltransferase reactions using mutated target RNAs as substrate which; in each, one of the target cytosines is mutated to another base, in parallel with the control experiment (using the wild-type RNA substrate). Comparison of methylation assays with wild-type and mutant substrates can possibly isolate the exact target sites of the enzyme. There are, however, at least two major challenges with this approach: (1) to study each one of the cytosines in a target RNA, separate RNA templates (wild type and mutated) should be made and tested. This is especially challenging if a given enzyme is targeting multiple sites within the same molecule, such as the case of the NSUN2 enzyme that targets multiple sites in the same RNA in some tRNAs and ncRNAs. (2) In some cases, mutating a non-target cytosine can decrease or abolish the target recognition or specificity of the enzyme due to conformation changes in the mutant RNA. This is going to be particularly misleading in identifying the true target cytosines. To overcome both of these issues, the bisulfite sequencing approach (see below) is a good alternative.

2. *Analysis by bisulfite sequencing*. Assembly of the methyltransferase reaction to be analyzed by bisulfite sequencing is identical to the aforementioned protocol above with the exception that instead of the hot ^{3}H-SAM now cold SAM (NEB) can be used, and the recovered RNA will be subjected to bisulfite treatment and sequencing as explained earlier in this chapter. Here, if the enzyme under study can methylate the substrate RNA, the exact target cytosine(s) will be identified immediately as they remain as cytosine(s) after bisulfite conversion of the target RNA.

REFERENCES

Abbasi-Moheb, L., Mertel, S., Gonsior, M., Nouri-Vahid, L., Kahrizi, K., Cirak, S., et al. (2012). Mutations in NSUN2 cause autosomal-recessive intellectual disability. *American Journal of Human Genetics*, *90*, 847–855.

Becker, M., Muller, S., Nellen, W., Jurkowski, T. P., Jeltsch, A., & Ehrenhofer-Murray, A. E. (2012). Pmt1, a Dnmt2 homolog in Schizosaccharomyces pombe, mediates tRNA methylation in response to nutrient signaling. *Nucleic Acids Research*, *40*, 11648–11658.

Blanco, S., Kurowski, A., Nichols, J., Watt, F. M., Benitah, S. A., & Frye, M. (2011). The RNA-methyltransferase Misu (NSun2) poises epidermal stem cells to differentiate. *PLoS Genetics*, *7*, e1002403.

Camara, Y., Asin-Cayuela, J., Park, C. B., Metodiev, M. D., Shi, Y., Ruzzenente, B., et al. (2011). MTERF4 regulates translation by targeting the methyltransferase NSUN4 to the mammalian mitochondrial ribosome. *Cell Metabolism*, *13*, 527–539.

Doll, A., & Grzeschik, K. H. (2001). Characterization of two novel genes, WBSCR20 and WBSCR22, deleted in Williams-Beuren syndrome. *Cytogenetics and Cell Genetics*, *95*, 20–27.

Dong, A., Yoder, J. A., Zhang, X., Zhou, L., Bestor, T. H., & Cheng, X. (2001). Structure of human DNMT2, an enigmatic DNA methyltransferase homolog that displays denaturant-resistant binding to DNA. *Nucleic Acids Research*, *29*, 439–448.

Durdevic, Z., Hanna, K., Gold, B., Pollex, T., Cherry, S., Lyko, F., et al. (2013). Efficient RNA virus control in Drosophila requires the RNA methyltransferase Dnmt2. *EMBO Reports*, *14*, 269–275.

Edelheit, S., Schwartz, S., Mumbach, M. R., Wurtzel, O., & Sorek, R. (2013). Transcriptome-wide mapping of 5-methylcytidine RNA modifications in bacteria, archaea, and yeast reveals m5C within archaeal mRNAs. *PLoS Genetics*, *9*, e1003602.

Fonagy, A., Swiderski, C., Wilson, A., Bolton, W., Kenyon, N., & Freeman, J. W. (1993). Cell cycle regulated expression of nucleolar antigen P120 in normal and transformed human fibroblasts. *Journal of Cellular Physiology*, *154*, 16–27.

Frye, M., Dragoni, I., Chin, S. F., Spiteri, I., Kurowski, A., Provenzano, E., et al. (2010). Genomic gain of 5p15 leads to over-expression of Misu (NSUN2) in breast cancer. *Cancer Letters*, *289*, 71–80.

Frye, M., & Watt, F. M. (2006). The RNA methyltransferase Misu (NSun2) mediates Myc-induced proliferation and is upregulated in tumors. *Current Biology: CB*, *16*, 971–981.

Goll, M. G., Kirpekar, F., Maggert, K. A., Yoder, J. A., Hsieh, C. L., Zhang, X., et al. (2006). Methylation of tRNAAsp by the DNA methyltransferase homolog Dnmt2. *Science*, *311*, 395–398.

Harris, T., Marquez, B., Suarez, S., & Schimenti, J. (2007). Sperm motility defects and infertility in male mice with a mutation in Nsun7, a member of the Sun domain-containing family of putative RNA methyltransferases. *Biology of Reproduction*, *77*, 376–382.

Hong, B., Wu, K., Brockenbrough, J. S., Wu, P., & Aris, J. P. (2001). Temperature sensitive nop2 alleles defective in synthesis of 25S rRNA and large ribosomal subunits in Saccharomyces cerevisiae. *Nucleic Acids Research*, *29*, 2927–2937.

Hussain, S., Sajini, A. A., Blanco, S., Dietmann, S., Lombard, P., Sugimoto, Y., et al. (2013). NSun2-mediated cytosine-5 methylation of vault noncoding RNA determines its processing into regulatory small RNAs. *Cell Reports*, *4*, 255–261.

Hussain, S., Tuorto, F., Menon, S., Blanco, S., Cox, C., Flores, J. V., et al. (2013). The mouse cytosine-5 RNA methyltransferase NSun2 is a component of the chromatoid body and required for testis differentiation. *Molecular and Cellular Biology*, *33*, 1561–1570.

Job, B., Bernheim, A., Beau-Faller, M., Camilleri-Broet, S., Girard, P., Hofman, P., et al. (2010). Genomic aberrations in lung adenocarcinoma in never smokers. *PLoS One*, *5*, e15145.

Khan, M. A., Rafiq, M. A., Noor, A., Hussain, S., Flores, J. V., Rupp, V., et al. (2012). Mutation in NSUN2, which encodes an RNA methyltransferase, causes autosomal-recessive intellectual disability. *American Journal of Human Genetics*, *90*, 856–863.

Khoddami, V., & Cairns, B. R. (2013). Identification of direct targets and modified bases of RNA cytosine methyltransferases. *Nature Biotechnology*, *31*, 458–464.

Khoddami, V., & Cairns, B. R. (2014). Transcriptome-wide target profiling of RNA cytosine methyltransferases using the mechanism-based enrichment procedure Aza-IP. *Nature Protocols*, *9*, 337–361.

Khosronezhad, N., Colagar, A. H., & Jorsarayi, S. G. (2014). T26248G-transversion mutation in exon7 of the putative methyltransferase Nsun7 gene causes a change in protein folding associated with reduced sperm motility in asthenospermic men. *Reproduction, Fertility, and Development*, *27*, 471–480.

Kiani, J., Grandjean, V., Liebers, R., Tuorto, F., Ghanbarian, H., Lyko, F., et al. (2013). RNA-mediated epigenetic heredity requires the cytosine methyltransferase Dnmt2. *PLoS Genetics*, *9*, e1003498.

Koboldt, D. C., Chen, K., Wylie, T., Larson, D. E., McLellan, M. D., Mardis, E. R., et al. (2009). VarScan: Variant detection in massively parallel sequencing of individual and pooled samples. *Bioinformatics*, *25*, 2283–2285.

Machnicka, M. A., Milanowska, K., Osman Oglou, O., Purta, E., Kurkowska, M., Olchowik, A., et al. (2013). MODOMICS: A database of RNA modification pathways—2013 update. *Nucleic Acids Research*, *41*, D262–D267.

Martinez, F. J., Lee, J. H., Lee, J. E., Blanco, S., Nickerson, E., Gabriel, S., et al. (2012). Whole exome sequencing identifies a splicing mutation in NSUN2 as a cause of a Dubowitz-like syndrome. *Journal of Medical Genetics*, *49*, 380–385.

Metodiev, M. D., Spahr, H., Loguercio Polosa, P., Meharg, C., Becker, C., Altmueller, J., et al. (2014). NSUN4 is a dual function mitochondrial protein required for both methylation of 12S rRNA and coordination of mitoribosomal assembly. *PLoS Genetics*, *10*, e1004110.

Motorin, Y., Lyko, F., & Helm, M. (2010). 5-Methylcytosine in RNA: Detection, enzymatic formation and biological functions. *Nucleic Acids Research*, *38*, 1415–1430.

Nix, D. A., Courdy, S. J., & Boucher, K. M. (2008). Empirical methods for controlling false positives and estimating confidence in ChIP-Seq peaks. *BMC Bioinformatics*, *9*, 523.

Okamoto, M., Hirata, S., Sato, S., Koga, S., Fujii, M., Qi, G., et al. (2012). Frequent increased gene copy number and high protein expression of tRNA (cytosine-5-)-methyltransferase (NSUN2) in human cancers. *DNA and Cell Biology*, *31*, 660–671.

Phalke, S., Nickel, O., Walluscheck, D., Hortig, F., Onorati, M. C., & Reuter, G. (2009). Retrotransposon silencing and telomere integrity in somatic cells of Drosophila depends on the cytosine-5 methyltransferase DNMT2. *Nature Genetics*, *41*, 696–702.

Rai, K., Chidester, S., Zavala, C. V., Manos, E. J., James, S. R., Karpf, A. R., et al. (2007). Dnmt2 functions in the cytoplasm to promote liver, brain, and retina development in zebrafish. *Genes & Development*, *21*, 261–266.

Robinson, J. T., Thorvaldsdottir, H., Winckler, W., Guttman, M., Lander, E. S., Getz, G., et al. (2011). Integrative genomics viewer. *Nature Biotechnology*, *29*, 24–26.

Schaefer, M., Pollex, T., Hanna, K., & Lyko, F. (2009). RNA cytosine methylation analysis by bisulfite sequencing. *Nucleic Acids Research*, *37*, e12.

Schaefer, M., Pollex, T., Hanna, K., Tuorto, F., Meusburger, M., Helm, M., et al. (2010). RNA methylation by Dnmt2 protects transfer RNAs against stress-induced cleavage. *Genes & Development*, *24*, 1590–1595.

Schosserer, M., Minois, N., Angerer, T. B., Amring, M., Dellago, H., Harreither, E., et al. (2015). Methylation of ribosomal RNA by NSUN5 is a conserved mechanism modulating organismal lifespan. *Nature Communications*, *6*, 6158.

Sharma, S., Yang, J., Watzinger, P., Kotter, P., & Entian, K. D. (2013). Yeast Nop2 and Rcm1 methylate C2870 and C2278 of the 25S rRNA, respectively. *Nucleic Acids Research, 41*, 9062–9076.

Spahr, H., Habermann, B., Gustafsson, C. M., Larsson, N. G., & Hallberg, B. M. (2012). Structure of the human MTERF4-NSUN4 protein complex that regulates mitochondrial ribosome biogenesis. *Proceedings of the National Academy of Sciences of the United States of America, 109*, 15253–15258.

Squires, J. E., Patel, H. R., Nousch, M., Sibbritt, T., Humphreys, D. T., Parker, B. J., et al. (2012). Widespread occurrence of 5-methylcytosine in human coding and non-coding RNA. *Nucleic Acids Research, 40*, 5023–5033.

Tuorto, F., Liebers, R., Musch, T., Schaefer, M., Hofmann, S., Kellner, S., et al. (2012). RNA cytosine methylation by Dnmt2 and NSun2 promotes tRNA stability and protein synthesis. *Nature Structural & Molecular Biology, 19*, 900–905.

Yadlapalli, S., & Yamashita, Y. M. (2013). Chromosome-specific nonrandom sister chromatid segregation during stem-cell division. *Nature, 498*, 251–254.

Yakubovskaya, E., Guja, K. E., Mejia, E., Castano, S., Hambardjieva, E., Choi, W. S., et al. (2012). Structure of the essential MTERF4:NSUN4 protein complex reveals how an MTERF protein collaborates to facilitate rRNA modification. *Structure, 20*, 1940–1947.

CHAPTER FOURTEEN

RNA 5-Methylcytosine Analysis by Bisulfite Sequencing

Matthias Schaefer[1]
Vienna Biocenter, Max F. Perutz Laboratories, Department of Biochemistry and Cell Biology, Universität Wien, Vienna, Austria
[1]Corresponding author: e-mail address: matthias.schaefer@univie.ac.at

Contents

Abstract

Cells have developed molecular machineries, which can chemically modify DNA and RNA nucleosides. One particular and chemically simple modification, (cytosine-5) methylation (m^5C), has been detected both in RNA and DNA suggesting universal use of m^5C for the function of these nucleotide polymers. m^5C can be reproducibly mapped to abundant noncoding RNAs (transfer RNA, tRNA and ribosomal RNA, rRNA), and recently, also nonabundant RNAs (including mRNAs) have been reported to carry this modification. Quantification of m^5C content in total RNA preparations indicates that a limited number of RNAs carry this modification and suggests specific functions for (cytosine-5) RNA methylation.

Methods in Enzymology, Volume 560
ISSN 0076-6879
http://dx.doi.org/10.1016/bs.mie.2015.03.007

What exactly is the biological function of m^5C in RNA? Before attempting to address this question, m^5C needs to be mapped specifically and reproducibly, preferably on a transcriptome-wide scale.

To facilitate the detection of m^5C in its sequence context, RNA bisulfite sequencing (RNA-BisSeq) has been developed. This method relies on the efficient chemical deamination of nonmethylated cytosine, which can be read out as single nucleotide polymorphism (nonmethylated cytosine as thymine vs. methylated cytosine as cytosine), when differentially comparing cDNA libraries to reference sequences after DNA sequencing. Here, the basic protocol of RNA-BisSeq, its current applications and limitations are described.

1. INTRODUCTION

In contrast to the limited number of known DNA modifications (<10), RNAs are "decorated" with (at least) 130 distinct posttranscriptional modifications (both at terminal and internal positions). The chemical nature of nucleoside modifications, especially in noncoding RNAs, can be diverse and very complex (Machnicka et al., 2013). Strikingly, various polynucleotide methylation systems have emerged independently and multiple times during evolution, indicating the conserved versatility and biological importance of methylated polynucleotides (Bujnicki, Droogmans, Grosjean, Purushothaman, & Lapeyre, 2004; Grosjean, 2009).

1.1 RNA (Cytosine-5) Methyltransferases

Among the various methylated nucleosides, m^5C holds a special position because it is found in DNA and the enzymatic machineries for m^5C synthesis in RNA are expressed in all kingdoms of life. Interestingly, the genomes of individual species encode many more (cytosine-5) RNA methyltransferases (RCMTs) than canonical DNA MTases (Goll & Bestor, 2005). Iterative database searches identified nine RCMT subfamilies (RCMT1–RCMT9, Table 1) (Bujnicki, Feder, Ayres, & Redman, 2004; Gustafsson, 1996; Pavlopoulou & Kossida, 2009; Reid, Greene, & Santi, 1999).

These include five subfamilies (RCMT1, RCMT3–6) that are only found in prokaryotes; three "strictly" eukaryotic subfamilies (RCMT2, RCMT7, and RCMT8), and one subfamily (RCMT9) encoded in some plants and Mycetozoa taxons.

RCMTs catalyze the transfer of a methyl group to the fifth carbon of the cytosine base in RNA sequences to produce C5-methylcytosine (m^5C)

Table 1 Known RCMTs, Subfamilies and RNA Substrates

Family	Name(s)	Organism	RNA Substrates	References
RCMT1	RsmB (Fmu/Fmv)	*E. coli*	16S rRNA (967)	Tscherne et al. (1999), Gu, Gustafsson, Ku, Yu, and Santi (1999), and Foster, Nunes, Greene, Moustakas, and Stroud (2003), **Schaefer, Pollex, Hanna, and Lyko (2009)**
RCMT1	PH0851	*Pyrococcus horikoshii*	16S rRNA (967)	Hikida, Kuratani, Bessho, Sekine, and Yokoyama (2010)
RCMT2	p120 (NSun1)	*Homo sapiens*	Unknown	Koonin (1994)
RCMT2	Nop2p	*Saccharomyces cerevisiae*	25S rRNA (2870)	Debeus et al. (1994) and Hong, Brockenbrough, Wu, and Aris (1997)
RCMT3	YebU	*Haloferax volcanii*	tRNA (39/40/48/49)	Grosjean, Gaspin, Marck, Decatur, and de Crécy-Lagard (2007)
RCMT3	YebU	*Pyrococcus abyssi*	tRNA (multiple)	Auxilien, Khadali El, Rasmussen, Douthwaite, and Grosjean (2007) and Machnicka et al. (2013)
RCMT7	aTrm4	*Pyrococcus horikoshii*	Unknown	Ishikawa et al. (2004)
RCMT7	RsmF (YebU)	*E. coli*	16S rRNA (1407)	Andersen and Douthwaite (2006), Hallberg et al. (2006), and **Schaefer et al. (2009)**

Continued

Table 1 Known RCMTs, Subfamilies and RNA Substrates—cont'd

Family	Name(s)	Organism	RNA Substrates	References
RCMT7	hTrm4 (NSUN2)	*Homo Sapiens*	pre-tRNA and tRNA (multiple); mRNA (variable); noncoding RNA (variable)	Brzezicha et al. (2006), Motorin and Helm (2011), Khan et al. (2012), **Squires et al. (2012)**, **Martinez et al. (2012)**, **Hussain, Sajini, et al. (2013)**, **Khoddami and Cairns (2014)**, and **Blanco et al. (2014)**
RCMT7	mTrm4 (NSUN2, Misu)	*Mus musculus*	pre-tRNA and tRNA (multiple); mRNA (variable); noncoding RNA (variable)	**Blanco et al. (2011)**, **Tuorto et al. (2012)**, **Hussain, Sajini, et al. (2013)**, and **Blanco et al. (2014)**
RCMT7	Trm4p (YBL024w, Ncl1)	*Saccharomyces cerevisiae*	pre-tRNA and tRNA (multiple)	Wu, Brockenbrough, Paddy, and Aris (1998), Motorin and Grosjean (1999), and Moon and Redman (2014)
RCMT7	trm4a + trm4b + (SPAC17D4.04; SPAC23C4.17)	*Schizosaccharomyces pombe*	tRNA-Asp^{GTC} (48,49,60-62)	**Becker et al. (2012)**
RCMT7	FLJ22609 (NSUN3)	*Homo Sapiens*	Unknown	Bujnicki, Feder, et al. (2004)
RCMT7	NSUN3	*Mus musculus*	Unknown	Chi and Delgado-Olguín (2013)
RCMT7	MGC22930 (NSUN4)	*Homo Sapiens*	Unknown	Bujnicki, Feder, et al. (2004)
RCMT7	(NSUN4)	*Mus musculus*	Mitochondrial 12S rRNA (911)	**Metodiev et al. (2014)**
RCMT8	WBSCR20A-C (NSUN5A-C)	*Homo Sapiens*	Unknown	Doll and Grzeschik (2001)

RCMT8	NSUN5 (Y53F4B_4)	*Caenorhabditis elegans*	28S rRNA (3381)	**Schosserer et al. (2015)**
RCMT8	WBSCR22	*Homo Sapiens*	Unknown	Doll and Grzeschik (2001)
RCMT8	NSUN7	*Homo Sapiens*	Unknown	
RCMT8	NSUN7	*Mus musculus*	Unknown	Harris, Marquez, Suarez, and Schimenti (2007)
RCMT8	Ynl022c (Rcm1)	*Homo Sapiens*	25S rRNA (2278)	Sharma, Yang, Watzinger, Kötter, and Entian (2013), **Gigova, Duggimpudi, Pollex, Schaefer, and Kos (2014)**, and **Schosserer et al. (2015)**
RlmI	RlmI (YccW)	*E. coli*	23S rRNA (1962)	Sunita et al. (2008) and Purta, O'Connor, Bujnicki, and Douthwaite (2008)
RlmI	RlmO	*Thermus thermophilus*	23S rRNA (1942)	Larsen, Rasmussen, Giessing, Jogl, and Kirpekar (2012)
NSUN6	NSUN6	*Mus musculus*	Unknown	Chi and Delgado-Olguín (2013)

Position of m^5C in individual RNAs in parentheses; application of RNA-BisSeq as bold references.

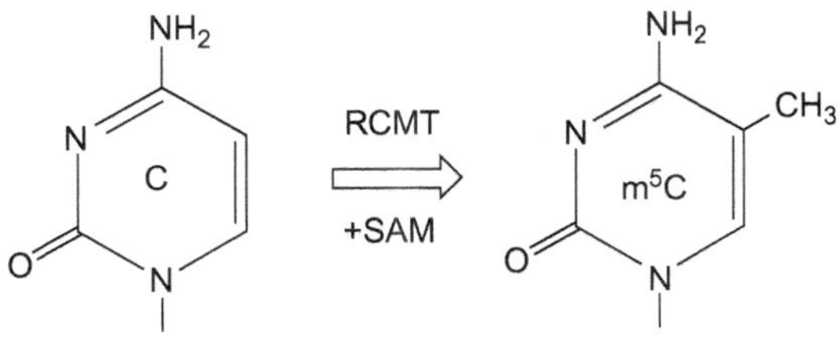

Figure 1 Cytosine is methylated to 5-methylcytosine by RNA (cytosine-5) methyltransferases (RCMTs). RCMTs catalyze the transfer of a methyl group (CH3) from S-adenosylmethionine (SAM) to the 5-carbon position of cytosine. C, cytosine; m^5C, 5-methylcytosine.

(Fig. 1). S-adenosyl-L-methionine acts as cofactor and methyl donor (Cheng & Roberts, 2001; Chiang et al., 1996). The catalytic mechanism of known RCMTs involves a cysteine thiolate attack at position 6 of the target cytosine base. This causes a covalent linkage between RCMT and RNA, and activates the methyl group transfer to C5 at the same time. A second cysteine residue is required to act as a general base in the following beta-elimination reaction, which restores the enzyme and releases the methylated product (King & Redman, 2002).

While (cytosine-5) methylation by RCMTs and DNA MTases involves the same chemical reaction mechanism (resulting in m^5C in RNA and DNA, respectively), these enzyme families have been historically distinguished because different motifs in their protein structures catalyze the methylation reaction (Liu & Santi, 2000). Of note, this distinction has been recently challenged through the addition of Dnmt2 enzymes to the RCMT family (Table 2); albeit as noncanonical members because Dnmt2 proteins have been historically annotated as the most widely conserved members of the eukaryotic (cytosine-5) DNA MTase protein family (Goll & Bestor, 2005; Jurkowski & Jeltsch, 2010). However, Dnmt2 enzymes are RCMTs with specificity toward the anticodon loop of tRNAs (Goll et al., 2006; Schaefer et al., 2010; Tuorto et al., 2012), a finding which has led to the renaming of Dnmt2 as tRNA aspartic acid methyltransferase 1 (Trdmt1). Intriguingly, Dnmt2/Trdmt1 enzymes methylate their substrates with the exact catalytic domain structure that was commonly used as basis for identifying and annotating (cytosine-5) DNA MTases (Jurkowski et al., 2008; Liu & Santi, 2000).

Table 2 Dnmt2/Trdmt1 Enzymes and RNA Substrates

Original Name (Other Names)	**Organism**	**RNA Substrates**	**References/Use of RNA-BisSeq**
Dnmt2 (Trdmt1)	*Homo Sapiens*	Multiple tRNAs (38)	Goll et al. (2006), **Squires et al. (2012)**, and **Khoddami and Cairns (2013)**
Dnmt2 (Trdmt1)	*Mus musculus*	Multiple tRNAs (38)	Goll et al. (2006), **Tuorto et al. (2012)**, and **Kiani et al. (2013)**
Dnmt2 (Mt2)	*Drosophila melanogaster*	Multiple tRNAs (38)	Goll et al. (2006) and **Schaefer et al. (2010)**
Dnmt2	*Danio rerio*	tRNA	Rai et al. (2007)
Dnmt2 (pmt1)	*Saccharomyces pombe*	Multiple tRNAs (38)	**Becker et al. (2012)**
Dnmt2 (DnmA)	*Dictyostelium discoideum*	tRNA-Asp^{GTC} (38)	**Müller et al. (2013)**
Ehmeth	*Entamoeba histolytica*	tRNA-Asp^{GTC} (38)	**Tovy, Tov, Gaentzsch, Helm, and Ankri (2010)**
Dnmt2 (GsDnmt)	*Geobacter sulfurreducens*	tRNA-Glu$^{CTC/TTC}$ (38)	**Shanmugam et al. (2013)**

Position of m^{5}C in individual RNAs in parentheses; application of RNA-BisSeq as bold references.

1.2 Biological Function of RNA (Cytosine-5) Methylation

Most of our knowledge about RCMT substrate choice has been derived from studying abundant noncoding RNAs in prokaryotes and simple eukaryotes. Initial work showed that the three founding members (Nop2, Ynl022, and Trm4) of eukaryotic RCMT subfamilies catalyze the formation of m^{5}C at specific positions in tRNA and rRNAs in yeast (Motorin & Grosjean, 1999; Motorin & Helm, 2011; Motorin, Lyko, & Helm, 2010; Sharma et al., 2013). However, it has now been established that orthologues of these RCMTs modify not only abundant (Blanco et al., 2011; Brzezicha et al., 2006; Tuorto et al., 2012) but also various nonabundant RNAs (i.e., 7SK and vault RNAs) in mammalian cells (Hussain, Sajini, et al., 2013; Khoddami & Cairns, 2013; Squires et al., 2012).

The analysis of genetic mutations in particular RCMT family members has provided first insights into the biological effects of aberrant (cytosine-5) RNA methylation. For instance, mutations in NCL1/TRM4 increased the sensitivity to antibiotics and affected ribosome decoding in yeast (Wu et al., 1998), indicating that m^5C is important for correct translation. This notion was supported by findings showing that m^5C at position 34 in $tRNA^{Leu}$ (catalyzed by NCL1/TRM4) modulated translational output in yeast (Strobel & Abelson, 1986). Importantly, the subsequent characterization of these RCMTs in multicellular organisms has revealed that their dysfunction does not necessarily cause strong phenotypes such as lethality (except in combination with mutations in other RCMTs) but rather tissue-specific as well as stress-induced effects such as decreased tRNA stability (Blanco et al., 2014; Schaefer et al., 2010; Tuorto et al., 2012). Deletion of NSun2 in *Drosophila* caused memory defects (Abbasi-Moheb et al., 2012) and mice with loss-of-function mutations in NSun2 showed impaired cellular differentiation of skin, testes, and brain (Blanco et al., 2011; Hussain, Tuorto, et al., 2013; Tuorto et al., 2012). Also, mutations in human NSUN2 have been linked to neurological disorders, which are associated with intellectual disability and Dubowitz-like syndrome (Abbasi-Moheb et al., 2012; Khan et al., 2012; Martinez et al., 2012). In addition, deletions in the human NSUN5A-C locus correlate with the manifestation of a complex developmental disorder, called Williams–Beuren syndrome, with multisystemic phenotypes including cognitive dysfunction (Doll & Grzeschik, 2001). Furthermore, a mutation in NSUN7 has been shown to cause reduced sperm motility and infertility in male mice (Harris et al., 2007). For most of these RCMTs, the respective RNAs that carry the m^5C modification have not been identified yet.

Of note, m^5C in RNA can reduce the activation of Toll-like receptors, whereas primary dendritic cells can be stimulated by m^5C-modified RNAs (Karikó, Buckstein, Ni, & Weissman, 2005) indicating that (cytosine-5) RNA methylation contributes to the distinction between self and nonself RNA (Dalpke & Helm, 2012; Ishii & Akira, 2005). Importantly, m^5C has also been detected in Y-RNAs (Hussain, Sajini, et al., 2013). Y-RNAs are small noncoding RNAs (80–130 nt) that bind Ro RNPs (consisting of 60 kDa protein Ro and La proteins), which are common autoantigens in patients with rheumatic disease syndromes such as systemic lupus erythematosus and Sjogren's syndrome (Lerner, Boyle, Hardin, & Steitz, 1981; Verhagen & Pruijn, 2011), suggesting that (cytosine-5) RNA methylation impacts on the function of the immune system.

In summary, the combined evidence from analysing *RCMT* mutations point toward an involvement of (cytosine-5) RNA methylation systems in RNA quality control, RNA stability, protein translation as well as immunity and consequently also in disease etiology. However, the molecular details of how the m^5C modification affects these diverse processes, especially under stress conditions, remain to be identified.

2. METHODS FOR IDENTIFYING m^5C-MODIFIED RNA IN SEQUENCE CONTEXT

Due to technological limitations, for a long-time RNA modifications could neither be detected nor quantified except when present in abundant RNAs. Recent developments, especially the application of sensitive mass spectrometry tools to nucleic acid chemistry (i.e., liquid chromatography combined with mass spectrometry, LC–MS) has considerably increased analytical sensitivity, opening up possibilities for quantifying also the modifications status of low-abundance RNAs (Kellner et al., 2014; Li et al., 2008; Russell & Limbach, 2013). Using LC–MS has provided unprecedented insights into the dynamic landscape of RNA modifications (including m^5C) in abundant RNAs, especially during stress conditions (Chan et al., 2010, 2012). However, all physicochemical methods include hydrolysis steps, which consequently lead to the loss of RNA sequence identity. Unless specific RNA species are pre-selected and purified prior to such LC–MS analysis (i.e., tRNAs; Gu, Begley, & Dedon, 2014) losing sequence information represents a considerable caveat for the identification of individual RNA molecules carrying specific modifications.

Therefore, methods allowing the detection of RNA modifications within their RNA sequence context are desirable and their development is now feasible, especially with the support of massive parallel high-throughput sequencing techniques. However, such methods are presently only available for the detection of a handful of RNA modifications (Carlile et al., 2014; Dominissini et al., 2012; Meyer et al., 2012; Motorin et al., 2010; Motorin, Muller, Behm-Ansmant, & Branlant, 2007; Schaefer et al., 2009; Schwartz et al., 2014).

In general, the relative position of modified nucleosides can be identified after affinity-based enrichment and sequencing of RNA fragments containing these nucleosides, a method which has been used for the detection of the relative position of methyl-6-adenosine residues in RNA (Dominissini et al., 2012). Recently, photo-crosslinking-assisted m^6A sequencing has been introduced, which improves the accuracy of the m^6A mapping procedure for RNA (Chen et al., 2015). Alternatively, the chemical reactivity of modified nucleosides can be exploited to reveal their exact position within an RNA sequence by comparing the reverse transcription (RT) profiles of cDNA libraries before and after chemical treatment of the RNA analyte.

Importantly, m^5C displays marked differential reactivity toward various chemical reagents if compared to cytosine (reviewed in Motorin et al. (2010)). For instance, the low reactivity of m^5C with sodium bisulfite has been exploited for almost 20 years by the DNA bisulfite sequencing technique (Frommer et al., 1992; Hayatsu, Negishi, Shiraishi, Tsuji, & Moriyama, 2007). Bisulfite treatment was adapted to RNA in combination with a primer extension approach to detect m^5C in a tRNA (Gu, Hurto, Hopper, Grayhack, & Phizicky, 2005) and later for the analysis of (cytosine-5) RNA methylation in high throughput format, which is now called RNA bisulfite sequencing (RNA-BisSeq) (Schaefer et al., 2009).

RNA-BisSeq is based on the comparatively low reactivity of m^5C with HSO_3^- ions under neutral or acidic pH conditions. Sodium bisulfite reacts with C residues in single-stranded RNA and chemically deaminates these to U residues. Incubation of denatured RNA with sodium bisulfite causes the formation of cytosine-sulfonate adducts, which are unstable. Rapid decomposition to uracil-sulfonate at basic pH conditions results in chemical deamination converting C-to-U (Fig. 2), which can be identified after cDNA synthesis, amplification, and sequencing as C-to-T changes. In contrast, any position that does not show such C-to-T changes might have existed as m^5C-modified in the input RNA.

After the initial development of RNA-BisSeq, the method has been commercialized. Various RNA bisulfite treatment kits are available (Zymosearch and Epigentek), providing the means to perform candidate

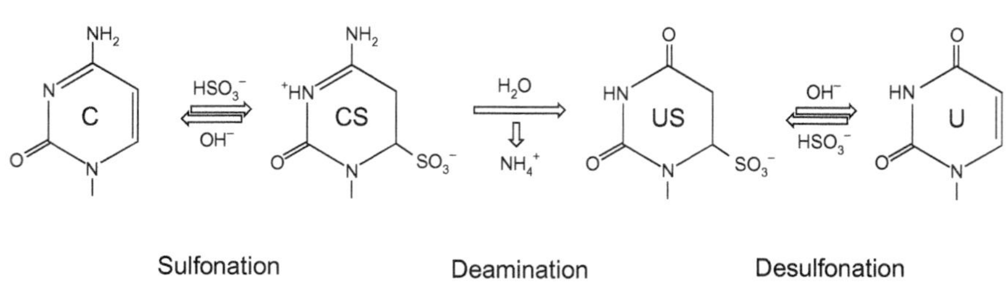

Figure 2 Sodium bisulfite reacts in acidic to neutral conditions with cytosine forming a sulfonate adduct, which rapidly decomposes to uracil-sulfonate at basic pH conditions resulting in the conversion of cytosine to uracil. C, cytosine; CS, cytosine-sulfonate; US, uridine sulfonate; U, uridine.

RNA-BisSeq experiments as well as transcriptome-wide identifications of m^5C in any RNA preparation.

First publications using RNA-BisSeq have now confirmed and quantified m^5C in known RCMT substrates (see Tables 1 and 2, and references therein) but have also reported previously unknown targets of RCMT activity (Edelheit, Schwartz, Mumbach, Wurtzel, & Sorek, 2013; Hussain, Aleksic, Blanco, Dietmann, & Frye, 2013; Squires et al., 2012). Importantly, among these m^5C-modified RNAs were mRNAs and various noncoding RNAs (Hussain, Sajini, et al., 2013; Squires et al., 2012). The existence of m^5C and other modifications in low-abundance RNAs such as mRNAs has drawn attention to the potential existence of an epitranscriptome (Meyer & Jaffrey, 2014, Saletore, Chen-Kiang, & Mason, 2013; Saletore et al., 2012). However, before the biological function of such an epitranscriptome, including an RNA methylome (Hussain, Aleksic, et al., 2013; Sibbritt, Patel, & Preiss, 2013) can be studied, it will be necessary to perform extensive sequencing experiments, to provide comprehensive information on the localisation of RNA modifications before tissue or even cell-specific RNA modification patterns can be mapped.

In addition to RNA-BisSeq, alternative approaches for the discovery of m^5C in RNA have been developed, especially for detecting low numbers of m^5C-modified RNA substrates. These techniques have been named 5-azacytidine-mediated RNA immunoprecipitation (Aza-IP) (Khoddami & Cairns, 2013), methylation-individual-nucleotide-resolution crosslinking and immunoprecipitation (miCLiP) (Hussain, Sajini, et al., 2013) and m^5C-RNA IP (Edelheit et al., 2013). These methods employ antibody-based purification of either methylated RNA or RNA methylation intermediates, which contain RCMTs that are "trapped" in the process of the methylation reaction (see Chapter 13 of this volume "Target identification of RNA cytosine methyltransferases"). Of note, studies applying these techniques primarily for m^5C discovery also used RNA-BisSeq when validating RNA methylation patterns, underscoring the importance of complementary technologies for the exact positional identification of m^5C in RNA. However, the number of (cytosine-5) methylation marks in individual RNA molecules appears to be low. For instance, m^5C in mRNA has been calculated as low as 0.18 sites/1000 nt or 0.45 sites per single copy mRNA (Sibbritt et al., 2013). Therefore, the total number of reported m^5C-modified RNAs remains presently very low. Only a systematic and reproducible interrogation of m^5C content in RNA preparations, preferably from different cellular contexts and in a transcriptome-wide as well as stress-dependent fashion, will establish if and under which circumstances RNA (cytosine-5) methylation is existent and physiologically relevant.

3. CONSIDERATIONS BEFORE RNA BISULFITE SEQUENCING (RNA-BisSeq)

RNA-BisSeq requires the purification of RNA before chemical deamination can be performed and therefore is solely an *in vitro* technology. Presently, the only variable parameter in the workflow of the method is the choice of input RNA. Such RNA preparations can come from *in vivo* as well

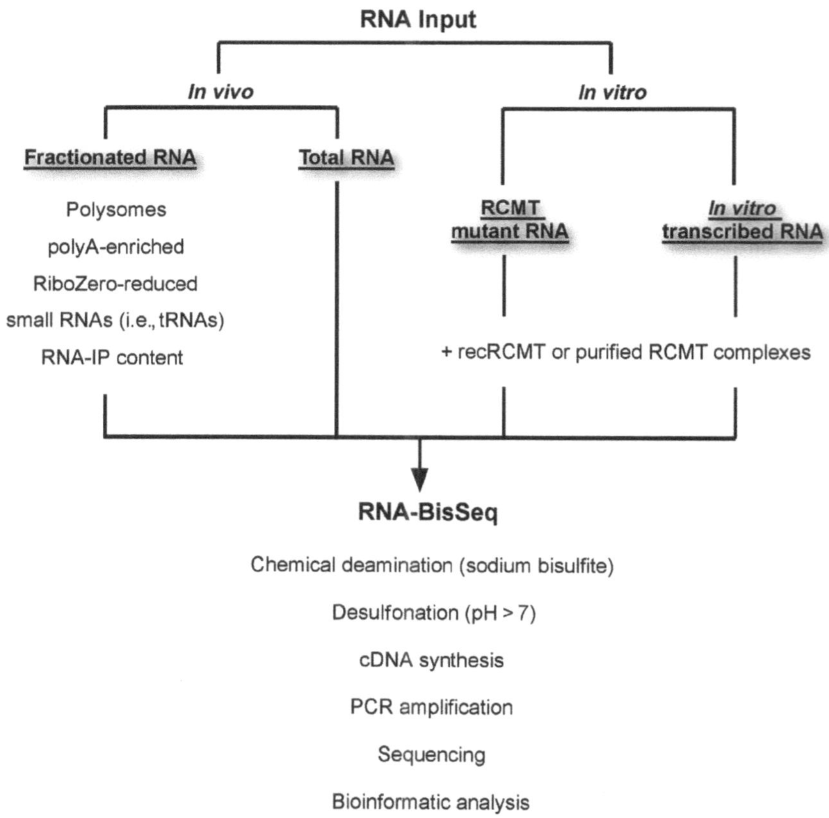

Figure 3 Sources of RNA input, which can be analyzed using the RNA-BisSeq protocol. *In vivo* RNA preparations should be prepared from wild type and *RCMT* mutant or *RCMT* knockdown cells in order to control for deamination artifacts and to allow stringent analysis of methylation patterns. *In vitro* methylation approaches require either recombinant RCMTs or purified RCMT complexes obtained by biochemical means (i.e., immunoprecipitation or density gradient fractionation).

as from *in vitro* sources. *In vivo* sources can be total or various fractionated RNAs (i.e., poly-A, tRNAs) that have been modified by endogenous RCMT activity. Alternatively, total or fractionated RNA from *RCMT* mutant cells can be used for *in vitro* methylation experiments using recombinant RCMTs or purified RCMT complexes. *In vitro*-methylated RNA could then be interrogated using RNA-BisSeq to obtain positional information on the introduced m^5C modifications (Fig. 3).

From the publication record on m^5C detection in RNA, it can be concluded that this modification is present mostly at specific positions within abundant noncoding RNAs (Motorin & Helm, 2011; Motorin et al., 2010) and in a few other RNAs including small noncoding RNAs and mRNAs (Dubin & Taylor, 1975; Dubin, Taylor, & Davenport, 1978; Hussain, Sajini, et al., 2013; Khoddami & Cairns, 2013; Squires et al., 2012). It follows that m^5C is neither ubiquitous nor a constitutive RNA modification (with the exception of its presence in some abundant noncoding RNAs), which should be taken into consideration when planning to use RNA-BisSeq for (cytosine-5) RNA methylation discovery. In addition, m^5C might mark RNA only under specific environmental or stress conditions (Becker et al., 2012; Bügl et al., 2000; Chan et al., 2010, 2012), which should guide the experimental design prior to employing RNA-BisSeq.

Although RNA-BisSeq can reveal the exact position of m^5C in a given RNA molecule, the use of this method is presently curtailed by the need to efficiently denature purified RNA preparations (a prerequisite for quantitative deamination) as well as by the harsh deamination conditions. Both parameters are largely incompatible with the stability of the RNA phosphodiester backbone leading to extensive RNA degradation, which is a serious caveat, especially when the identification of m^5C in low-copy number RNAs is the aim of the experiment.

3.1 Mapping of m^5C Positions in Previously Identified RNAs

The method to map m^5C in RNAs has been described in the original RNA-BisSeq protocol (Pollex, Hanna, & Schaefer, 2010) and is summarized in brief in Section 5.

This approach is applicable for the validation of known or predicted m^5C marks in candidate RNAs. Such RNAs might have been predicted by RNA enrichment experiments that are based on antibody-mediated purification of either the modified RNA itself (m^5C-RNA IP) or various

RNA–enzyme intermediates (miCLiP and AZA-IP). Importantly, these methods do not suffer from chemically induced RNA degradation, which potentially allows enriching for even very low amounts of m^5C-modified RNA from a large pool of RNAs. Here, the strength of RNA-BisSeq lies in mapping the exact position of the m^5C residues in the identified candidate RNA regions.

3.2 Discovery of m^5C in RNAs

Owing to recent technical developments, especially the increasing cost-effectiveness of high-throughput sequencing, RNA-BisSeq could be used for the systematic and transcriptome-wide discovery of (cytosine-5) methylation in RNA. To date, only a few publications have reported the use of RNA-BisSeq in such a context.

Transcriptome-wide as well as mRNA-enriched RNA-BisSeq has been performed (Edelheit et al., 2013; Hussain, Sajini, et al., 2013; Squires et al., 2012) albeit by using the original methodology that was adapted from DNA bisulfite sequencing (Pollex et al., 2010; Schaefer et al., 2009). The published protocols of these experiments follow standard RNA sequencing workflows involving cDNA synthesis (either from ribosomal depleted or from total RNAs), PCR amplification of adapter-ligated libraries and the use of high-throughput sequencing platforms. However, nontargeted and exploratory RNA-BisSeq approaches using total RNA preparations might fail to uncover low-copy number m^5C marks in RNA because these cannot be detected without prior enrichment of specific classes of RNAs, an efficient cDNA library amplification protocol, as well as the inclusion of meaningful controls.

3.3 Quantification of m^5C in Candidate RNAs

Once the exact position of m^5C in individual RNA molecules has been identified and validated (see Tables 1 and 2), RNA-BisSeq can be used to quantify the extent and/or changes of (cytosine-5) RNA methylation at that position in different cells, tissues or organs, and under different experimental or environmental conditions (i.e., stress exposure). For quantification, such a sequencing approaches should involve the use of barcode identifiers that allow measuring actual cDNA molecule numbers, which have been copied from individual input RNA molecules (Fig. 4). Use of barcode identifiers has been pioneered by the recent development of single-cell sequencing technologies that push the limits of RNA and DNA identification from

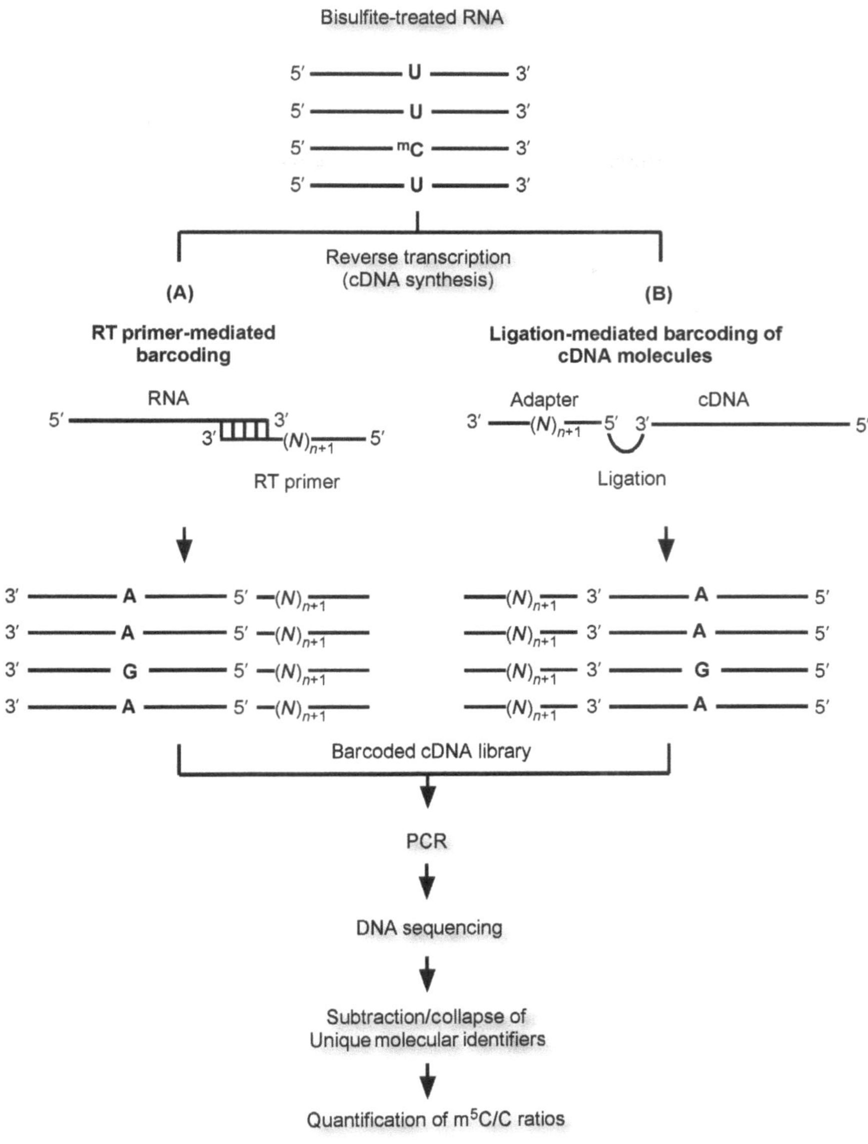

Figure 4 Two possibilities for (random) cDNA library barcoding that allow calculating original RNA molecules that have entered the cDNA synthesis and PCR-based amplification. Barcoding cDNA molecules with unique molecular identifiers (UMIs) will allow determining the clonal cDNA origin of the obtained sequence reads. (A) RNAs are barcoded during the RT reaction with RT primers that contain a stretch of random code (N_{n+1}). (B) cDNA synthesis is performed normally using any set of RT primers followed by ligation of DNA adapters containing a stretch of random code (N_{n+1}). Random barcoding allows tagging of 4^N different cDNA molecules that originated from 4^N different RNAs (i.e., $N=6$, 4096 cDNA molecules). For the determination of actual methylation ratios (m^5C/C) at a specific position of an RNA sequence, reads containing the same barcode (UMI) should be collapsed into one single read during the bioinformatics analysis of obtained sequencing reads.

ultra-low amounts of input material (Dey, Kester, Spanjaard, Bienko, & van Oudenaarden, 2015; Kivioja et al., 2011; Shiroguchi, Jia, Sims, & Xie, 2012; Zong, Lu, Chapman, & Xie, 2012). Low input represents a major challenge when employing RNA-BisSeq for the detection of m^5C in low-copy number RNAs. Using unique molecular identifiers (UMIs), such as random barcodes in adapters (included either during or after the RT reaction) allow approximating the number of RNA molecules that have entered the PCR-based amplification cycles (Fig. 4). This information will be crucial when determining the ratio of unmethylated/methylated cytosine at a given position in a pool of RNAs with the same sequence.

3.4 Controls for RNA-BisSeq

Sodium bisulfite does not efficiently deaminate cytosine to uracil unless cytosine is present in single-stranded regions of the analyte RNA. Although RNA preparations are denatured during the RNA-BisSeq protocol by using moderately high temperatures (60–70 °C), such treatment has to be kept as short as possible to avoid excessive RNA degradation. On the other hand, incomplete denaturation of RNA will cause nonquantitative deamination resulting in false-positive signals, which will indicate m^5C at positions where double-stranded regions impaired the deamination process. In addition, any peptide–RNA adducts, which can be the result of incomplete denaturation or removal of RNA binding proteins (Danilevskaya, Slot, Traverse, Hogan, & Pardue, 1994) might also affect the deamination process. It follows that it is almost impossible to completely rule out the occurrence of deamination artifacts when performing RNA-BisSeq. Therefore, sequencing reads containing deamination artifacts have to be distinguished from reads with true m^5C-derived signals by using either experimental control RNA samples that have been processed in parallel and/or a bioinformatics pipeline that employs stringent postexperimental data analysis.

Experimental controls such as sodium bisulfite treatment of specific (candidate) RNAs (synthesized by *in vitro* transcription) are one option that may allow deducing the extent of the deamination resistance of a specific RNA sequence. *In vitro*-transcribed RNA controls have been used for the analysis of the m^5C status of long noncoding RNAs in human cells (Amort et al., 2013) and as spike-in RNAs during the mapping of m^5C in an archeal transcriptome (Edelheit et al., 2013). However, the deamination efficiency of an *in vitro*-synthesized RNA might differ considerably from an RNA that has been extracted from cellular material because various (unknown) parameters could interfere with the deamination of cytosines such as preexisting modifications in cellular RNA or remaining traces of RNA binding proteins (see above). These variables might affect the deamination process of endogenous RNAs but not of *in vitro*-synthesized RNAs. Therefore, the most stringent approach to control for deamination artifacts is the use of RNA preparations from material harboring genetic mutations in the respective (cytosine-5) RNA methyltransferase gene, which render the enzyme catalytically inactive. Obtaining such material depends first on the availability of *RCMT* mutant cells or tissues and secondly on the viability of the cells harboring these *RCMT* mutations. Genetic mutations have been successfully used when analyzing RNA substrates of various RCMTs (Becker et al., 2012; Blanco et al., 2014; Gigova et al., 2014; Hussain, Sajini, et al., 2013; Kiani et al., 2013; Schaefer et al., 2009, 2010; Tuorto et al., 2012). Alternatively, individual RCMT expression can be reduced using RNA interference (RNAi)-mediated knockdown. Although this can be cumbersome in tissues or whole animals (with the exception of worm and fly systems), RNAi is most easily achieved in cell culture and has been successfully used to validate m^5C positions in miCLiP- and Aza-IP-identified NSun2 substrates (Khoddami & Cairns, 2013). Nevertheless, the efficiency of individual RCMT knockdowns should be tested using Western blotting before extracting RNA for RNA-BisSeq. In addition, it should be kept in mind that even after successful knockdown of individual RCMTs the half-lives of m^5C-modified RNAs could be extensive (such as reported for some tRNAs; Phizicky & Hopper, 2010), which could interfere with correctly distinguishing RCMT-dependent m^5C signals from deamination artifacts.

In summary, it is advisable considering the following points before embarking on RNA-BisSeq projects to ensure meaningful outcomes of the experiments:

For discovery of m^5C-modified RNAs

- Is the expression pattern of specific RCMTs known and can a specific cell type, tissue, or organ provide the RNA input material?
- Is it necessary to introduce stress conditions before extracting RNA?
- Can RNA be enriched from subcellular structures (i.e., nucleus vs. cytoplasm, mitochondria vs. cytoplasm) before treatment with sodium bisulfite?
- Can specific RNA classes be enriched (i.e., poly-A vs. small RNAs) before treatment with sodium bisulfite?
- How can random barcodes be introduced into the workflow?

- Which controls should be included to subtract deamination artifacts/false-positive signals?

For validation and quantification of m^{5}C positions in specific RNAs

- Is enrichment of candidate RNAs possible? (see above)
- How can random barcodes be introduced into the workflow?
- Which controls can be included to subtract deamination artifacts/false-positive signals?

Considerations such as these are crucially important for the successful outcome of any RNA-BisSeq experiment.

4. GENERAL PROTOCOL FOR RNA-BisSeq

The original protocols used self-made sodium bisulfite solution (Gu et al., 2005) or adapted a DNA bisulfite conversion kit (Pollex et al., 2010; Schaefer et al., 2009) to the treatment of RNA. Three commercial products are now available that offer specific RNA bisulfite treatment options:

- http://www.zymoresearch.com/epigenetics/rna-methylation/bisulfite-conversion/ez-rna-methylation-kit
- http://www.epigentek.com/catalog/methylamp-rna-bisulfite-conversion-kit-p-3644.html
- http://www.epigentek.com/catalog/epinext-rna-bisulfite-seq-kit-illumina-p-3696.html

These products claim to have addressed the RNA degradation issue by including undisclosed components that protect RNA from hydrolysis. The manuals state that even low amounts of input RNA (5 ng) can be successfully treated with sodium bisulfite and quantitatively recovered for subsequent cDNA synthesis. However, at the time of this writing, only one academic study used one of these commercial products, albeit for targeted RNA-BisSeq on long noncoding RNAs (Amort et al., 2013). Therefore, it is presently not possible to objectively judge the quality of the RNA deamination process or the yield of bisulfite-converted RNA that can be recovered when using these kits.

4.1 RNA Extraction

RNA isolated from any tissue, cell population or RNA that was recovered from protein complexes, had been size fractionated (i.e., urea-PAGE) or subtracted from the total RNA pool by oligonucleotide-mediated hybridization (i.e., oligo-dT enrichment of poly-adenylated RNAs) can be used as starting material. In general, the higher the amount of input RNA the more likely it is to recover sufficient numbers of individual and intact RNA molecules after sodium bisulfite treatment. In some instances, it might be even advisable to include Proteinase K treatment to reduce RNA loss due to protein-RNA adducts (see above) when using organic extraction of RNAs. When using total RNA preparations as starting material it should also be kept in mind that RNA from most cell types contains >90% ribosomal RNAs and 2–4% transfer RNAs, leaving candidate RNAs from single copy genes significantly underrepresented in such input material RNA.

Therefore, depending on the RNA class or species one wishes to interrogate for m^5C content and position, sufficient material should be prepared to guarantee at least 1 μg of input RNA (or less when enrichment of specific RNAs can be performed).

Note: Use RNAse-free materials and equipment throughout the procedure. Be careful with DEPC-treated materials and solutions when downstream processes require enzymatic activities because DEPC might not only inhibit RNAses but also other enzymes (Wolf, Lesnaw, & Reichmann, 1970). See also technical notes "RNase and DEPC Treatment: Fact or Laboratory Myth" at LifeTechnologies.

4.2 Treatment of RNA with DNAse

Most RNA preparations contain traces of genomic DNA. This DNA will not be efficiently denatured due to the application of only moderately high temperatures (60–70 °C) during RNA denaturation instead of >94 °C that is necessary to denature double-stranded DNA. Not removing these DNA traces will certainly add nondeaminated sequences to the downstream PCR-based amplification cycles and therefore will add reads containing deamination artifacts to the analysis.

Materials

- DNA-*free*™ DNA Removal Kit (Invitrogen)
- Purified RNA

Protocol

1. Treat RNA according to manufacturer's instructions.
2. Adjust DNAse-treated RNA to a volume of 10 μl.

4.3 Sodium Bisulfite Treatment of RNA

The protocol described in this paragraph is an adaption of the original method that was developed for ≥1 μg of input RNA and use with the EpiTect Kit (Qiagen) for deamination of RNA (Pollex et al., 2010; Schaefer et al., 2009).

Note: If other commercially available RNA deamination kits are employed follow the respective manufacturer's instructions.

Materials

- EpiTect Kit (Qiagen)
- DNAse-treated RNA preparation
- PCR thermal cycler
- Micro Bio-spin 6 columns (BioRad)
- GlycoBlue coprecipitant (LifeTechnologies)
- Cooled microcentrifuge

Note: Keep sodium bisulfite solutions at room temperature because highly concentrated sodium bisulfite precipitates at low temperatures.

Protocol

1. Transfer DNAseI-treated RNA solution into a PCR tube.
2. Add 42.5 μl freshly made bisulfite solution (EpiTect kit) and mix thoroughly by pipetting.
3. Add 17.5 μl DNA stabilization solution (EpiTect kit) and mix thoroughly by pipetting.
4. Place samples in thermal cycler and run the following program to deaminate RNAs:
   - **i.** 70 °C: 5 min
   - **ii.** 60 °C: 60 min
   - **iii.** Repeat steps (i) and (ii) two to six times (depending on target RNA abundance and predicted secondary structures)
   - **iv.** 25 °C—hold
5. Desalt reaction by passage through a Micro Bio-spin 6 column according to manufacturer's instructions.
6. Add 1 volume (70 μl) of 1 *M* Tris–HCl (pH 9.0) and incubate at 37 ° C—30 min for desulfonation.
7. Precipitate RNA by adding sodium acetate (pH 5.2) to 0.3 *M*, 15 μg GlycoBlue coprecipitant and 3 volumes of 100% ethanol and incubate at −80 °C for ≥1 h.
8. Pellet RNA by centrifugation at 4 °C for ≥30 min at top speed.
9. Wash RNA pellet with 75% ethanol.
10. Air-dry ≤5 min at room temperature.
11. Store deaminated RNA pellet until further use at −80 °C.

4.4 Synthesis of cDNA from Sodium Bisulfite-Treated RNA

The protocol described in this paragraph is an adaption of the original method that was developed for the detection of m^5C positions in abundant candidate RNAs such as tRNAs and rRNAs (Pollex et al., 2010; Schaefer et al., 2009).

Note: If a discovery approach of m^5C positions on a transcriptome-wide level is the aim of the experiment follow the materials and methods section of the respective publication (Blanco et al., 2014; Edelheit et al., 2013; Hussain, Sajini, et al., 2013; Squires et al., 2012).

Materials

- SuperScript III kit (Invitrogen)
- RNAseOUT (Invitrogen)
- Deaminated RNA preparation
- PCR thermal cycler

Protocol

1. Set up a cDNA synthesis reaction:
   $\geq$200 ng deaminated RNA resuspended in RNAse-free water
   2 μl 10 m*M* dNTPs
   50 n*M* RT primers
   RNAse-free water to 13 μl

   Note: RT primers can be random or specific primers (i.e., stem-loop structured for priming small RNAs or 18- to 24-mers for other RNAs). Importantly, RT primers should not contain Gs to avoid RT of RNA that was not efficiently deaminated. In addition, when quantification of m^5C positions is the aim of the experiment, consider using primers with a stretch of random barcodes (Fig. 4).
2. Place sample in a thermal cycler and incubate at 70 °C for 5 min.
3. Put reaction on ice before adding the following components:
   4 μl 5 × RT buffer
   1 μl 0.1 *M* DTT
   1 μl RNaseOUT (40 units)
   1 μl SuperScript III (200 units)
4. Place samples in thermal cycler and run the following program:
   1. 25 °C—5 min
   2. 50 °C—50 min
   3. 75 °C—15 min
   4. 4 °C—hold

4.5 PCR Amplification

Deamination of the majority of cytosines to uracils in the analyte RNA pool, followed by RT of these positions into adenines will lead to stretches of A-mononucleotide polymers in cDNA preparations, which might cause conventional DNA polymerases to stutter or slip. It can therefore be of advantage to use high-performance proofreading DNA polymerases during PCR amplification steps (Fazekas, Steeves, & Newmaster, 2010).

Materials

- Proofreading DNA polymerase and reaction buffers (various suppliers)
- 10 m*M* dNTPs
- PCR thermal cycler
- QIAEX II Gel Extraction Kit (Qiagen)

Protocol

1. Set up PCR reaction:

   2.5 μl cDNA
   Proofreading DNA polymerase reaction buffer to 1×
   1 μl 10 m*M* dNTPs
   200 n*M* specific forward and reverse primers
   Proofreading DNA polymerase according to manufacturer's instructions
   RNAse-free water to 25 μl

   Note: PCR primers can be specific primers that hybridize to deaminated candidate cDNA sequences (forward C-less, reverse G-less) either without (for subsequent TA cloning) or with specific adapters (for high-throughput sequencing).
2. Place samples in thermal cycler and run the following program:
   1. 98 °C—5 min
   2. 94 °C—30 s
   3. 50 °C—30 s
   4. 68 °C—30 s
   5. Repeat steps 2 to 4 a further 30 times
   6. 68 °C—5 min
   7. 10 °C—hold
3. Visualize PCR product using agarose gel electrophoresis (2–3%).
4. Excise PCR product and clean up using QIAEX II Gel Extraction Kit according to manufacturer's instructions.

4.6 Analysis of PCR Products

Depending on the scale and purpose of the RNA methylation analysis approach (validation or quantification), purified PCR fragments can be directly cloned using TA cloning followed by Sanger sequencing or can be further processed for subsequent sequencing on high-throughput platforms (i.e., 454 Roche or Illumina sequencers).

4.6.1 Sanger Sequencing of PCR Products

Proofreading DNA polymerases produce PCR products with blunt ends. In order to perform direct cloning of such PCR products into plasmids with manufactured T-overhangs, nontemplate-directed addition of an adenine residue to the 3′ ends of the PCR product has to be performed using *Taq* DNA polymerase.

Materials

- Purified PCR product
- dATP
- *Taq* DNA Polymerase with reaction buffer
- Chemically competent bacteria
- TA cloning kit
- Miniprep kit

Protocol

1. Set up an adenylation reaction:
   Purified PCR product (10–100 ng/100 bp length)
   1 × *Taq* DNA polymerase reaction buffer (with $MgCl_2$)
   200 n*M* dATP
   1 unit *Taq* DNA polymerase
   Water to 50 µl
2. Incubate reaction at 72 °C for 20 min.
3. Proceed immediately with TA cloning according to manufacturer's instructions.
   Note: Since 3′A overhangs will gradually be lost during storage, it is advisable to immediately proceed with TA cloning after the adenylation reaction.
4. Transform 2 µl of the TA cloning reaction into chemically competent *Escherichia coli* using standard procedures and spread onto kanamycin plate that is prepared for blue/white screening. Incubate overnight at 37 °C.
5. Select white colonies for culturing in selective media.
6. Miniprep samples using a standard plasmid kit according to manufacturer's instructions.
7. Send a representative number of plasmids for Sanger sequencing using suitable primers for the used plasmid.

4.6.2 Sequencing of PCR Products on High-Throughput Platforms

Various sequencing platforms provide the means to create massive numbers of reads with short read lengths (<100 bp). Alternatively, pyrosequencing on a 454-platform allows obtaining read lengths >100 bp albeit with limited sequencing depths. In addition, using high-throughput sequencing platforms allows multiplexing of PCR amplicons (i.e., technical and biological replicates, *RCMT* mutant or knockdown controls) through the introduction of standard sequencing barcodes during library preparation. However, the chemical treatment of RNA with sodium bisulfite and the desulfonation step at basic pH causes random RNA degradation, which reduces the likelihood of amplifying high numbers of long cDNA molecules (>200 bp) that are derived from low-abundance RNAs. Therefore, depending on the read numbers that are necessary to deduce statistically relevant information about the m^5C status of a candidate RNA region and the length distribution of the prepared PCR amplicons the appropriate sequencing platform should be chosen. Preparation of sequencing libraries should be performed according to the manufacturer's instructions of high-throughput sequencing kits.

5. RNA-BisSeq DATA ANALYSIS

Independently of the used sequencing platform, the existence of cytosine calls has to be assessed by comparing the obtained reads to reference sequences. These can be specific sequences (candidate regions) or sequence information from whole genomes and transcriptomes. Preferably, reference sequences should be *in silico* converted (all C-to-T) in order to provide a reference data set with 100% deamination at all cytosine positions, which allows identifying nondeaminated cytosines as potential m^5C marks in the experimental samples.

For targeted RNA-BisSeq by Sanger sequencing, the analysis of the limited number of individual reads can be represented in the form of circle diagrams. For targeted high-throughput RNA-BisSeq, no specific pipeline or program has presently been made public for the analysis of the sequencing data. However, a number of open source DNA methylation analysis programs such as Bismarck, Methy-Pipe or BisQC have been developed over the last years (Chen et al., 2013; Jiang et al., 2013; Krueger & Andrews, 2011), which could be used as entry points for the analysis of RNA-BisSeq data. In addition, most RNA-BisSeq publications have described individual approaches for data processing, mapping, and visualization of m^5C positions in RNA.

The general processing workflow for high-throughput data analysis consists of demultiplexing of pooled libraries, removal of adapters, and sequencing barcodes, followed by aligning reads representing candidate target regions (amplicons), to the respective converted reference sequence. Multiple reads originating from PCR amplification of the same cDNA molecule can be identified by the existence of identical barcodes (UMIs) and should be reduced to one read per barcode to determine exact cDNA counts. Thresholds for conversion rates (C-to-T) should be set either globally (>99%) or specifically in such way that stretches of nonconverted Cs can be filtered out (i.e., reads with >three Cs in a row). This will exclude reads with deamination artifacts from the analysis. Cytosine calls from experimental data sets should be compared to those in control data sets (i.e., *RCMT* mutant, RCMT knockdown material, or spike-in controls), followed by statistical tests for significance (including a measure for false-discovery rates). Different publications have either used custom-made or open source heat map generation programs (i.e., matrix2png) to visualize methylated cytosines in RNA.

6. LIMITATIONS OF RNA-BisSeq

While RNA-BisSeq has been repeatedly used to validate m^5C at either known RNA positions or at RNA regions that have been predicted to contain m^5C the use of the method for m^5C discovery has not been firmly established. The reason for this is most likely that m^5C in RNA is neither ubiquitous nor constitutive (with the exception of already identified m^5C positions in abundant noncoding RNAs), but rather a dynamic modification whose establishment depends on various environmental inputs (Becker et al., 2012; Bügl et al., 2000; Chan et al., 2010, 2012). Because the amount of m^5C at a given position in a specific RNA can vary considerably within a pool of these RNAs (i.e., Dnmt2-mediated tRNA-Asp^{GTC} methylation at C38 ranges between 30% and 100% depending on the cell type, unpublished observations), the detection of very low m^5C levels might not be possible using RNA-BisSeq. It is therefore advisable to perform a combination of Aza-IP or miCLiP and RNA-BisSeq for the discovery and validation of m^5C in presently unknown RNAs.

The main confounding factor that precludes mapping of m^5C in low-copy RNAs with a potentially low methylation ratio is the hydrolytic degradation of RNAs during the denaturation, treatment with sodium bisulfite and the desulfonation steps. This methodological caveat needs to be addressed in the future to improve recovery and inclusion of most input RNA sequence in the RNA-BisSeq output. In general, RNA autohydrolysis could be reduced by stably modifying the 2′-OH on the ribose moieties. Of note, patents have been filed that describe the modification of 2′-position in RNAs prior to cDNA synthesis (US Patent No. 6867290, "Modified polynucleotides and uses thereof"). It remains to be tested whether such modifications are stable enough to be maintained throughout the sodium bisulfite treatment and desulfonation procedure to increase the yield of reads from low-copy RNAs.

Additional technical issues exist that might cause the inability to detect low levels of m^5C in RNA. It has been reported that also m^5C can be deaminated (C-to-T) albeit at slower rates than C is deaminated to U (Hayatsu, 2008; Hayatsu & Shiragami, 1979; Hayatsu, Wataya, Kai, & Iida, 1970). Given that the concentration of sodium bisulfite in commercially available kits has been optimized to guarantee high deamination rates, the possibility to "over-deaminate" and thereby lose m^5C residues in RNA is likely, which will negatively impact both on the discovery of low m^5C levels, especially in presently unknown RNAs as well as on the quantification of m^5C at known positions. In addition, also hm^5C, an oxidation product of m^5C, has been detected in RNA (Fu et al., 2014; Huber et al., 2015; Rácz, Juhász, Király, & Lásztity, 1982). Although the amount of hm^5C in RNA is very low, its presence in specific RNAs could cause cytosine calls after RNA-BisSeq because bisulfite sequencing does not allow distinguishing m^5C from hm^5C, unless performed under specific conditions (Booth et al., 2012).

Despite these limitations, the use of RNA-BisSeq might be expanded to studying other cytosine base modifications. The chemistry of the deamination process indicates that also other chemical groups can interfere with this reaction. Among these modifications are 3-methylcytidine (m3C), which has been detected in human 18S and 28S rRNA (Iwanami & Brown, 1968), N4-acetylcytidine (ac4C) (Thomas, Gordon, & Rogg, 1978), and N4-methylcytidine (4mC), found in mitochondrial 13S rRNA (Dubin et al., 1978). For instance, N4, 2′-*O*-dimethylcytidine (m4mC), a derivative of 4mC in *E. coli*, has been detected by RNA-BisSeq at position 1402 of bacterial 16S rRNA (Schaefer et al., 2009).

ACKNOWLEDGMENTS

I am grateful to Tim Pollex, Katharina Hanna, and Frank Lyko for their contribution to the development of RNA-BisSeq. The development of the method was supported by a grant from the DFG (FOR1082).

REFERENCES

Abbasi-Moheb, L., Mertel, S., Gonsior, M., Nouri-Vahid, L., Kahrizi, K., Cirak, S., et al. (2012). Mutations in NSUN2 cause autosomal-recessive intellectual disability. *American Journal of Human Genetics*, *90*, 847–855.

Amort, T., Soulière, M. F., Wille, A., Jia, X.-Y., Fiegl, H., Wörle, H., et al. (2013). Long non-coding RNAs as targets for cytosine methylation. *RNA Biology*, *10*, 1003–1009.

Andersen, N. M., & Douthwaite, S. (2006). YebU is a m5C methyltransferase specific for 16 S rRNA nucleotide 1407. *Journal of Molecular Biology*, *359*, 777–786.

Auxilien, S., Khadali El, F., Rasmussen, A., Douthwaite, S., & Grosjean, H. (2007). Archease from Pyrococcus abyssi improves substrate specificity and solubility of a tRNA m5C methyltransferase. *The Journal of Biological Chemistry*, *282*, 18711–18721.

Becker, M. M., Müller, S. S., Nellen, W. W., Jurkowski, T. P. T., Jeltsch, A. A., & Ehrenhofer-Murray, A. E. A. (2012). Pmt1, a Dnmt2 homolog in Schizosaccharomyces pombe, mediates tRNA methylation in response to nutrient signaling. *Nucleic Acids Research*, *40*, 11648–11658.

Blanco, S., Dietmann, S., Flores, J. V., Hussain, S., Kutter, C., Humphreys, P., et al. (2014). Aberrant methylation of tRNAs links cellular stress to neuro-developmental disorders. *The EMBO Journal*, *33*, 2020–2039.

Blanco, S., Kurowski, A., Nichols, J., Watt, F. M., Benitah, S. A., & Frye, M. (2011). The RNA-methyltransferase Misu (NSun2) poises epidermal stem cells to differentiate. *PLoS Genetics*, 7, e1002403.

Booth, M. J., Branco, M. R., Ficz, G., Oxley, D., Krueger, F., Reik, W., et al. (2012). Quantitative sequencing of 5-methylcytosine and 5-hydroxymethylcytosine at single-base resolution. *Science (New York, N.Y.)*, *336*, 934–937.

Brzezicha, B., Schmidt, M., Makalowska, I., Jarmolowski, A., Pienkowska, J., & Szweykowska-Kulinska, Z. (2006). Identification of human tRNA:m5C methyltransferase catalysing intron-dependent m5C formation in the first position of the anticodon of the pre-tRNA Leu (CAA). *Nucleic Acids Research*, *34*, 6034–6043.

Bügl, H., Fauman, E. B., Staker, B. L., Zheng, F., Kushner, S. R., Saper, M. A., et al. (2000). RNA methylation under heat shock control. *Molecular Cell*, *6*, 349–360.

Bujnicki, J. M., Droogmans, L., Grosjean, H., Purushothaman, S. K., & Lapeyre, B. (2004). *Bioinformatics-guided identification and experimental characterization of novel RNA methyltransferas. Nucleic acids and molecular biology*. Berlin, Heidelberg: Springer Berlin Heidelberg, 139–168.

Bujnicki, J. M., Feder, M., Ayres, C. L., & Redman, K. L. (2004). Sequence-structure-function studies of tRNA:m5C methyltransferase Trm4p and its relationship to DNA:m5C and RNA:m5U methyltransferases. *Nucleic Acids Research*, *32*, 2453–2463.

Carlile, T. M., Rojas-Duran, M. F., Zinshteyn, B., Shin, H., Bartoli, K. M., & Gilbert, W. V. (2014). Pseudouridine profiling reveals regulated mRNA pseudouridylation in yeast and human cells. *Nature*, *515*, 143–146.

Chan, C. T. Y., Dyavaiah, M., DeMott, M. S., Taghizadeh, K., Dedon, P. C., & Begley, T. J. (2010). A quantitative systems approach reveals dynamic control of tRNA modifications during cellular stress. *PLoS Genetics*, *6*, e1001247.

Chan, C. T. Y., Pang, Y. L. J., Deng, W., Babu, I. R., Dyavaiah, M., Begley, T. J., et al. (2012). Reprogramming of tRNA modifications controls the oxidative stress response by codon-biased translation of proteins. *Nature Communications*, *3*, 937.

Chen, G. G., Diallo, A. B., Poujol, R., Nagy, C., Staffa, A., Vaillancourt, K., et al. (2013). BisQC: An operational pipeline for multiplexed bisulfite sequencing. *BMC Genomics*, *15*, 290.

Chen, K., Lu, Z., Wang, X., Fu, Y., Luo, G. Z., Liu, N., et al. (2015). High-resolution N6-Methyladenosine (m6A) map using photo-crosslinking-assisted m6A sequencing. *Angewandte Chemie, International Edition, 54*, 1587–1590.

Cheng, X., & Roberts, R. J. (2001). AdoMet-dependent methylation, DNA methyltransferases and base flipping. *Nucleic Acids Research, 29*, 3784–3795.

Chi, L., & Delgado-Olguín, P. (2013). Expression of NOL1/NOP2/sun domain (Nsun) RNA methyltransferase family genes in early mouse embryogenesis. *Gene Expression Patterns, 13*, 319–327.

Chiang, P. K., Gordon, R. K., Tal, J., Zeng, G. C., Doctor, B. P., Pardhasaradhi, K., et al. (1996). S-Adenosylmethionine and methylation. *FASEB Journal: Official Publication of the Federation of American Societies for Experimental Biology, 10*, 471–480.

Dalpke, A., & Helm, M. (2012). RNA mediated Toll-like receptor stimulation in health and disease. *RNA Biology, 9*, 828–842.

Danilevskaya, O. N., Slot, F., Traverse, K. L., Hogan, N. C., & Pardue, M. L. (1994). Drosophila telomere transposon HeT-A produces a transcript with tightly bound protein. *Proceedings of the National Academy of Sciences of the United States of America, 91*, 6679–6682.

Debeus, E., Brockenbrough, J. S., Hong, B., & Aris, J. P. (1994). Yeast Nop2 encodes an essential nucleolar protein with homology to a human proliferation marker. *The Journal of Cell Biology, 127*, 1799–1813.

Dey, S. S., Kester, L., Spanjaard, B., Bienko, M., & van Oudenaarden, A. (2015). Integrated genome and transcriptome sequencing of the same cell. *Nature Biotechnology, 33*(3), 285–289.

Doll, A., & Grzeschik, K. H. (2001). Characterization of two novel genes, WBSCR20 and WBSCR22, deleted in Williams-Beuren syndrome. *Cytogenetic and Genome Research, 95*, 20–27.

Dominissini, D., Moshitch-Moshkovitz, S., Schwartz, S., Salmon-Divon, M., Ungar, L., Osenberg, S., et al. (2012). Topology of the human and mouse m(6)A RNA methylomes revealed by m(6)A-seq. *Nature, 485*, 201–U84.

Dubin, D. T., & Taylor, R. H. (1975). The methylation state of poly A-containing messenger RNA from cultured hamster cells. *Nucleic Acids Research, 2*, 1653–1668.

Dubin, D. T., Taylor, R. H., & Davenport, L. W. (1978). Methylation status of 13S ribosomal RNA from hamster mitochondria: The presence of a novel riboside, N4-methylcytidine. *Nucleic Acids Research, 5*, 4385–4397.

Edelheit, S., Schwartz, S., Mumbach, M. R., Wurtzel, O., & Sorek, R. (2013). Transcriptome-wide mapping of 5-methylcytidine RNA modifications in bacteria, archaea, and yeast reveals m5C within archaeal mRNAs. *PLoS Genetics, 9*, e1003602.

Fazekas, A., Steeves, R., & Newmaster, S. (2010). Improving sequencing quality from PCR products containing long mononucleotide repeats. *BioTechniques, 48*, 277–285.

Foster, P. G., Nunes, C. R., Greene, P., Moustakas, D., & Stroud, R. M. (2003). The first structure of an RNA m5C methyltransferase, Fmu, provides insight into catalytic mechanism and specific binding of RNA substrate. *Structure (London, England: 1993), 11*, 1609–1620.

Frommer, M., McDonald, L. E., Millar, D. S., Collis, C. M., Watt, F., Grigg, G. W., et al. (1992). A genomic sequencing protocol that yields a positive display of 5-methylcytosine residues in individual DNA strands. *Proceedings of the National Academy of Sciences of the United States of America, 89*, 1827–1831.

Fu, L., Guerrero, C. R., Zhong, N., Amato, N. J., Liu, Y., Liu, S., et al. (2014). Tet-mediated formation of 5-hydroxymethylcytosine in RNA. *Journal of the American Chemical Society, 136*, 11582–11585.

Gigova, A., Duggimpudi, S., Pollex, T., Schaefer, M., & Kos, M. (2014). A cluster of methylations in the domain IV of 25S rRNA is required for ribosome stability. *RNA (New York, N.Y.), 20*, 1632–1644.

Goll, M. G., & Bestor, T. H. (2005). Eukaryotic cytosine methyltransferases. *Annual Review of Biochemistry*, *74*, 481–514.
Goll, M. G., Kirpekar, F., Maggert, K. A., Yoder, J. A., Hsieh, C.-L., Zhang, X., et al. (2006). Methylation of tRNAAsp by the DNA methyltransferase homolog Dnmt2. *Science (New York, NY)*, *311*, 395–398.
Grosjean, H. (2009). *DNA and RNA modification enzymes*. Austin, Texas: Landes Bioscience.
Grosjean, H., Gaspin, C., Marck, C., Decatur, W. A., & de Crécy-Lagard, V. (2007). RNomics and Modomics in the halophilic archaea Haloferax volcanii: Identification of RNA modification genes. *BMC Genomics*, *9*, 470.
Gu, C., Begley, T. J., & Dedon, P. C. (2014). tRNA modifications regulate translation during cellular stress. *FEBS Letters*, *588*, 4287–4296.
Gu, X. R., Gustafsson, C., Ku, J., Yu, M., & Santi, D. V. (1999). Identification of the 16S rRNA m5C967 methyltransferase from Escherichia coli. *Biochemistry*, *38*, 4053–4057.
Gu, W., Hurto, R. L., Hopper, A. K., Grayhack, E. J., & Phizicky, E. M. (2005). Depletion of Saccharomyces cerevisiae tRNA(His) guanylyltransferase Thg1p leads to uncharged tRNAHis with additional m(5)C. *Molecular and Cellular Biology*, *25*, 8191–8201.
Gustafsson, C. (1996). Identification of new RNA modifying enzymes by iterative genome search using known modifying enzymes as probes. *Nucleic Acids Research*, *24*, 3756–3762.
Hallberg, B. M., Ericsson, U. B., Johnson, K. A., Andersen, N. M., Douthwaite, S., Nordlund, P., et al. (2006). The structure of the RNA m5C methyltransferase YebU from Escherichia coli reveals a C-terminal RNA-recruiting PUA domain. *Journal of Molecular Biology*, *360*, 774–787.
Harris, T., Marquez, B., Suarez, S., & Schimenti, J. (2007). Sperm motility defects and infertility in male mice with a mutation in Nsun7, a member of the Sun domain-containing family of putative RNA methyltransferases. *Biology of Reproduction*, *77*, 376–382.
Hayatsu, H. (2008). Discovery of bisulfite-mediated cytosine conversion to uracil, the key reaction for DNA methylation analysis—A personal account. *Proceedings of the Japan Academy. Series B, Physical and Biological Sciences*, *84*, 321–330.
Hayatsu, H., Negishi, K., Shiraishi, M., Tsuji, K., & Moriyama, K. (2007). Chemistry of bisulfite genomic sequencing; advances and issues. *Nucleic Acids Symposium Series*, *2004*, 47–48.
Hayatsu, H., & Shiragami, M. (1979). Reaction of bisulfite with the 5-hydroxymethyl group in pyrimidines and in phage DNAs. *Biochemistry*, *18*, 632–637.
Hayatsu, H., Wataya, Y., Kai, K., & Iida, S. (1970). Reaction of sodium bisulfite with uracil, cytosine, and their derivatives. *Biochemistry*, *9*, 2858–2865.
Hikida, Y., Kuratani, M., Bessho, Y., Sekine, S.-I., & Yokoyama, S. (2010). Structure of an archaeal homologue of the bacterial Fmu/RsmB/RrmB rRNA cytosine 5-methyltransferase. *Acta Crystallographica. Section D, Biological Crystallography*, *66*, 1301–1307.
Hong, B., Brockenbrough, J. S., Wu, P., & Aris, J. P. (1997). Nop2p is required for pre-rRNA processing and 60S ribosome subunit synthesis in yeast. *Molecular and Cellular Biology*, *17*, 378–388.
Huber, S. M., van Delft, P., Mendil, L., Bachman, M., Smollett, K., Werner, F., et al. (2015). Formation and abundance of 5-hydroxymethylcytosine in RNA. *ChemBioChem*, *16*(5), 752–755. http://dx.doi.org/10.1002/cbic.201500013.
Hussain, S., Aleksic, J., Blanco, S., Dietmann, S., & Frye, M. (2013). Characterizing 5-methylcytosine in the mammalian epitranscriptome. *Genome Biology*, *14*, 215.
Hussain, S., Sajini, A. A., Blanco, S., Dietmann, S., Lombard, P., Sugimoto, Y., et al. (2013). NSun2-mediated cytosine-5 methylation of vault noncoding RNA determines its processing into regulatory small RNAs. *Cell Reports*, *4*, 255–261.
Hussain, S., Tuorto, F., Menon, S., Blanco, S., Cox, C., Flores, J. V., et al. (2013). The mouse cytosine-5 RNA methyltransferase NSun2 is a component of the chromatoid body and required for testis differentiation. *Molecular and Cellular Biology*, *33*, 1561–1570.

Ishii, K. J., & Akira, S. (2005). TLR ignores methylated RNA? *Immunity*, *23*, 111–113.
Ishikawa, I., Sakai, N., Tamura, T., Yao, M., Watanabe, N., & Tanaka, I. (2004). Crystal structure of human p120 homologue protein PH1374 from Pyrococcus horikoshii. *Proteins: Structure, Function, and Bioinformatics*, *54*, 814–816.
Iwanami, Y., & Brown, G. M. (1968). Methylated bases of ribosomal ribonucleic acid from HeLa cells. *Archives of Biochemistry and Biophysics*, *126*, 8–15.
Jiang, P., Sun, K., Lun, F. M. F., Guo, A. M., Wang, H., Chan, K. C. A., et al. (2013). Methy-Pipe: An integrated bioinformatics pipeline for whole genome bisulfite sequencing data analysis. *PloS One*, *9*, e100360.
Jurkowski, T. P. T., & Jeltsch, A. A. (2010). On the evolutionary origin of eukaryotic DNA methyltransferases and Dnmt2. *PloS One*, *6*, e28104.
Jurkowski, T. P., Meusburger, M., Phalke, S., Helm, M., Nellen, W., Reuter, G., et al. (2008). Human DNMT2 methylates tRNA(Asp) molecules using a DNA methyltransferase-like catalytic mechanism. *RNA (New York, N.Y.)*, *14*, 1663–1670.
Karikó, K., Buckstein, M., Ni, H., & Weissman, D. (2005). Suppression of RNA recognition by Toll-like receptors: The impact of nucleoside modification and the evolutionary origin of RNA. *Immunity*, *23*, 165–175.
Kellner, S., Ochel, A., Thüring, K., Spenkuch, F., Neumann, J., Sharma, S., et al. (2014). Absolute and relative quantification of RNA modifications via biosynthetic isotopomers. *Nucleic Acids Research*, *42*(18), e142.
Khan, M. A., Rafiq, M. A., Noor, A., Hussain, S., Flores, J. V., Rupp, V., et al. (2012). Mutation in NSUN2, which encodes an RNA methyltransferase, causes autosomal-recessive intellectual disability. *American Journal of Human Genetics*, *90*, 856–863.
Khoddami, V., & Cairns, B. R. (2013). Identification of direct targets and modified bases of RNA cytosine methyltransferases. *Nature Biotechnology*, *31*, 458–464.
Khoddami, V., & Cairns, B. R. (2014). Transcriptome-wide target profiling of RNA cytosine methyltransferases using the mechanism-based enrichment procedure Aza-IP. *Nature Protocols*, *9*, 337–361.
Kiani, J., Grandjean, V., Liebers, R., Tuorto, F., Ghanbarian, H., Lyko, F., et al. (2013). RNA-mediated epigenetic heredity requires the cytosine methyltransferase Dnmt2. *PLoS Genetics*, *9*, e1003498.
King, M. Y., & Redman, K. L. (2002). RNA methyltransferases utilize two cysteine residues in the formation of 5-methylcytosine. *Biochemistry*, *41*, 11218–11225.
Kivioja, T., Vähärautio, A., Karlsson, K., Bonke, M., Enge, M., Linnarsson, S., et al. (2011). Counting absolute numbers of molecules using unique molecular identifiers. *Nature Methods*, *9*, 72–74.
Koonin, E. V. (1994). Prediction of an rRNA methyltransferase domain in human tumor-specific nucleolar protein P120. *Nucleic Acids Research*, *22*, 2476–2478.
Krueger, F., & Andrews, S. R. (2011). Bismark: A flexible aligner and methylation caller for Bisulfite-Seq applications. *Journal of Gerontology*, *27*, 1571–1572.
Larsen, L. H. G., Rasmussen, A., Giessing, A. M. B., Jogl, G., & Kirpekar, F. (2012). Identification and characterization of the Thermus thermophilus 5-methylcytidine (m5C) methyltransferase modifying 23S ribosomal RNA (rRNA) base C1942. *Journal of Biological Chemistry*, *287*, 27593–27600.
Lerner, M. R., Boyle, J. A., Hardin, J. A., & Steitz, J. A. (1981). Two novel classes of small ribonucleoproteins detected by antibodies associated with lupus erythematosus. *Science (New York, N.Y.)*, *211*, 400–402.
Li, H.-Y. H., Wang, S.-M. S., Liu, H.-M. H., Li, J. J., Han, D. D., Bu, S.-S. S., et al. (2008). Analysis of modified nucleosides in the urine of patients with malignant cancer by liquid chromatography/electrospray ionization mass spectrometry. *Rapid Communications in Mass Spectrometry*, *22*, 3161–3171.

Liu, Y., & Santi, D. V. (2000). m5C RNA and m5C DNA methyl transferases use different cysteine residues as catalysts. *Proceedings of the National Academy of Sciences of the United States of America*, *97*, 8263–8265.

Machnicka, M. A., Milanowska, K., Osman Oglou, O., Purta, E., Kurkowska, M., Olchowik, A., et al. (2013). MODOMICS: A database of RNA modification pathways–2013 update. *Nucleic Acids Research*, *41*, D262–D267.

Martinez, F. J., Lee, J. H., Lee, J. E., Blanco, S., Nickerson, E., Gabriel, S., et al. (2012). Whole exome sequencing identifies a splicing mutation in NSUN2 as a cause of a Dubowitz-like syndrome. *Journal of Medical Genetics*, *49*, 380–385.

Metodiev, M. D., Spåhr, H., Polosa, P. L., Meharg, C., Becker, C., Altmueller, J., et al. (2014). NSUN4 is a dual function mitochondrial protein required for both methylation of 12S rRNA and coordination of mitoribosomal assembly. *PLoS Genetics*, *10*, e1004110.

Meyer, K. D., & Jaffrey, S. R. (2014). The dynamic epitranscriptome: N6-methyladenosine and gene expression control. *Nature Reviews. Molecular Cell Biology*, *15*, 313–326.

Meyer, K. D. K., Saletore, Y. Y., Zumbo, P. P., Elemento, O. O., Mason, C. E. C., & Jaffrey, S. R. S. (2012). Comprehensive analysis of mRNA methylation reveals enrichment in 3' UTRs and near stop codons. *Cell*, *149*, 1635–1646.

Moon, H. J., & Redman, K. L. (2014). Trm4 and Nsun2 RNA:m(5)C methyltransferases form metabolite-dependent, covalent adducts with previously methylated RNA. *Biochemistry*, *53*(45), 7132–7144.

Motorin, Y., & Grosjean, H. (1999). Multisite-specific tRNA:m5C-methyltransferase (Trm4) in yeast Saccharomyces cerevisiae: Identification of the gene and substrate specificity of the enzyme. *RNA (New York, N.Y.)*, *5*, 1105–1118.

Motorin, Y., & Helm, M. (2011). RNA nucleotide methylation. *Wiley Interdisciplinary Reviews. RNA*, *2*, 611–631.

Motorin, Y., Lyko, F., & Helm, M. (2010). 5-methylcytosine in RNA: Detection, enzymatic formation and biological functions. *Nucleic Acids Research*, *38*, 1415–1430.

Motorin, Y., Muller, S., Behm-Ansmant, I., & Branlant, C. (2007). Identification of modified residues in RNAs by reverse transcription-based methods. *Methods in Enzymology*, *425*, 21–53.

Müller, S., Windhof, I. M., Maximov, V., Jurkowski, T., Jeltsch, A., Förstner, K. U., et al. (2013). Target recognition, RNA methylation activity and transcriptional regulation of the Dictyostelium discoideum Dnmt2-homologue (DnmA). *Nucleic Acids Research*, *41*, 8615–8627.

Pavlopoulou, A., & Kossida, S. (2009). Phylogenetic analysis of the eukaryotic RNA (cytosine-5)-methyltransferases. *Genomics*, *93*, 350–357.

Phizicky, E. M., & Hopper, A. K. (2010). tRNA biology charges to the front. *Genes and Development*, *24*, 1832–1860.

Pollex, T., Hanna, K., & Schaefer, M. (2010). Detection of cytosine methylation in RNA using bisulfite sequencing. *Cold Spring Harbor Protocols*, *2010*. pdb.prot5505.

Purta, E., O'Connor, M., Bujnicki, J. M., & Douthwaite, S. (2008). YccW is the m5C methyltransferase specific for 23S rRNA nucleotide 1962. *Journal of Molecular Biology*, *383*, 641–651.

Rácz, I., Juhász, A., Király, I., & Lásztity, D. (1982). Changes in the content of modified nucleotides of total transfer RNA of wheat seedlings during greening. *Planta*, *154*, 397–401.

Rai, K., Chidester, S., Zavala, C. V., Manos, E. J., James, S. R., Karpf, A. R., et al. (2007). Dnmt2 functions in the cytoplasm to promote liver, brain, and retina development in zebrafish. *Genes and Development*, *21*, 261–266.

Reid, R., Greene, P. J., & Santi, D. V. (1999). Exposition of a family of RNA m(5)C methyltransferases from searching genomic and proteomic sequences. *Nucleic Acids Research*, *27*, 3138–3145.

Russell, S. P., & Limbach, P. A. (2013). Evaluating the reproducibility of quantifying modified nucleosides from ribonucleic acids by LC–UV–MS. *Journal of Chromatography B, 923–924*, 74–82.

Saletore, Y., Chen-Kiang, S., & Mason, C. E. (2013). Novel RNA regulatory mechanisms revealed in the epitranscriptome. *RNA Biology, 10*, 342–346.

Saletore, Y., Meyer, K., Korlach, J., Vilfan, I. D., Jaffrey, S., & Mason, C. E. (2012). The birth of the Epitranscriptome: Deciphering the function of RNA modifications. *Genome Biology, 13*, 175.

Schaefer, M., Pollex, T., Hanna, K., & Lyko, F. (2009). RNA cytosine methylation analysis by bisulfite sequencing. *Nucleic Acids Research, 37*, e12.

Schaefer, M., Pollex, T., Hanna, K., Tuorto, F., Meusburger, M., Helm, M., et al. (2010). RNA methylation by Dnmt2 protects transfer RNAs against stress-induced cleavage. *Genes and Development, 24*, 1590–1595.

Schosserer, M., Minois, N., Angerer, T. B., Amring, M., Dellago, H., Harreither, E., et al. (2015). Methylation of ribosomal RNA by NSUN5 is a conserved mechanism modulating organismal lifespan. *Nature Communications, 6*, 6158.

Schwartz, S., Bernstein, D. A., Mumbach, M. R., Jovanovic, M., Herbst, R. H., León-Ricardo, B. X., et al. (2014). Transcriptome-wide mapping reveals widespread dynamic-regulated pseudouridylation of ncRNA and mRNA. *Cell, 159*, 148–162.

Shanmugam, R., Aklujkar, M., Schäfer, M., Reinhardt, R., Nickel, O., Reuter, G., et al. (2013). The Dnmt2 RNA methyltransferase homolog of Geobacter sulfurreducens specifically methylates tRNA-Glu. *Nucleic Acids Research, 42*, 6487–6496.

Sharma, S., Yang, J., Watzinger, P., Kötter, P., & Entian, K.-D. (2013). Yeast Nop2 and Rcm1 methylate C2870 and C2278 of the 25S rRNA, respectively. *Nucleic Acids Research, 41*, 9062–9076.

Shiroguchi, K., Jia, T. Z., Sims, P. A., & Xie, X. S. (2012). Digital RNA sequencing minimizes sequence-dependent bias and amplification noise with optimized single-molecule barcodes. *Proceedings of the National Academy of Sciences of the United States of America, 109*, 1347–1352.

Sibbritt, T., Patel, H. R., & Preiss, T. (2013). Mapping and significance of the mRNA methylome. *Wiley Interdisciplinary Reviews. RNA, 4*, 397–422.

Squires, J. E. J., Patel, H. R. H., Nousch, M. M., Sibbritt, T. T., Humphreys, D. T. D., Parker, B. J. B., et al. (2012). Widespread occurrence of 5-methylcytosine in human coding and non-coding RNA. *Nucleic Acids Research, 40*, 5023–5033.

Strobel, M. C., & Abelson, J. (1986). Effect of intron mutations on processing and function of Saccharomyces cerevisiae SUP53 tRNA in vitro and in vivo. *Molecular and Cellular Biology, 6*, 2663–2673.

Sunita, S., Tkaczuk, K. L., Purta, E., Kasprzak, J. M., Douthwaite, S., Bujnicki, J. M., et al. (2008). Crystal structure of the Escherichia coli 23S rRNA:m(5)C Methyltransferase RlmI (YccW) reveals evolutionary links between RNA modification Enzymes. *Journal of Molecular Biology, 383*, 652–666.

Thomas, G., Gordon, J., & Rogg, H. (1978). N4-Acetylcytidine. A previously unidentified labile component of the small subunit of eukaryotic ribosomes. *The Journal of Biological Chemistry, 253*, 1101–1105.

Tovy, A., Tov, R. S., Gaentzsch, R., Helm, M., & Ankri, S. (2010). A new nuclear function of the Entamoeba histolytica glycolytic enzyme enolase: The metabolic regulation of cytosine-5 methyltransferase 2 (Dnmt2) activity. *PLoS Pathogens, 6*, e1000775.

Tscherne, J. S., Nurse, K., Popienick, P., Michel, H., Sochacki, M., & Ofengand, J. (1999). Purification, cloning, and characterization of the 16S RNA m 5C967 methyltransferase from Escherichia coli. *Biochemistry, 38*, 1884–1892.

Tuorto, F. F., Liebers, R. R., Musch, T. T., Schaefer, M. M., Hofmann, S. S., Kellner, S. S., et al. (2012). RNA cytosine methylation by Dnmt2 and NSun2 promotes tRNA stability and protein synthesis. *Nature Structural & Molecular Biology*, *19*, 900–905.

Verhagen, A. P. M. A., & Pruijn, G. J. M. G. (2011). Are the Ro RNP-associated Y RNAs concealing microRNAs? Y RNA-derived miRNAs may be involved in autoimmunity. *BioEssays: News and Reviews in Molecular, Cellular and Developmental Biology*, *33*, 674–682.

Wolf, B., Lesnaw, J. A., & Reichmann, M. E. (1970). A mechanism of the irreversible inactivation of bovine pancreatic ribonuclease by diethylpyrocarbonate. A general reaction of diethylpyrocarbonate. A general reaction of diethylpyrocarbonate with proteins. *European Journal of Biochemistry/FEBS*, *13*, 519–525.

Wu, P., Brockenbrough, J. S., Paddy, M. R., & Aris, J. P. (1998). NCL1, a novel gene for a non-essential nuclear protein in Saccharomyces cerevisiae. *Gene*, *220*, 109–117.

Zong, C., Lu, S., Chapman, A. R., & Xie, X. S. (2012). Genome-wide detection of single-nucleotide and copy-number variations of a single human cell. *Science (New York, N.Y.)*, *338*, 1622–1626.

CHAPTER FIFTEEN

Biochemical and Transcriptome-Wide Identification of A-to-I RNA Editing Sites by ICE-Seq

Shunpei Okada, Masayuki Sakurai, Hiroki Ueda, Tsutomu Suzuki[1]
Department of Chemistry and Biotechnology, Graduate School of Engineering, University of Tokyo, Bunkyo-ku, Tokyo, Japan
[1]Corresponding author: e-mail address: ts@chembio.t.u-tokyo.ac.jp

Contents

Abstract

Inosine (I) is a modified adenosine (A) in RNA. In Metazoa, I is generated by hydrolytic deamination of A, catalyzed by adenosine deaminase acting RNA (ADAR) in a process called A-to-I RNA editing. A-to-I RNA editing affects various biological processes by modulating gene expression. In addition, dysregulation of A-to-I RNA editing results

Methods in Enzymology, Volume 560
ISSN 0076-6879
http://dx.doi.org/10.1016/bs.mie.2015.03.014

in pathological consequences. I on RNA strands is converted to guanosine (G) during cDNA synthesis by reverse transcription. Thus, the conventional method used to identify A-to-I RNA editing sites compares cDNA sequences with their corresponding genomic sequences. Combined with deep sequencing, this method has been applied to transcriptome-wide screening of A-to-I RNA editing sites. This approach, however, produces a large number of false positives mainly owing to mapping errors. To address this issue, we developed a biochemical method called inosine chemical erasing (ICE) to reliably identify genuine A-to-I RNA editing sites. In addition, we applied the ICE method combined with RNA-seq, referred to as ICE-seq, to identify transcriptome-wide A-to-I RNA editing sites. In this chapter, we describe the detailed protocol for ICE-seq, which can be applied to various sources and taxa.

1. INTRODUCTION

RNA molecules contain various types of chemical modifications that are introduced posttranscriptionally. More than one hundred chemical modification types have been reported in various RNAs from all domains of life (Machnicka et al., 2013). RNA modifications confer chemical diversity resulting in a variety of biological functions. Inosine (I) is a modified adenosine (A) but has chemical similarity to guanosine (G). In fact, I is able to base pair with cytidine (C) instead of uridine (U). In RNA strands, I is produced by the hydrolytic deamination of A mediated by adenosine deaminase acting on RNA (ADAR) in a process called A-to-I RNA editing (Bass, 2002; Nishikura, 2006). In metazoans, I is frequently found in double-stranded RNA (dsRNA) (Bass, 2002; Nishikura, 2010). Although three ADAR family members (ADAR1–3) are present in mammalian genomes (Nishikura, 2010), only ADAR1 and ADAR2 exhibit enzymatic activity for A-to-I RNA editing (Chen et al., 2000; Melcher et al., 1996). Both *Adar1* and *Adar2* are essential genes in mice. *Adar1* null mice exhibit embryonic lethal phenotypes accompanied by widespread apoptosis, defective hematopoiesis, and an upregulated interferon response (Hartner, Walkley, Lu, & Orkin, 2009; Wang, Khillan, Gadue, & Nishikura, 2000; Wang et al., 2004). A recent study showed that embryonic lethality of *Adar1* null mice is rescued by the depletion of *Mavs*, which plays a central role in the recognition of RNA viruses in infected cells and the activation of interferon signaling, indicating that an aberrant innate immune system caused by the loss of ADAR1 enzymatic activity might severely affect mouse development (Mannion et al., 2014). *Adar2* null mice show postnatal seizures and lethality (Higuchi et al., 2000), but this phenotype is partially rescued by the introduction of an edited *Gria2* allele in which Gln at the Q/R editing site is mutated to Arg. This strongly suggests that the Q/R editing in *Gria2* mRNA is a functional target of ADAR2 in mice. An aberrant level of A-to-I RNA editing is associated with amyotrophic lateral sclerosis (Kawahara et al., 2004) and psychiatric disorders such as depression and bipolar disorder (Lerer et al., 2001; Massat et al., 2007). Moreover, more than a hundred pathogenic mutations in human ADAR1 have been reported in patients with dyschromatosis symmetrica hereditaria (Hayashi & Suzuki, 2013; Slotkin & Nishikura, 2013; Tojo et al., 2006) or Aicardi-Goutieres syndrome (Rice et al., 2012; Slotkin & Nishikura, 2013), indicating that the decreased level of A-to-I RNA editing mediated by ADAR1 might be involved in the pathogenesis of these diseases.

A-to-I RNA editing can cause amino acid alterations (Burns et al., 1997; Higuchi et al., 1993; Hoopengardner, Bhalla, Staber, & Reenan, 2003; Ohlson, Pedersen, Haussler, & Ohman, 2007; Sakurai et al., 2014), alternative splicing (Rueter, Dawson, & Emeson, 1999), nuclear retention (Chen, DeCerbo, & Carmichael, 2008), prevention of aberrant exonization (Sakurai, Yano, Kawabata, Ueda, & Suzuki, 2010), nonsense-mediated mRNA decay (Agranat, Raitskin, Sperling, & Sperling, 2008), regulation of circular RNA biogenesis (Ivanov et al., 2015), RNA interference (Bass, 2006), altered translation (Hundley, Krauchuk, & Bass, 2008), and miRNA-mediated translational regulation (Borchert et al., 2009; Wang et al., 2013). However, the functional roles of most A-to-I RNA editing sites are unknown. To determine the biochemical and physiological roles of A-to-I RNA editing and molecular pathogenesis of human diseases associated with the dysregulation of RNA editing, it is important to comprehensively characterize A-to-I RNA editing sites.

A conventional method used to identify I sites is to compare a cDNA sequence with its corresponding genomic sequence (Kawahara, 2012; Levanon et al., 2004; Paz et al., 2007). Since I base pairs with C, it is converted to G at the corresponding site in the cDNA strand by reverse transcription and subsequent PCR amplification. An A in the genomic sequence that is partially or completely converted to G at the corresponding cDNA position indicates a candidate A-to-I RNA editing site. Using this method, vast numbers of such RNA and DNA differences (RDD) in the human transcriptome have been screened using next-generation sequencing technologies (Bahn et al., 2012; Li et al., 2011; Peng et al., 2012). However, it is reported that these sites include many false positives (Kleinman & Majewski, 2012; Lin, Piskol, Tan, & Li, 2012; Pickrell, Gilad, & Pritchard, 2012;

Piskol, Peng, Wang, & Li, 2013). The high error rate using the RDD method is attributed to the difficulty in differentiating between a G derived from a base I and a G from an unannotated SNP, sequencing error, or inevitable mapping error.

To solve these problems in identifying A-to-I RNA editing sites, we developed a highly reliable biochemical method called inosine chemical erasing (ICE) (Sakurai & Suzuki, 2011; Sakurai et al., 2010). The ICE method is based on the cyanoethylation of I bases with acrylonitrile to form N^1-cyanoethylinosine (ce^1I). Because the N^1-cyanoethyl group prevents base-pairing with C, ce^1I inhibits first-strand cDNA extension, resulting in fewer G peaks in cDNA sequence chromatograms after PCR amplification. By contrast, no reduction in G peaks in cDNA is observed at false positive sites, i.e., G bases on the RNA strand that do not react with acrylonitrile. Thus, the ICE method can be used to differentiate between a G that originated from an I site and a G that arose from a SNP or methodological error. However, the ICE method can only be applied to I sites in specific regions of interest. This limitation produces a considerable bias in the detection of novel A-to-I RNA editing sites. To obtain an unbiased transcriptome-wide landscape of A-to-I RNA editing sites, we developed the ICE method using RNA-seq technology, and termed the method ICE-seq (Sakurai et al., 2014). ICE-seq detects true I sites based on the reduction of sequence reads containing G at the RNA position corresponding to an I site upon cyanoethylation. Using ICE-seq, we have successfully identified 19,791 novel editing sites in the human adult brain transcriptome (Sakurai et al., 2014). In this chapter, we describe the detailed ICE-seq protocol using poly(A)$^+$ RNA from the human adult brain, including the cyanoethylation of RNA, preparation of cDNA, and data processing using our original software ICEBreaker.

2. OUTLINE OF THE ICE-seq PROTOCOL

First, poly(A)$^+$ RNA or rRNA-depleted RNA is treated with acrylonitrile to cyanoethylate RNA I sites (CE+ condition) (Fig. 1A, right column in Fig. 1B). As a reference, untreated RNA is also prepared (CE− condition, left column in Fig. 1B). Then, both RNA samples are subjected to cDNA library preparation for RNA-seq. Since ce^1I prevents base-pairing with C and inhibits reverse transcription, the cDNAs containing G derived from I are not produced in the CE+ condition. After second-strand synthesis, both ends of the cDNA fragments are repaired and capped with adaptors.

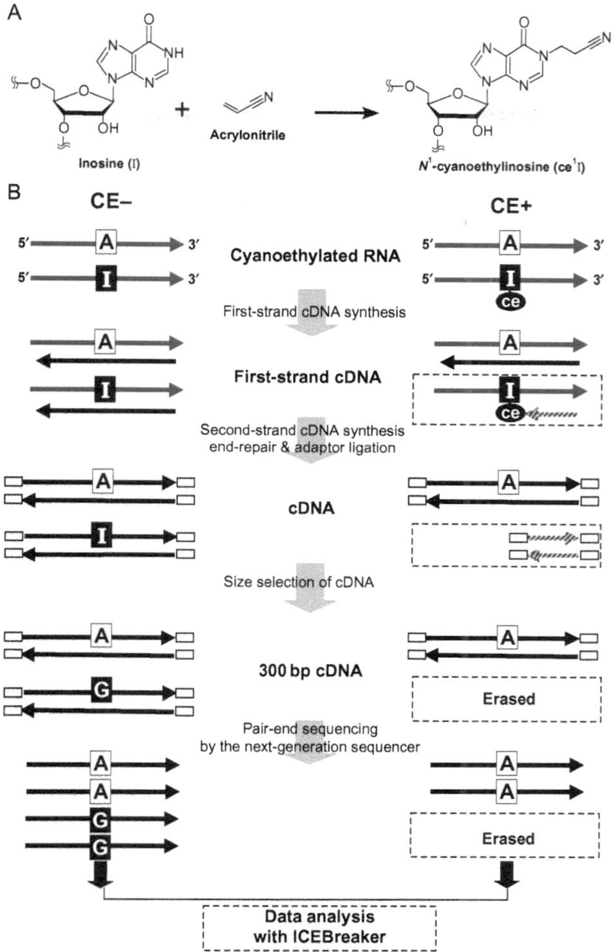

Figure 1 ICE chemistry and outline of the ICE-seq protocol. (A) Chemistry of inosine cyanoethylation. Inosine (I) on the RNA strand is cyanoethylated with acrylonitrile to form N^1-cyanoethylinosine (ce^{1}I). (B) Outline of ICE-seq. Schemes without (CE− condition) or with (CE+ condition) cyanoethylation of RNA are shown on the left and right, respectively. RNA and cDNA are indicated by gray and black arrows, respectively. The I in the RNA strand (black box) is specifically cyanoethylated to form ce^{1}I (CE+). In both conditions, RNA bearing an A at the editing site (white box) is converted to a T in the cDNA during first-strand synthesis. In the CE− condition, RNA bearing I is transcribed to C in the cDNA. In the CE+ condition, first-strand cDNA extension is arrested at the ce^{1}I site (striped arrow). Second strands of cDNA are synthesized to obtain double-stranded cDNA that is then subjected to the end-repair reaction and adaptor ligation. The size selection and subsequent amplification of cDNAs enrich 300 bp cDNAs. The cDNAs for the CE− and CE+ conditions are subjected to deep sequencing with the paired-end read option. After data analysis of sequence reads with our original integrated software, ICEBreaker, novel A-to-I RNA editing sites are output based on a reduction of G-containing reads upon cyanoethylation.

Then, cDNA fragments of approximately 300 base pairs (bp) are isolated, enriched by PCR amplification, and subjected to deep sequencing. The sequencing reads are mapped to the reference sequence for data analysis. The G bases originating from I bases found in the CE− condition are eliminated or decreased in the CE+ condition. If there is no difference in A-to-G ratio or read number between the CE− and CE+ conditions, the particular G base reflects a SNP, pseudogene, sequencing error, or mapping error.

3. INOSINE CYANOETHYLATION WITH ACRYLONITRILE

ICE-seq can be performed using RNA samples from any organism, various tissue types, and clinical specimens. The quality of the RNA sample is critical for ICE-seq analysis. Genomic DNA contamination affects the accuracy of A-to-I editing frequency estimates and results in poor sensitivity with respect to editing site detection. To eliminate genomic DNA contamination, RNA samples should be treated with RNase-free DNase prior to use.

The efficiency of cyanoethylation is sensitive to pH and buffer conditions; therefore, RNA cleanup kits or ethanol precipitation should be used prior to cyanoethylation. In the current ICE-seq protocol, we use poly(A)$^+$ RNA, but rRNA-depleted RNA is also suitable. Insufficient rRNA removal can affect the sequence read mapping efficiency. RNA quality should be assessed using Agilent 2100 Bioanalyzer to confirm the removal of rRNAs.

In our previous study (Sakurai et al., 2014), we cyanoethylated RNA samples in two different conditions (CE+, mild, and CE++, strong) for ICE-seq analysis. However, we recommend preparing duplicate or triplicate samples for a single cyanoethylation condition based on the high reproducibility of the reaction. When cyanoethylation is performed at 70 °C, most I bases in structured regions are exposed, resulting in efficient cyanoethylation.

3.1 Materials

- Poly(A)$^+$ RNA purified from human adult brain RNA
- Acrylonitrile (15.2 *M*; Tokyo Chemical Industry Co. Ltd., cat. no. A0146; Tokyo, Japan). Store at 4 °C in the dark. A fresh lot should be used for efficient cyanoethylation.
- Triethylamine (Nacalai Tesque, cat. no. 34804–85; Kyoto, Japan)

- Ethanol
- Acetic acid
- RNeasy MinElute Cleanup Kit (QIAGEN, cat. no. 74204)
- CE buffer: 50% ethanol and 1.1 *M* triethylammonium acetate. Mix 5 ml of 100% ethanol and 1.53 ml of 7.19 *M* trimethylamine, adjust the pH to 8.6 with acetic acid, and add RNase-free water to a volume of 10 ml. Divide small aliquots into tightly-sealed tubes and store at 4 °C. Note that the pH of this buffer may change. Check the pH of the buffer prior to use.
- PCR primers: hsCDK13 Q/R site 1st RT-PCR Fwd, 5′-GGCCTCC TCCTCTTGCTTC-3′; hsCDK13 Q/R site 1st RT-PCR Rev, 5′-GACTGGGAGCTCACATCCTC-3′; hsCDK13 Q/R site 2nd PCR Fwd, 5′-GCCTCCTCCTCTTGCTTCAG-3′; hsCDK13 Q/R site 2nd PCR Rev, 5′-GCTCACATCCTCGTATTCCAC-3′.

For a detailed list of additional reagents required for direct sequencing, see our previous papers Sakurai and Suzuki (2011) and Sakurai et al. (2010).

3.2 Procedures

Mix 4 μl of poly(A)$^+$ RNA (0.5 μg), 4 μl of 15.2 *M* acrylonitrile, and 30 μl of CE buffer in a 1.5 ml microcentrifuge tube (38 μl in total volume) to obtain a final concentration of 1.6 *M* acrylonitrile. Seal the tube tightly with parafilm and incubate it at 70 °C for 15 min (CE+) or 30 min (CE++). For the CE− condition, prepare the control RNA by replacing the acrylonitrile with RNase-free water, and incubate the solution at 70 °C for 15 min. After incubation, cool the tube immediately on ice and quench the reaction by adding 162 μl of RNase-free water. Purify the cyanoethylated RNA using the RNeasy MinElute Cleanup Kit according to the manufacturer's protocol.

Confirm the cyanoethylation efficiency by direct sequencing, as described previously (Sakurai & Suzuki, 2011; Sakurai et al., 2010). To verify the cyanoethylation efficiency, use the Q/R editing site of *CDK13* mRNA, which is ubiquitously expressed in various human tissues and culture cells (Marques et al., 2000). This site has a more than 90% editing frequency in human brain (Sakurai et al., 2014) and exhibits relatively high editing ratios (50–90%) in other tissues and culture cells (unpublished data). If the RNA sample is successfully cyanoethylated, a decrease in the G signal derived from I will be observed in the sequence chromatogram, as shown in Fig. 2.

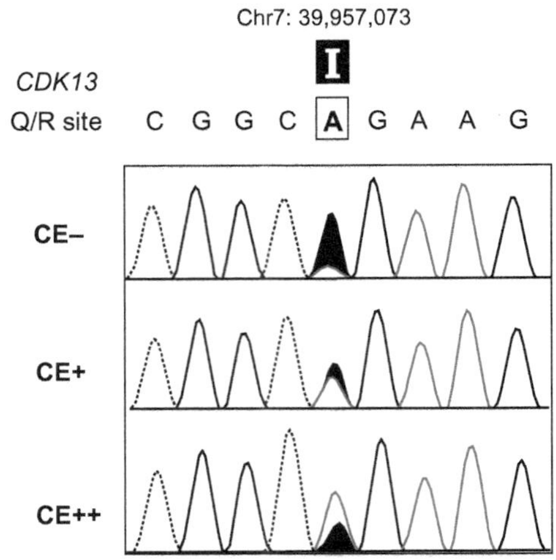

Figure 2 Evaluation of cyanoethylation using the ICE method. The sequence chromatogram of the Q/R site in *CDK13* mRNA obtained by direct sequencing. The top, middle, and bottom panels show the sequence chromatograms of cDNAs in the CE−, CE+, and CE++ conditions, respectively. The G signal derived from I at the Q/R site is decreased in response to cyanoethylation. Traces for A, G, and C are shown as a gray line, black line, and black dashed line, respectively. The G peaks at the Q/R site are filled with a dark color.

4. cDNA LIBRARY PREPARATION AND DEEP SEQUENCING

Here, we describe the cDNA preparation procedure using the mRNA-seq Sample Prep Kit (Illumina, San Diego, CA, USA). If an alternative TruSeq RNA Sample Preparation v2 Kit (Illumina, cat. no. RS-122-2001) is used, see Table 1 for a summary of procedural differences. Because A-to-I RNA editing occurs in double-stranded regions, reverse transcription should be performed at temperatures of 50 °C or higher using a thermostable reverse transcriptase such as SuperScript III.

In primates, the majority of A-to-I RNA editing sites reside in *Alu* repetitive elements, which are a type of short-intersperse repetitive element (SINE) (Levanon et al., 2004; Paz-Yaacov et al., 2010; Sakurai et al., 2014). To identify A-to-I RNA editing sites in *Alu* elements, accurate sequence read mapping is required. Based on the average length of *Alu* elements, prepare cDNA segments of approximately 300 bp to ensure accurate mapping of one of the cDNA ends to the regions adjacent to *Alu* elements. If TruSeq RNA Sample Preparation v2 Kit is used, cDNA fragments should be prepared around 400–450 bp considering that the total length of adaptor sequence ligated to both ends of cDNA is 122 bp.

Table 1 Procedures for the TruSeq RNA Sample Preparation v2 Kit Compared with the Original Protocol

Original Protocol	TruSeq RNA Sample Preparation v2 Kit
Fragmentation of RNA	Purify and Fragment mRNA: Incubate RFP
First-strand cDNA synthesis	Synthesize First-Strand cDNA: Make CDP (replace SuperScript II with SuperScript III), Incubate 1 CDP
Second-strand cDNA synthesis	Synthesize Second-Strand cDNA: Add SSM, Incubate 2 CDP, Purify CDP
End repair	Perform End Repair: Make IMP, Incubate IMP, Cleanup IMP
3′ end adenylation	Adenylate 3′ Ends: Add ATL, Incubate 1 ALP,
Adaptor ligation	Ligate Adapters: Add LIG, Incubate 2 ALP, Add STL (Replace "Cleanup ALP" with purification by MinElute Cleanup Kit)
First agarose gel purification	(Perform the same procedure described in our original protocol, but change the size of excised cDNA fragment from 300 bp to 500–550 bp. The position corresponding to the target cDNA size plus 100 bp should be excised in this step due to the Y-shape conformation of illumina's adaptor oligo)
PCR amplification	Enrich DNA Fragments: Make PCR, Amp PCR (Replace "Cleanup ALP" with purification by MinElute Cleanup Kit)
Second agarose gel purification	(Perform the same procedure described in our original protocol, but change the size of excised cDNA fragment from 300 to 400–450 bp)

In our study of the human transcriptome, we found that 250–300 million sequence reads mapped to the reference are sufficient to detect I sites with high accuracy (Sakurai et al., 2014)[1]. Read number is positively correlated with the detection of sites with a low-frequency A-to-I RNA editing.

[1] For application to other organisms, the RNA sequencing depth should be determined based on the general criteria for RNA-seq for each organism. The number of reads should be determined by referring to the transcriptome size and examining whether known editing sites are detected in pilot sequencing of cDNA in the CE− condition.

4.1 Materials

4.1.1 Reagents

- mRNA-seq Sample Prep Kit (Illumina, cat. no. 1004814). This kit includes fragmentation buffer, fragmentation stop solution, random primers, 5 × first-strand buffer, 100 m*M* DTT, 25 m*M* dNTPs, 40 U/μl RNaseOUT, 10 × second-strand buffer, 2 U/μl RNase H, 10 U/μl DNA polymerase I, 10 × end-repair buffer, 3 U/μl T4 DNA polymerase, 5 U/μl Klenow DNA polymerase, A-tailing buffer, 1 m*M* dNTPs, 5 U/μl Klenow fragment (3′ to 5′ exo-), DNA ligase buffer, PE adaptor oligo mix, 600 U/μl DNA ligase, 5 × cloned Phu Buffer, PCR primer 1.0, PCR primer 2.0, and Phu polymerase.
- SuperScript III RT (200 U/μl; Life Technologies, cat. no. 18080–044; Carlsbad, CA, USA)
- OptiKinase (10 U/μl; USB, cat. no. 78334Y; Cleveland, OH, USA)
- NuSieve GTG agarose (Lonza, cat. no. 50080)
- 1 × TAE
- SYBR Gold Nucleic Acid Gel Stain (Life Technologies, cat. no. S-11494)
- β-agarase (1 U/μl; Lonza, cat. no. 58001)
- 10 *M* ammonium acetate or 3 *M* sodium acetate
- Ethanol
- Glycogen (20 mg/ml; Roche, cat. no. 901393; Basel, Switzerland)
- QIAquick PCR Purification Kit (QIAGEN, cat. no. 28104)
- MinElute PCR Purification Kit (QIAGEN, cat. no. 28001)
- EB buffer. 1 × EB buffer is supplied in the QIAquick and MinElute PCR Purification Kits. Dilute EB buffer with an equal volume of RNase-free water to yield 0.5 × EB buffer.
- Agilent DNA 1000 Kit (Agilent Technologies, cat. no. 5067–1504; Santa Clara, CA, USA)

4.1.2 Equipment

- Agilent 2100 Bioanalyzer (Agilent Technologies, cat. no. G2939A)
- Genome Analyzer II (Illumina)

4.2 RNA Fragmentation

Perform fragmentation of 100 ng poly(A)$^+$ RNA treated with or without cyanoethylation in 20 μl of 1 × fragmentation buffer at 95 °C for 5 min[2].

[2] RNA quality can be reduced to 10 ng when using the TruSeq RNA Sample Preparation v2 Kit.

Add 2 μl of the fragmentation stop solution to the mixture. Precipitate the RNA with ethanol and dissolve the RNA pellets in 11.1 μl of RNase-free water.

4.3 First- and Second-Strand cDNA Synthesis

Mix each RNA solution with 1 μl (3 μg) of random primer. Incubate the solution at 65 °C for 5 min and chill it on ice. Add 4 μl of 5 × first-strand buffer, 2 μl of 100 m*M* DTT, 0.4 μl of 25 m*M* dNTPs, and 0.5 μl of 40 U/μl RNaseOUT. Mix the solution well by pipetting and incubate it at 25 °C for 2 min. Add 1 μl of 200 U/μl SuperScript III to the mixture (20 μl in total volume) and incubate it at 25 °C for 10 min, 50 °C for 50 min, and 70 °C for 15 min. Next, add 10 μl of 10 × second-strand buffer, 1.2 μl of 25 m*M* dNTPs, and 62.8 μl of RNase-free water to the mixture, mix it well by pipetting and leave it on ice for 5 min. Add 1 μl of 2 U/μl RNase H and 5 μl of 10 U/μl DNA polymerase I to the mixture (100 μl in total volume). Mix the solution well by pipetting and incubate the solution at 16 °C for 2.5 h. Purify the cDNA using the QIAquick PCR Purification Kit and elute the cDNA with 50 μl of 1 × EB buffer.

4.4 End Repairing, 3′ End Adenylation, and Adaptor Ligation

For end repairing, mix 50 μl of the cDNA with 10 μl of 10 × end-repair buffer, 1.6 μl of 25 m*M* dNTP, 5 μl of 3 U/μl T4 DNA polymerase, 1 μl of 5 U/μl Klenow DNA polymerase, 5 μl of 10 U/μl OptiKinase, and 27.4 μl of RNase-free water (100 μl in total volume). Mix the solution well by pipetting and incubate it at 20 °C for 30 min. Purify the cDNA using the QIAquick PCR Purification Kit and elute it with 50 μl of 0.5 × EB buffer. Dry the eluate with a vacuum centrifuge evaporator and dissolve the end-repaired cDNA with 32 μl of RNase-free water.

For adenylation, mix 32 μl of the end-repaired cDNA with 5 μl of A-tailing buffer, 10 μl of 1 m*M* dNTP, and 3 μl of 5 U/μl Klenow fragment (3′ to 5′ exo-) (50 μl in total volume). Mix the solution well by pipetting and incubate it at 37 °C for 30 min, purify the cDNA using the QIAquick PCR Purification Kit, and elute it with 50 μl of 0.5 × EB buffer. Dry the eluate with a vacuum centrifuge evaporator, and dissolve the A-tailed cDNA with 25 μl of RNase-free water.

For adaptor ligation, mix 25 μl of the A-tailed cDNA with 25 μl of DNA ligase buffer, 1 μl of PE adaptor oligo mix, and 1 μl of 600 U/μl DNA ligase (52 μl in total volume). Mix the solution well by pipetting and incubate it at

25 °C for 15 min. Purify the cDNA using the MinElute PCR Purification Kit, and elute it with 10 μl of 0.5 × EB buffer. Dry the eluate with a vacuum centrifuge evaporator and dissolve the adaptor-ligated cDNA with 5 μl of RNase-free water.

4.5 First Agarose Gel Purification

Mix 5 μl of adaptor-ligated cDNA with 1 μl of 6 × loading dye and run the sample on a 3.5% NuSieve GTG Agarose Gel (10 cm × 5–10 cm × 1 cm) with 1 × TAE running buffer at 100 V and 4 °C for approximately 100 min until the Bromophenol blue (BPB) reaches a distance of approximately 1/5 of the gel length from the bottom. Stain the gel with SYBR Gold in 1 × TAE buffer, and excise the region corresponding to approximately 300 bp of cDNA (500–550 bp, if TruSeq RNA Sample Preparation v2 Kit is used. The cDNAs ligated with adaptor oligos used in the TruSeq kit retarded its migration in agarose gel electrophoresis with about 100 bp than the expected size due to the unique conformation of the adator oligos.). Weigh the gel piece, estimate its volume, and melt it at 70 °C for 10 min. Add 2 μl of the 50 × β-agarase buffer (supplied with β-agarase) per 100 μl of the melted gel solution. Cool the solution to 45 °C. Add 4 μl of 1 U/μl β-agarase per 100 μl of the melted gel. Incubate the solution at 45 °C for 60 min. Chill the solution at −20 °C for 15 min and centrifuge it at 16,000 × *g* for 15 min at 4 °C. Transfer the supernatants to a new tube carefully without carrying over undigested agarose gels[3]. Add ammonium acetate (or sodium acetate) to the supernatant to a final concentration of 2.5 *M* (or 0.3 *M* for sodium acetate). Mix the supernatant with 1 μl of 20 mg/ml glycogen and a 2.8-fold volume of 100% ethanol, and then precipitate the 300 bp cDNA with ethanol. Dissolve the cDNA with 50 μl of RNase-free water, purify it using the MinElute PCR Purification Kit, and elute it with 10 μl of 1 × EB buffer.

4.6 PCR Amplification

Add 27 μl of RNase-free water to 10 μl of the 300 bp cDNA, and mix it with 10 μl of 5 × cloned Phu Buffer, 1 μl of PCR primer 1.0, 1 μl of PCR primer 2.0, 0.5 μl of 25 m*M* dNTPs, and 0.5 μl of Phu polymerase (50 μl in total volume), and mix the solution well by pipetting. Perform PCR amplification using the following program: 1 cycle at 98 °C for 30 s, 20 cycles at 98 °C for 10 s, 65 °C for 30 s, and 72 °C for 30 s, and 1 cycle at 72 °C for 5 min. Purify the amplified cDNA using the MinElute PCR Purification Kit and

[3] Agarose gel digestion with β-agarase is replaced with other protocols like the QIAquick Gel Extraction kit protocol (QIAGEN).

elute it with 10 μl of 0.5 × EB buffer. Dry the eluate with a vacuum centrifuge evaporator and dissolve it with 5 μl of RNase-free water.

4.7 Second Agarose Gel Purification

Repeat the procedure for the isolation of 300 bp cDNA (400–450 bp in the case of TruSeq RNA Sample v2 Kit) as described in Section 4.5. Check the quality of the purified cDNA using an Agilent 2100 Bioanalyzer and the Agilent DNA 1000 Kit.

4.8 Deep Sequencing

Perform deep sequencing with cDNAs prepared for the three conditions (CE −, CE +, and CE++) using the Genome Analyzer II according to the manufacturer's protocol applying the paired-end read option.

5. DATA ANALYSIS

After deep sequencing, map the sequence reads to both the genome reference and transcriptome reference with BWA (Li & Durbin, 2009). We do not recommend splice-aware type mapping software like TopHat (Trapnell, Pachter, & Salzberg, 2009) or SORPsplice (Huang et al., 2011)[4]. After merging the mapping results for both references, perform a local alignment of unmapped read pairs to the A/G-masked reference. Initial A/G mismatch sites can be extracted from the mapping results. Then, real A-to-I RNA editing sites can be discriminated from false positives resulting from SNPs, pseudogenes, mapping errors, etc.

To automate these data processes, we developed an integrated software package called ICEBreaker (http://sourceforge.net/projects/fastpassngs), which maps sequence reads to the reference, takes a mapping file in BAM format as an input, piles up sequence reads at A-to-G mismatch positions, and compares normalized G base counts, and read depths between CE conditions. It calculates the ICE score based on normalized decreases in the frequency of G base counts upon cyanoethylation. Moreover, it filters possible artificial false positives around error-prone regions containing indels, non-A-to-G mismatches, and strand directional bias of mapped reads, and removes candidate sites matched with SNPs reported in the Single Nucleotide Polymorphism

[4] Splice-aware mapping software like TopHat (Trapnell et al., 2009) or SORPsplice (Huang et al., 2011) have mapping errors, especially frequent misalignment against pseudogenes (Garber, Grabherr, Guttman, & Trapnell, 2011). These problems are caused by their mapping approach, which splits sequence reads and maps each part of split reads to a reference independently (Garber et al., 2011).

Database (dbSNP) (Sherry et al., 2001). These processes result in a list of candidate sites and associated ICE scores generated by ICEBreaker.

5.1 Software

- ICEBreaker (http://sourceforge.net/projects/fastpassngs)
- SAMtools program (http://samtools.sourceforge.net/) for manipulating BAM-formatted files (Li et al., 2009)
- JAVA runtime environment (http://java.com/ja/download/)
- BWA (http://bio-bwa.sourceforge.net/) (Li & Durbin, 2009)
- ANNOVAR (http://www.openbioinformatics.org/annovar/) (Wang, Li, & Hakonarson, 2010)

5.2 Preparation of the Genome and Transcriptome References

For the human genome reference, download the "hg18.2bit" file from http://hgdownload.cse.ucsc.edu/goldenPath/hg18/bigZips/. If using a genomic reference for another taxon, download it via http://hgdownload.cse.ucsc.edu/downloads.html. Create the BWA index with the genomic reference according to the BWA manual. Prepare an hg18.size file that contains chromosome names and sizes. For the transcriptome reference, prepare a gene coordinate-sorted file that is a tab-delimited text composed of the transcript ID, chromosome, strand, exonStarts, and exonEnds. We implemented this protocol with an hg18 knownGene track retrieved from the UCSC Genome Bioinformatics site. It can be downloaded from the UCSC table browser (http://genome.ucsc.edu/cgi-bin/hgTables?command=start). For example, Edit the file name to "knowngenes.coord." The latest version of genomic and transcriptome sequence files should be used as a reference in your study.

5.3 Read Mapping

Trim the read length if the averaged quality value (QV) at the end of each read is less than 10. Use BWA to map reads for each CE condition to the transcriptome reference and the genomic reference, allowing five and four mismatches, respectively. For example, output files are named "genomemap.bam" (genomic reference) and "cDNAmap.bam" (transcriptome reference). Do not use the Smith–Waterman option in this step. Convert the coordinates of the mapping results against the transcriptome reference to the genomic coordinates using the following ICEBreaker command:

```
./icebreaker1.2.jar changeCoord -in /path/to/cDNAmap.bam -out
/path/to/cDNAmapGenomeCoord.bam -coordinate /path/to/knowngenes.coord
-sizeRef /path/to/hg18.size
```

Be sure to use the gene coordinated-sorted file, when generating the transcriptome reference. Merge the transcriptome-mapped BAM file and the genome-mapped BAM file. If a sequence read is mapped to different locations in the two references, choose the mapped site bearing fewer mismatches using the following ICEBreaker command:

```
./icebreaker1.2.jar takeOptimal -inSam1 /path/to/genomemap.bam -inSam1 /path/to/cDNAmapGenomeCoord.bam -out /path/to/optimal/optimal.bam
```

Perform a local alignment of unmapped read pairs with the A/G-masked reference using the Smith–Waterman algorithm (Smith & Waterman, 1981) by executing the following command in ICEBreaker:

```
./icebreaker1.2.jar unmapResque -in /path/to/optimal.bam -out /path/to/optimalremap.bam -ref /path/to/hg18.2bit
```

Sort and index the BAM files using SAMtools (Li et al., 2009). Prepare BAM files for each CE condition (CE−, CE+, and CE++).

5.4 Data Process by ICEBreaker

ICEBreaker processes BAM files and extracts A-to-I RNA editing candidates as follows:

(a) Binomial test for base call quality

The base call QV for each editing candidate is averaged and converted to an error probability (p) for each site according to the equation $QV = -10\log_{10}(p)$. The binomial (random chance) probability ($P1$) of k mismatches at each site where n total reads are mapped is calculated according to the equation $P1 = p^k \cdot (1-p)^k$. To simplify the calculation, the binomial probability is approximated by the normal distribution (Z value) as follows:

$$Z = \frac{k-\mu}{\sqrt{\sigma^2}}$$

The mean and standard deviation are described as $\mu = np$ and $\sigma^2 = np(1-p)$, respectively.

Candidates with Z values less than 10 are filtered out.

(b) Filtering process based on read depth, supporting read count, and base call quality

For candidate sites, the editing frequency should be higher than 10%, the read depth should be greater than 15, and the minimal number of G reads should be 6 in the CE− condition. Read depths of greater than 5 and 3 are required in the CE+ and CE++ conditions, respectively. Any base bearing a Phred base call QV of less than 10 is not counted when piling up bases.

Initial A/G mismatches extracted immediately after the mapping process include a large number of false positives arising from low sequence depth and/or low editing frequency. To minimize these artifacts, the thresholds for sequence depth, G read count, and editing frequency should be independently optimized for each analysis.

(c) ICE score calculation

The ICE score measures the decrease in the frequency of G bases upon cyanoethylation. First, calculate ΔNg, which is defined as the logarithmic value that represents the decreased ratio of Ng to total sites in the CE+ or CE++ condition compared to the CE− condition, as follows:

$$\Delta Ng(\mathrm{CE}+) = \log_2\left(\frac{Ng(\mathrm{CE}+)+1}{Ng(\mathrm{CE}-)+1}\cdot R_{\mathrm{adj}}(\mathrm{CE}+)\right)$$

$$\Delta Ng(\mathrm{CE}++) = \log_2\left(\frac{Ng(\mathrm{CE}++)+1}{Ng(\mathrm{CE}-)+1}\cdot R_{\mathrm{adj}}(\mathrm{CE}++)\right)$$

where the R_{adj} value is a coefficient that normalizes mapped reads for different CE conditions, calculated as $R_{\mathrm{adj}}(\mathrm{CE}+) = Nb(\mathrm{CE}-)/Nb(\mathrm{CE}+)$ and $R_{\mathrm{adj}}(\mathrm{CE}++) = Nb(\mathrm{CE}-)/Nb(\mathrm{CE}++)$. Nb represents the total number of mapped bases.

The ICE score is calculated as follows:

$$\mathrm{ICE\ score} = \sqrt{\Delta Ng(\mathrm{CE}+)^2 + \Delta Ng(\mathrm{CE}++)^2}$$

For duplicate or triplicate samples in a single CE condition, ΔNg values of all A-to-G mismatch sites for each experiment are plotted in two-dimensional or three-dimensional graphs, respectively. Then, calculate the distance of each point from the origin to determine the ICE score.

(d) Final filtering process using the quality of supporting reads and dbSNP

Supporting reads that contain a G at candidate sites are examined using a heuristic approach[5]. Supporting reads are removed if they meet at least one of following four criteria:

[5] The method for removing artifacts using a heuristic algorithm has been studied to detect somatic single nucleotide variations (SNV), especially in cancer genome sequencing (Cibulskis et al., 2013; Koboldt et al., 2012).

1. They contain an indel call within a distance of 5 nt from the candidate position;
2. They mapped only in a single direction (+ or – strand);
3. They contain mismatches other than an A-to-G substitution within a distance of 75 nt from the candidate position; or
4. They have non-A-to-G base mismatches at the candidate position with a mismatch ratio of greater than 0.1.

The distance from the candidate position in the third criteria should correspond to the read length analyzed by the next-generation sequencer used in each experiment.

In addition, candidate sites that overlap with dbSNP sites are removed.

5.5 Detection of A-to-I RNA Editing Sites with ICEBreaker

Execute the following command in ICEBreaker:

```
./icebreaker1.2.jar analysis -icem /path/to/icem.bam -icep /path/to/icep.bam -icepp /path/to/icepp.bam
-out /path/to/out/dir -ref /path/to/twobit/ref -dbSNP /path/to/dbSNP.txt -knownEditing /path/to/knownEditing
```

The candidate list of A-to-I RNA editing sites including ICE scores and neighboring sequences in FASTQ format is output by ICEBreaker based on the process described in Section 5.4.

Mapped reads are displayed in the regions bearing known editing sites in the *CDK13* CDS and *PAICS* 3′ UTR (Fig. 3). Around the Q/R editing site of *CDK13*, we observed a reduction in read depth. We also detected a specific decrease in the G base count relative to the A base count at the site depending on the strength of cyanoethylation (Fig. 3A). We observed similar changes in the A/G ratio at I sites in the sense and antisense strands of *Alu* elements in the *PAICS* 3′ UTR (Fig. 3B), indicating a high reliability of the method for detecting A-to-I editing clusters in *Alu* elements.

The candidate sites for A-to-I RNA editing detected in our previous study (Sakurai et al., 2014) had ICE scores ranging from 0 to 6 (black line), as shown in the histogram in Fig. 4. By contrast, A-to-G SNP sites had low ICE scores (gray line) since cyanoethylation does not affect the *Ng* of SNP sites. Only 241 SNP sites exceeded the ICE scores threshold of 1.75 among the 25,285 sites detected by ICE-seq. Thus, the *P*-value of SNP sites is 0.01, indicating that this threshold can be applied to discriminate authentic A-to-I RNA editing sites from false positives. By filtering

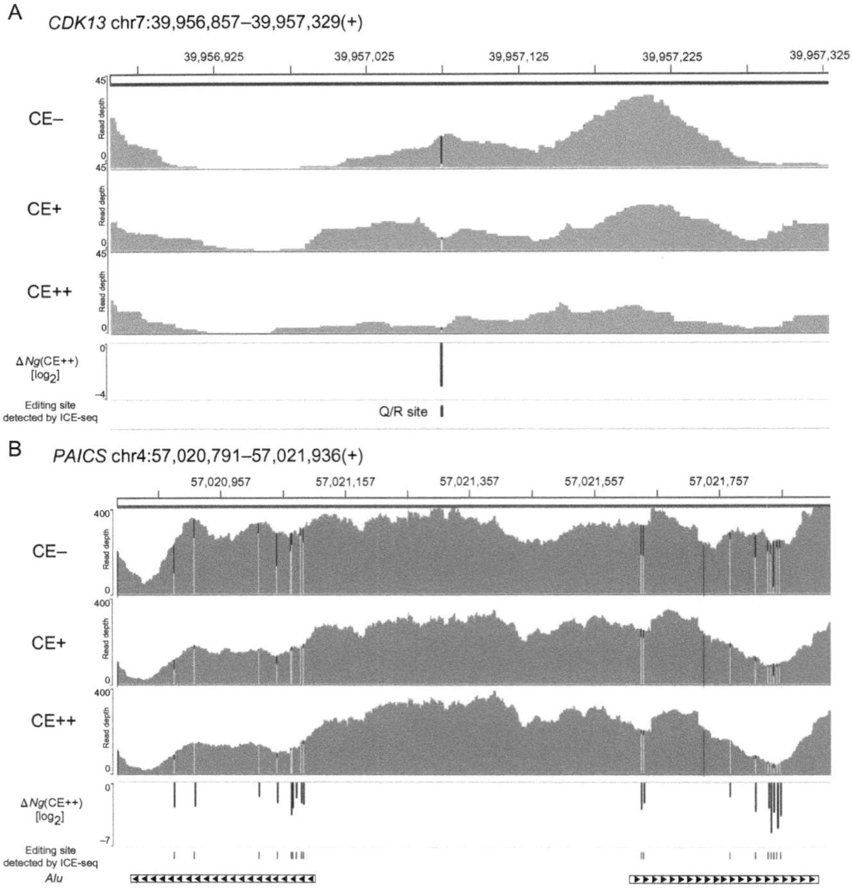

Figure 3 Schematic views of the mapped regions containing known editing sites. (A) The first exon of *CDK13* containing the Q/R site. (B) 3′ UTR of *PAICS* containing editing clusters in *Alu* elements. In each gene, the scale bar at the top of a diagram indicates chromosomal position, and the following three panels show histograms of mapped reads in the CE−, CE+, and CE++ conditions in which A or G read counts at editing positions are indicated as white or black bar graphs, respectively. The fourth and fifth panels show a bar graph of ΔNg (CE++) and the positions of editing sites detected by ICE-seq. The regions containing *Alu* elements are displayed in the *PAICS* 3′ UTR.

the candidate sites with ICE scores of greater than 1.75, the final list of A-to-I RNA editing sites is obtained. In addition, genes bearing A-to-I RNA editing sites can be annotated with a program such as ANNOVAR (Wang et al., 2010).

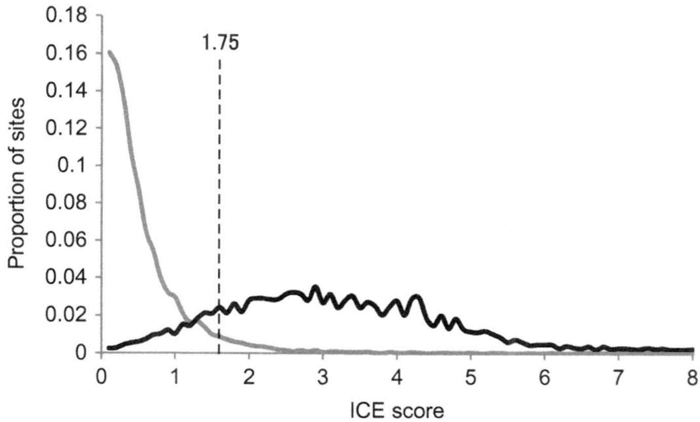

Figure 4 ICE score filtering. Histograms of the ICE scores for annotated A-to-G SNP sites (gray) and the candidates and known editing sites identified by ICE-seq (black) (Sakurai et al., 2014). An ICE score of 1.75 is indicated by a dashed line.

6. EVALUATION OF ICE-Seq

In our previous study (Sakurai et al., 2014), we detected 14,225 RNA editing sites in non-CDS mRNA regions. To estimate the accuracy of A-to-I RNA editing site detection by ICE-seq, we validated 931 sites using the ICE method and confirmed 902 as authentic A-to-I RNA editing sites, demonstrating a high accuracy rate (96.9%) (Sakurai et al., 2014).

The editing rate at each site can be estimated by its *Gr* value (G reads/ [A reads + G reads]), and there is a strong correlation between *Gr* and the G peak ratio in the Sanger sequencing chromatogram ($R^2 = 0.77$), indicating that ICE-seq can be used to calculate the editing frequency at each site with high accuracy (Sakurai et al., 2014).

A single A-to-I RNA editing site tends to have a low ICE score compared to multiple editing sites because the ΔNg value at each site is affected by the pausing efficiency of reverse transcription at cyanoethylated I sites. In general, most A-to-I RNA editing sites in CDS regions are single sites. Therefore, careful analysis is required to detect novel editing sites in CDS regions. For this purpose, we recommend the identification of candidate sites in CDS regions by setting lower ICE score thresholds (e.g., 1.45 in our previous study) (Sakurai et al., 2014). However, lower ICE score thresholds decrease overall accuracy (39.6% in our analysis); therefore, validation by ICE method for all candidates in CDS regions is required (Sakurai et al., 2014).

Since ICE-seq is used to identify A-to-I RNA editing sites based on decreased G read counts upon cyanoethylation, it is difficult to discriminate sites with a low frequency of editing from false positives. To ensure the accuracy of detection, we set the threshold for editing frequency to 10% in this protocol. However, the detection sensitivity for sites with low editing frequency (<10%) will be improved by increasing the read depth and the quality of deep sequencing (Bazak et al., 2014). Alternatively, the addition of a group of synthetic spike-ins with variable frequencies of I sites can also be used to effectively generate an editing frequency calibration curve and to normalize RNA-seq data across various CE conditions (Loven et al., 2012; Schwartz et al., 2014). This may facilitate the detection of sites with a low frequency of A-to-I editing.

ACKNOWLEDGMENTS

We thank members of the Suzuki laboratory for experimental assistance and fruitful discussion regarding this study. This work was supported by Grants-in-Aid for Scientific Research on Priority Areas from the Ministry of Education, Science, Sports, and Culture of Japan (to T.S.), and by a grant from the New Energy and Industrial Technology Development Organization (NEDO) (to T.S.).

REFERENCES

Agranat, L., Raitskin, O., Sperling, J., & Sperling, R. (2008). The editing enzyme ADAR1 and the mRNA surveillance protein hUpf1 interact in the cell nucleus. *Proceedings of the National Academy of Sciences of the United States of America, 105*(13), 5028–5033.

Bahn, J. H., Lee, J. H., Li, G., Greer, C., Peng, G., & Xiao, X. (2012). Accurate identification of A-to-I RNA editing in human by transcriptome sequencing. *Genome Research, 22*(1), 142–150.

Bass, B. L. (2002). RNA editing by adenosine deaminases that act on RNA. *Annual Review of Biochemistry, 71*, 817–846.

Bass, B. L. (2006). How does RNA editing affect dsRNA-mediated gene silencing? *Cold Spring Harbor Symposia on Quantitative Biology, 71*, 285–292.

Bazak, L., Haviv, A., Barak, M., Jacob-Hirsch, J., Deng, P., Zhang, R., et al. (2014). A-to-I RNA editing occurs at over a hundred million genomic sites, located in a majority of human genes. *Genome Research, 24*(3), 365–376.

Borchert, G. M., Gilmore, B. L., Spengler, R. M., Xing, Y., Lanier, W., Bhattacharya, D., et al. (2009). Adenosine deamination in human transcripts generates novel microRNA binding sites. *Human Molecular Genetics, 18*(24), 4801–4807.

Burns, C. M., Chu, H., Rueter, S. M., Hutchinson, L. K., Canton, H., Sanders-Bush, E., et al. (1997). Regulation of serotonin-2C receptor G-protein coupling by RNA editing. *Nature, 387*(6630), 303–308.

Chen, C. X., Cho, D. S., Wang, Q., Lai, F., Carter, K. C., & Nishikura, K. (2000). A third member of the RNA-specific adenosine deaminase gene family, ADAR3, contains both single- and double-stranded RNA binding domains. *RNA, 6*(5), 755–767.

Chen, L. L., DeCerbo, J. N., & Carmichael, G. G. (2008). Alu element-mediated gene silencing. *EMBO Journal, 27*(12), 1694–1705.
Cibulskis, K., Lawrence, M. S., Carter, S. L., Sivachenko, A., Jaffe, D., Sougnez, C., et al. (2013). Sensitive detection of somatic point mutations in impure and heterogeneous cancer samples. *Nature Biotechnology, 31*(3), 213–219.
Garber, M., Grabherr, M. G., Guttman, M., & Trapnell, C. (2011). Computational methods for transcriptome annotation and quantification using RNA-seq. *Nature Methods, 8*(6), 469–477.
Hartner, J. C., Walkley, C. R., Lu, J., & Orkin, S. H. (2009). ADAR1 is essential for the maintenance of hematopoiesis and suppression of interferon signaling. *Nature Immunology, 10*(1), 109–115.
Hayashi, M., & Suzuki, T. (2013). Dyschromatosis symmetrica hereditaria. *Journal of Dermatology, 40*(5), 336–343.
Higuchi, M., Maas, S., Single, F. N., Hartner, J., Rozov, A., Burnashev, N., et al. (2000). Point mutation in an AMPA receptor gene rescues lethality in mice deficient in the RNA-editing enzyme ADAR2. *Nature, 406*(6791), 78–81.
Higuchi, M., Single, F. N., Kohler, M., Sommer, B., Sprengel, R., & Seeburg, P. H. (1993). RNA editing of AMPA receptor subunit GluR-B: A base-paired intron-exon structure determines position and efficiency. *Cell, 75*(7), 1361–1370.
Hoopengardner, B., Bhalla, T., Staber, C., & Reenan, R. (2003). Nervous system targets of RNA editing identified by comparative genomics. *Science, 301*(5634), 832–836.
Huang, S., Zhang, J., Li, R., Zhang, W., He, Z., Lam, T. W., et al. (2011). SOAPsplice: Genome-wide ab initio detection of splice junctions from RNA-seq data. *Frontiers in Genetics, 2*, 46.
Hundley, H. A., Krauchuk, A. A., & Bass, B. L. (2008). C. elegans and H. sapiens mRNAs with edited 3' UTRs are present on polysomes. *RNA, 14*(10), 2050–2060.
Ivanov, A., Memczak, S., Wyler, E., Torti, F., Porath, H. T., Orejuela, M. R., et al. (2015). Analysis of intron sequences reveals hallmarks of circular RNA biogenesis in animals. *Cell Reports, 10*(2), 170–177.
Kawahara, Y. (2012). Quantification of adenosine-to-inosine editing of microRNAs using a conventional method. *Nature Protocols*, 7(7), 1426–1437.
Kawahara, Y., Ito, K., Sun, H., Aizawa, H., Kanazawa, I., & Kwak, S. (2004). Glutamate receptors: RNA editing and death of motor neurons. *Nature, 427*(6977), 801.
Kleinman, C. L., & Majewski, J. (2012). Comment on "Widespread RNA and DNA sequence differences in the human transcriptome" *Science, 335*(6074), 1302, author reply 1302.
Koboldt, D. C., Zhang, Q., Larson, D. E., Shen, D., McLellan, M. D., Lin, L., et al. (2012). VarScan 2: Somatic mutation and copy number alteration discovery in cancer by exome sequencing. *Genome Research, 22*(3), 568–576.
Lerer, B., Macciardi, F., Segman, R. H., Adolfsson, R., Blackwood, D., Blairy, S., et al. (2001). Variability of 5-HT2C receptor cys23ser polymorphism among European populations and vulnerability to affective disorder. *Molecular Psychiatry, 6*(5), 579–585.
Levanon, E. Y., Eisenberg, E., Yelin, R., Nemzer, S., Hallegger, M., Shemesh, R., et al. (2004). Systematic identification of abundant A-to-I editing sites in the human transcriptome. *Nature Biotechnology, 22*(8), 1001–1005.
Li, H., & Durbin, R. (2009). Fast and accurate short read alignment with Burrows-Wheeler transform. *Bioinformatics, 25*(14), 1754–1760.
Li, H., Handsaker, B., Wysoker, A., Fennell, T., Ruan, J., Homer, N., et al. (2009). The sequence alignment/map format and SAMtools. *Bioinformatics, 25*(16), 2078–2079.
Li, M., Wang, I. X., Li, Y., Bruzel, A., Richards, A. L., Toung, J. M., et al. (2011). Widespread RNA and DNA sequence differences in the human transcriptome. *Science, 333*(6038), 53–58.

Lin, W., Piskol, R., Tan, M. H., & Li, J. B. (2012). Comment on "Widespread RNA and DNA sequence differences in the human transcriptome" *Science*, *335*(6074), 1302, author reply 1302.

Loven, J., Orlando, D. A., Sigova, A. A., Lin, C. Y., Rahl, P. B., Burge, C. B., et al. (2012). Revisiting global gene expression analysis. *Cell*, *151*(3), 476–482.

Machnicka, M. A., Milanowska, K., Oglou, O. O., Purta, E., Kurkowska, M., Olchowik, A., et al. (2013). MODOMICS: A database of RNA modification pathways-2013 update. *Nucleic Acids Research*, *41*(D1), D262–D267.

Mannion, N. M., Greenwood, S. M., Young, R., Cox, S., Brindle, J., Read, D., et al. (2014). The RNA-editing enzyme ADAR1 controls innate immune responses to RNA. *Cell Reports*, *9*(4), 1482–1494.

Marques, F., Moreau, J. L., Peaucellier, G., Lozano, J. C., Schatt, P., Picard, A., et al. (2000). A new subfamily of high molecular mass CDC2-related kinases with PITAI/VRE motifs. *Biochemical and Biophysical Research Communications*, *279*(3), 832–837.

Massat, I., Lerer, B., Souery, D., Blackwood, D., Muir, W., Kaneva, R., et al. (2007). HTR2C (cys23ser) polymorphism influences early onset in bipolar patients in a large European multicenter association study. *Molecular Psychiatry*, *12*(9), 797–798.

Melcher, T., Maas, S., Herb, A., Sprengel, R., Seeburg, P. H., & Higuchi, M. (1996). A mammalian RNA editing enzyme. *Nature*, *379*(6564), 460–464.

Nishikura, K. (2006). Editor meets silencer: Crosstalk between RNA editing and RNA interference. *Nature Reviews Molecular Cell Biology*, 7(12), 919–931.

Nishikura, K. (2010). Functions and regulation of RNA editing by ADAR deaminases. *Annual Review of Biochemistry*, *79*, 321–349.

Ohlson, J., Pedersen, J. S., Haussler, D., & Ohman, M. (2007). Editing modifies the GABA(A) receptor subunit alpha3. *RNA*, *13*(5), 698–703.

Paz, N., Levanon, E. Y., Amariglio, N., Heimberger, A. B., Ram, Z., Constantini, S., et al. (2007). Altered adenosine-to-inosine RNA editing in human cancer. *Genome Research*, *17*(11), 1586–1595.

Paz-Yaacov, N., Levanon, E. Y., Nevo, E., Kinar, Y., Harmelin, A., Jacob-Hirsch, J., et al. (2010). Adenosine-to-inosine RNA editing shapes transcriptome diversity in primates. *Proceedings of the National Academy of Sciences of the United States of America*, *107*(27), 12174–12179.

Peng, Z., Cheng, Y., Tan, B. C., Kang, L., Tian, Z., Zhu, Y., et al. (2012). Comprehensive analysis of RNA-Seq data reveals extensive RNA editing in a human transcriptome. *Nature Biotechnology*, *30*(3), 253–260.

Pickrell, J. K., Gilad, Y., & Pritchard, J. K. (2012). Comment on "Widespread RNA and DNA sequence differences in the human transcriptome" *Science*, *335*(6074), 1302, author reply 1302.

Piskol, R., Peng, Z., Wang, J., & Li, J. B. (2013). Lack of evidence for existence of noncanonical RNA editing. *Nature Biotechnology*, *31*(1), 19–20.

Rice, G. I., Kasher, P. R., Forte, G. M., Mannion, N. M., Greenwood, S. M., Szynkiewicz, M., et al. (2012). Mutations in ADAR1 cause Aicardi-Goutieres syndrome associated with a type I interferon signature. *Nature Genetics*, *44*(11), 1243–1248.

Rueter, S. M., Dawson, T. R., & Emeson, R. B. (1999). Regulation of alternative splicing by RNA editing. *Nature*, *399*(6731), 75–80.

Sakurai, M., & Suzuki, T. (2011). Biochemical identification of A-to-I RNA editing sites by the inosine chemical erasing (ICE) method. *Methods in Molecular Biology*, *718*, 89–99.

Sakurai, M., Ueda, H., Yano, T., Okada, S., Terajima, H., Mitsuyama, T., et al. (2014). A biochemical landscape of A-to-I RNA editing in the human brain transcriptome. *Genome Research*, *24*(3), 522–534.

Sakurai, M., Yano, T., Kawabata, H., Ueda, H., & Suzuki, T. (2010). Inosine cyanoethylation identifies A-to-I RNA editing sites in the human transcriptome. *Nature Chemical Biology*, *6*(10), 733–740.

Schwartz, S., Bernstein, D. A., Mumbach, M. R., Jovanovic, M., Herbst, R. H., Leon-Ricardo, B. X., et al. (2014). Transcriptome-wide mapping reveals widespread dynamic-regulated pseudouridylation of ncRNA and mRNA. *Cell*, *159*(1), 148–162.

Sherry, S. T., Ward, M. H., Kholodov, M., Baker, J., Phan, L., Smigielski, E. M., et al. (2001). dbSNP: The NCBI database of genetic variation. *Nucleic Acids Research*, *29*(1), 308–311.

Slotkin, W., & Nishikura, K. (2013). Adenosine-to-inosine RNA editing and human disease. *Genome Medicine*, *5*(11), 105.

Smith, T. F., & Waterman, M. S. (1981). Identification of common molecular subsequences. *Journal of Molecular Biology*, *147*(1), 195–197.

Tojo, K., Sekijima, Y., Suzuki, T., Suzuki, N., Tomita, Y., Yoshida, K., et al. (2006). Dystonia, mental deterioration, and dyschromatosis symmetrica hereditaria in a family with ADAR1 mutation. *Movement Disorders*, *21*(9), 1510–1513.

Trapnell, C., Pachter, L., & Salzberg, S. L. (2009). TopHat: Discovering splice junctions with RNA-Seq. *Bioinformatics*, *25*(9), 1105–1111.

Wang, Q., Hui, H., Guo, Z., Zhang, W., Hu, Y., He, T., et al. (2013). ADAR1 regulates ARHGAP26 gene expression through RNA editing by disrupting miR-30b-3p and miR-573 binding. *RNA*, *19*(11), 1525–1536.

Wang, Q., Khillan, J., Gadue, P., & Nishikura, K. (2000). Requirement of the RNA editing deaminase ADAR1 gene for embryonic erythropoiesis. *Science*, *290*(5497), 1765–1768.

Wang, K., Li, M., & Hakonarson, H. (2010). ANNOVAR: Functional annotation of genetic variants from high-throughput sequencing data. *Nucleic Acids Research*, *38*(16), e164.

Wang, Q., Miyakoda, M., Yang, W., Khillan, J., Stachura, D. L., Weiss, M. J., et al. (2004). Stress-induced apoptosis associated with null mutation of ADAR1 RNA editing deaminase gene. *Journal of Biological Chemistry*, *279*(6), 4952–4961.

CHAPTER SIXTEEN

Radical SAM-Mediated Methylation of Ribosomal RNA

Vanja Stojković*, Danica Galonić Fujimori*,†,1

*Department of Cellular and Molecular Pharmacology, University of California San Francisco, San Francisco, California, USA

†Department of Pharmaceutical Chemistry, University of California San Francisco, San Francisco, California, USA

[1]Corresponding author: e-mail address: danica.fujimori@ucsf.edu

Contents

Methods in Enzymology, Volume 560
ISSN 0076-6879
http://dx.doi.org/10.1016/bs.mie.2015.03.002

1. INTRODUCTION

Post-transcriptional modifications of nucleotides play an important and ubiquitous, yet incompletely understood role in proper RNA function. To date, more than 100 different types of modifications have been reported for various RNA molecules, including ribosomal RNA (rRNA), transfer RNA (tRNA), messenger RNA (mRNA), small nuclear (snRNA), and microRNA (miRNA) (Machnicka et al., 2013). These modifications are chemically diverse and are comprised of methylation, hydroxylation, acetylation, deamination, isomerization, selenylation, reduction, and cyclization, to name a few (Agris, 1996). From this wide variety of modifications, methylation of nucleotides, both at nucleobases and at the ribose moiety, are one of the most common and simplest modifications, and the most prevalent modification in rRNA (Sergeeva, Bogdanov, & Sergiev, 2014). Most of the methylation sites are located in the functional centers of the ribosome: in the decoding and peptidyl transferase centers, as well as on the interface of ribosomal subunits. The roles of such modifications are not completely understood; however, their conservation across bacterial species implies their functional importance. It is believed that their roles span from structural stabilization of substrate RNA to modulation of the function of the ribosome (e.g., maintenance of translational fidelity) (Li & Mason, 2014; Motorin & Helm, 2011; Sergeeva et al., 2014; Sergiev et al., 2011). Additionally, post-transcriptional modification of rRNA can also confer antibiotic resistance toward ribosome-targeting antibiotics (Leclercq & Courvalin, 1991; Long, Poehlsgaard, Kehrenberg, Schwarz, & Vester, 2006; McCusker & Fujimori, 2012; Poehlsgaard & Douthwaite, 2005; Vester & Long, 2009; Wilson, 2014).

Several classes of RNA methylating enzymes perform methylation of rRNA. These enzymes are mechanistically diverse and their mode of methylation depends on the electronic demands of the substrate (Motorin & Helm, 2011). Nucleophilic sites in nucleotides are modified by methyltransferases that transfer methyl group from *S*-adenosylmethionine (SAM) to nucleophilic atoms (e.g., heteroatoms of nucleobases and hydroxyl groups in the ribose) via an S_N2 displacement mechanism (e.g., Dunkle et al., 2014; Hansen, Kirpekar, & Douthwaite, 2001; Lövgren & Wikström, 2001; Tscherne, Nurse, Popienick, & Ofengand, 1999). Another mechanistically distinct type of RNA methyltransferases is enzymes that perform methylation of cytosine and uridine at the C5 carbon. These enzymes activate their substrates via Michael addition of the catalytic Cys to C6 carbon on the pyrimidine ring. The formed adduct activates the C5 position for a nucleophilic attack on the methyl group of SAM. Subsequent elimination of the Cys moiety leads to the formation of C5-methylated cytosines and uridines (e.g., Agarwalla, Stroud, & Gaffney, 2004; King & Redman, 2002). Unique among rRNA methylating enzymes are those that methylate the C2 and C8 amidine carbons of adenosine 2503 (A2503). Recent studies by our lab and by others (Boal et al., 2011; Grove et al., 2011, 2013; McCusker et al., 2012; Silakov et al., 2014; Yan & Fujimori, 2011; Yan et al., 2010) have unraveled a mechanistically novel type of RNA methylation, catalyzed by radical SAM methylsynthases RlmN and Cfr. Unlike methyltransferases, these enzymes incorporate only a methylene fragment, ultimately derived from the methyl group of SAM, into the methyladenosine product. In this chapter, we describe in detail several procedures used in our lab to detect the activity of RlmN and Cfr both *in vivo* and *in vitro* (Fig. 1). Although we will use RlmN/Cfr as our model systems, the methodology presented herein can be applied to study various types of

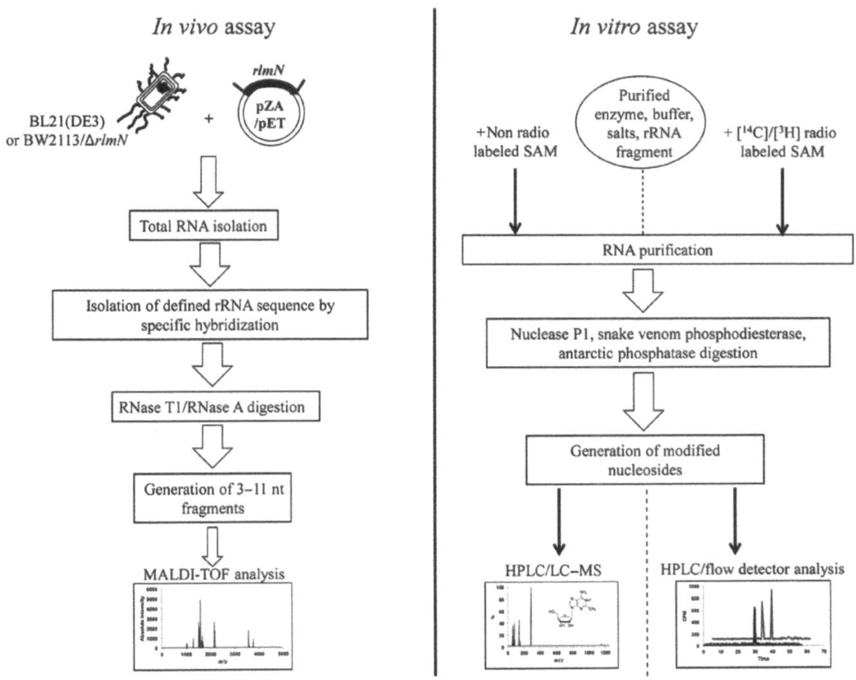

Figure 1 Outline of *in vivo* and *in vitro* strategies used to characterize methylation status of specific nucleotide in 23S rRNA by radical SAM methylating enzymes RlmN and Cfr.

RNA methylations. As 23S rRNA is a shared substrate for both RlmN and Cfr, this chapter will focus on methods implemented in our lab to examine the activity of these enzymes toward 23S rRNA.

2. MECHANISM OF RNA METHYLATION BY RlmN AND Cfr

RlmN and Cfr belong to the radical SAM superfamily. A hallmark of this family is characteristic and highly conserved CX_3CX_2C motif that ligates the [4Fe–4S] iron–sulfur cluster (Sofia, Chen, Hetzler, Reyes-Spindola, & Miller, 2001). This iron–sulfur cluster serves as a cofactor and its reduction is required for initiation of the reaction. Both RlmN and Cfr assemble a methyl group on their products from a hydrogen atom initially present on the substrate and a methylene fragment ultimately derived from SAM. Experimental evidence suggests a shared mechanism among these enzymes (Fig. 2) (Boal et al., 2011; Grove et al., 2011, 2013; McCusker et al., 2012; Silakov et al., 2014; Yan & Fujimori, 2011). Their mutual substrate is a single adenosine nucleotide—A2503 of 23S rRNA (*Escherichia coli* numbering). RlmN from *E. coli* is known to also methylate a subset of tRNAs *in vivo* (Benitez-Paez, Villarroya, & Armengod, 2012). Several lines of evidence indicate that their reaction requires two SAM molecules. In the first-step of the reaction, one of the conserved Cys residues (C355, *E. coli* RlmN numbering) is methylated by SAM in

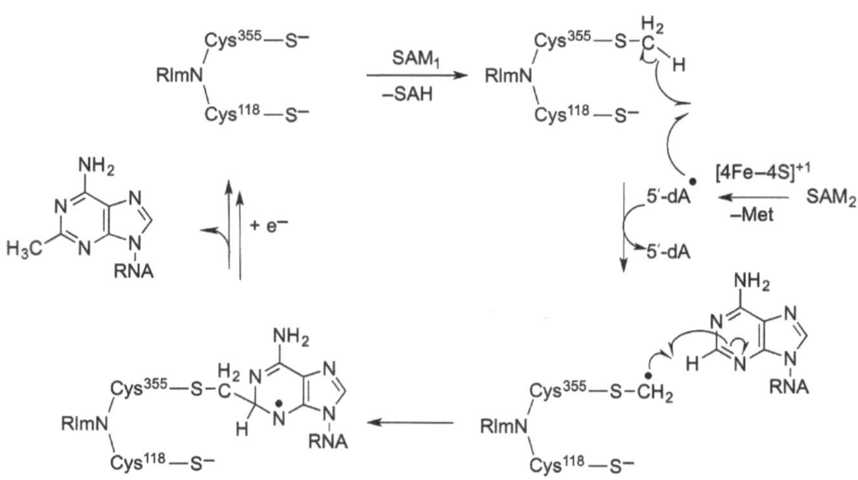

Figure 2 Proposed mechanism of RNA methylation by the radical SAM enzymes RlmN and Cfr.

an S_N2 displacement reaction. Following methylation of C355, the reduced [4Fe–4S] cluster catalyzes the reductive cleavage of the second SAM molecule producing a 5′-deoxyadenosyl (5′-dA•) radical. The 5′-dA• abstracts a hydrogen from a methyl group on C355 producing a cysteine-bound methylene radical. In the subsequent step, this methylene radical adds to the amidine carbon on the adenine ring (C2 position for RlmN; C8 for Cfr), resulting in an enzyme–RNA adduct. This adduct has both been isolated by mutagenesis (McCusker et al., 2012) and characterized spectroscopically (Silakov et al., 2014). The adduct is resolved by a second conserved cysteine, C118, which initiates the fragmentation of the enzyme–RNA crosslink to release the enzyme and to form the methylated RNA product (Boal et al., 2011; Grove et al., 2011; McCusker et al., 2012; Silakov et al., 2014).

3. EXPRESSION OF RlmN AND Cfr

3.1 Protocol for Expression of His_6-Tagged RlmN and Cfr Proteins in LB Medium

The *E. coli* strain BL21(DE3) is cotransformed with pET21a-RlmN (or pET15b-Cfr) with pDB1282, a plasmid that encodes proteins required for the biogenesis of the iron–sulfur clusters and their incorporation into target enzymes. This bacterial transformation is plated on LB agar containing 100 μg/ml ampicillin and 50 μg/ml kanamycin, and incubated at 37 °C. A single colony is used to inoculate 50 ml of LB medium containing antibiotics, which is then incubated overnight at 37 °C, 180 rpm. Four flasks with 1 l of LB medium containing 100 μg/ml ampicillin and 50 μg/ml kanamycin are inoculated with 10 ml of the overnight culture and cultivated at 37 °C until they reach OD_{600} of 0.3–0.4. At this point, expression of the *isc* operon from pDB1282 is initiated via addition of 20% (w/v) arabinose solution to a final concentration of 0.2%. Next, $FeCl_3$ (200 m*M*) and L-cysteine (200 m*M*) are added drop-wise to a final concentration of 200 μ*M* of each. Flasks are cooled on ice while the temperature of the incubator is reduced to 18 °C. When the content is cooled to ~20 °C, cultures are returned to an incubator and left to shake until OD_{600} reaches 0.6–0.8. At that point, the production of RlmN is induced by the addition of isopropyl β-D-1-thiogalactopyranoside (IPTG) to a final concentration of 200 μ*M*, and the bacterial culture is incubated at 18 °C for an additional 24 h. Cells are harvested by centrifugation at 4000 rpm for 20 min at 4 °C. Due to the presence of Fe^{3+} and S^{2-} in the LB medium during the expression, as well as enzymes responsible for the biogenesis of [4Fe–4S] cluster, RlmN overexpressed by this protocol is methylated at C355 (C338 in *Staphylococcus aureus* Cfr) (Boal et al., 2011; Grove et al., 2011, 2013; McCusker et al., 2012).

3.2 Protocol for the Expression of apo His_6-Tagged RlmN and Cfr Proteins in Minimal Medium

A single colony of the BL21(DE3) *E. coli* cells containing pET21a-RlmN or pET15b-Cfr is used to inoculate 50 ml of LB medium containing 100 μg/ml ampicillin, which is left to incubate overnight at 37 °C with shaking at 180 rpm. The following day, each of four 1.35 l of M9 minimal medium (Sigma–Aldrich) containing 150 μl of 1 *M* $CaCl_2$, 3 ml of 1 *M* $MgSO_4$, 30 ml of 1 *M* glucose, and 1.5 ml of 100 mg/ml Amp is inoculated with 15 ml of the overnight culture. Flasks are incubated at 37 °C, 180 rpm until OD_{600} reaches 0.3, at which point 150 ml of 5xEZ mix amino acids (Teknova) is added. At an OD_{600} of ~0.6, cultures are cooled on ice to ~20 °C, while the temperature of the incubator is reduced to 18 °C. Fifteen milliliters of 7.5 m*M* 1,10-phenanthroline (Sigma–Aldrich) and 1.5 ml of 200 m*M* IPTG are added into each flask, followed by the incubation at 18 °C for 24 h. Cells are harvested by centrifugation at 4000 rpm for 20 min at 4 °C. Then cells are washed and repelleted twice with 1 × PBS (pH 7.2) cold solution to remove any residual phenanthroline. This protocol allows for the production of apo RlmN and Cfr. As Fe^{3+} and S^{2-} are absent under these conditions, the [4Fe–4S] cluster is not assembled. As a result C355 (*E. coli* RlmN) and corresponding C338 (*S. aureus* Cfr) are unmodified.

4. PURIFICATION OF RlmN AND Cfr

Due to the high sensitivity of the [4Fe–4S] cluster to oxygen, purification of these proteins is carried out under strict anaerobic conditions. Anaerobic conditions are maintained during purification of both iron–sulfur containing and apo-RlmN and Cfr in order to prevent any oxygen contamination in downstream applications. The temperature of the anaerobic chamber (MBraun) is kept at ~10 °C, and oxygen levels are maintained below 1.8 ppm.

Lysis buffer (50 m*M* HEPES pH 7.5, 300 m*M* KCl, 2 m*M* imidazole, 10% glycerol, 10 m*M* β-mercaptoethanol), wash buffer (50 m*M* HEPES pH 7.5, 300 m*M* KCl, 10 m*M* imidazole, 10% glycerol, 10 m*M* β-mercaptoethanol), elution buffer (50 m*M* HEPES pH 7.5, 300 m*M* KCl, 250 m*M* imidazole, 10% glycerol, 10 m*M* β-mercaptoethanol), 2 × reconstitution buffer (200 m*M* HEPES, pH 7.5, 1 *M* KCl, 20% glycerol), and gel-filtration buffer (10 m*M* HEPES, pH 7.5, 500 m*M* KCl, 10% glycerol, 5 m*M* DTT) are all prepared in advance and deoxygenated by bubbling under argon for at least 1 h prior to being introduced into the anaerobic chamber, where they are left to equilibrate overnight.

For purification, 10–15 g of cell pellets are brought into the anaerobic chamber and resuspended in 35–40 ml of the lysis buffer. Cells are then transferred into an air-tight sonicator cell and lysed outside of the anaerobic chamber. Sonication is performed on ice at 50% duty cycle using a Branson 450 Sonifier sonic dismembrator (Fisher). Power output is set to 5–7, with 1 min duration of each pulse followed by sufficient cooling of the sonicator cell between each sonic disruption. The sonicator cell is then returned to the glove box, where the lysate is poured into centrifuge tubes. The cells are centrifuged at 35,000 rpm, 4 °C, for 45 min. Meanwhile, ~7–8 ml of 50% slurry of Talon cobalt resin (Clontech) is prepared by washing the resin with distilled deionized water (ddI H_2O) 3 times, resuspension in ~10 ml of ddI H_2O and deoxygenation under argon for ~45 min. After centrifugation, the centrifuge tubes together with the Talon resin and 5 g of desalting P6 resin (BioRad—BioGel-P6DG Gel) are reintroduced into the glove box where the supernatant is carefully decanted into a clean Erlenmeyer flask. Talon resin is added to the supernatant and this suspension is left to equilibrate for 1 h with gentle stirring. After 1 h, the supernatant-resin suspension is applied to a column, after which the column content is washed with 50 ml of the lysis buffer followed by 25 ml of the wash buffer. The enzyme is eluted with 10 ml of elution buffer and collected in 1 ml fractions. Typically, coexpression of the enzymes with pDB1282 results in fractions that contain significant brown color, while the enzymes expressed in the minimal medium are colorless. Protein-containing fractions, as identified by the Bradford method, are pooled and desalted on manually prepared P6 desalting column, and protein is eluted with 1 × reconstitution buffer. After desalting, protein concentration is adjusted to ~50–60 μ*M*, 1 *M* DTT is added to a final concentration of 200 m*M*, and enzyme is left to equilibrate for 1 h. The reconstitution of the [4Fe–4S] cluster is then carried out by addition of $FeCl_3 \cdot 6H_2O$ and $Na_2S \cdot 9H_2O$ (both at 200 m*M* stock concentration). Iron solution is added drop-wise to a slowly stirred enzyme solution over a 30 min period, to a final concentration of 500 μ*M*. The resulting solution is stirred for an additional 10 min prior to the drop-wise addition of the sulfide solution over 90 min, to a final concentration of 600 μ*M*. The resulting dark brown solution is gently stirred for an additional 15 h. Following incubation, the reconstitution mixture is transferred to an ultracentrifuge tube, sealed inside of a glove box, and centrifuged at 30,000 rpm, 4 °C for 60 min. The tube is then brought into the glove box, and supernatant is desalted on the P6 resin (manually poured) in the gel filtration buffer. After desalting, the reconstituted enzyme is concentrated and flash frozen in liquid N_2, before being stored at −80 °C. To remove adventitiously bound iron and sulfide as well as aggregated protein from chemically reconstituted RlmN and Cfr, enzymes are further purified by gel filtration on S200 column (GE Healthcare) in gel filtration buffer. If needed, the purity of the enzyme can be further increased by anion exchange chromatography using Hi-TrapQ column (GE Healthcare) (Yan et al., 2010). It is important to note that the iron–sulfur cluster reconstitution step is performed not only for the apo enzymes, but also for enzymes coexpressed with iron–sulfur cluster assembly machinery in the LB medium—conditions that typically result in only partial reconstitution of the cluster.

5. *IN VITRO* METHYLATION ASSAYS FOR RlmN AND Cfr

5.1 General Overview of *In Vitro* Methylation Assays

The standard *in vitro* methylation assay for RNA methylating enzymes measures the incorporation of radioactively labeled methyl group (e.g., ^{14}C or ^{3}H) from SAM into the RNA product. The resulting RNA product is isolated by phenol/chloroform extractions or by using a commercial kit, and radioactivity incorporation is measured by liquid scintillation counter (LSC). Alternatively, upon quenching of the reaction, reaction mixture is spotted onto a filter paper, unreacted radiolabeled SAM removed by washing the paper with trichloroacetic acid (TCA, 5%), and the filter paper subsequently transferred into vials containing scintillation liquid. The amount of transferred methyl group is quantified with LSC. Evaluation of the radioactivity allows for determination of molar amount of modified nucleotide within the RNA molecule. Additionally, a time course analysis allows for determination of the rate of a reaction (Chervin, Kittendorf, & Garcia, 2007).

A choice between ^{14}C and ^{3}H is solely on the researcher. In comparison to ^{14}C, tritium offers ~1000-fold higher specific radioactivity allowing for higher sensitivity of detection in LSC. In addition, due to a higher specific radioactivity, a broader range of final SAM concentrations can be used in the experiments. In contrast, ^{14}C has higher counting efficiency than tritium, and events such as chemiluminescence do not significantly interfere with counting. Both compounds, [^{14}C-methyl]-SAM and [^{3}H-methyl]-SAM, are readily commercially available and can be purchased with a range of specific radioactivities (Perkin-Elmer).

In an alternative approach, activity of an RNA methylating enzyme can be assessed by the use of nonradioactive SAM, and subsequent detection of the product by liquid chromatography/mass spectrometry (LC/MS). In short, following the incubation of RNA with an enzyme and SAM, methylated RNA is isolated from the reaction mixture and digested to single nucleosides. The resulting nucleosides are then separated on HPLC and analyzed by LC/MS. While this approach allows for isolation of the product and analysis of its molecular weight, a disadvantage of this method is its lower sensitivity as compared to the radioactivity-incorporation-based method.

Regardless of method used for product analysis, it is important to indicate that one of the byproducts of the reaction, *S*-adenosylhomocysteine (SAH) is known to act as a competitive inhibitor of many methyltransferases (Clarke & Banfield, 2001). However, in most assays, including those utilized in our lab, SAM is present in a large excess over the substrate and therefore the amount of SAH produced during the course of the reaction should not cause significant inhibition. Nonetheless, as with any enzymatic assay, once the activity of RNA methylating enzyme is determined *in vitro*, it is important to optimize the experimental conditions prior to more systematic study of the mechanism and specificity of the enzyme. These include optimization of the reductant, type, and concentration of specific metal ions (e.g., Mg^{2+} for RNA substrates), the pH and the composition of the reaction buffer, etc. Moreover, strict nuclease-free conditions must be maintained in the experiment, and in some instances addition of RNase inhibitors is required. In our group, both RNase inhibitor from human placenta (New England BioLabs) and RNasin (Promega) have been used with equal success. Lastly, the stability of SAM needs to be taken into account, since the rate of SAM hydrolysis increases with an increase in pH and temperature.

5.2 Protocol for Substrate Preparation

When isolated from wild-type *E. coli* cells, naturally occurring RNAs are fully modified and as such are not suitable for testing *in vitro* activities of RNA methylating enzymes. While this challenge can be overcome by using knockout strains (e.g., BW2113/Δ*rlmN* strain, see below; Yan et al., 2010), an alternative approach relies on *in vitro* transcription of RNA fragments. When working with smaller RNAs, such as tRNAs that are ~75-nt long, the whole biomolecule is usually transcribed. However, when working with longer substrates such as 23S rRNA, transcription of smaller fragments is preferred, as they are easier to prepare and handle than the full 23S rRNA. As the common substrate for RlmN and Cfr is 23S rRNA, here we describe *in vitro* transcription of 23S rRNA fragments that encompass the substrate nucleotide, A2503. In our earlier investigations, we have determined substrate requirements (i.e., length of the transcribed rRNA fragment) for the methylation of *in vitro* transcribed rRNA by RlmN and Cfr (Yan et al., 2010).

The transcription reactions for any fragment of rRNA can be carried out by following this protocol where the only difference is changing the primers used in the PCR amplification step. *In vitro* transcription reactions of 23S rRNA fragments are carried out in the following manner: in the first step, a desired fragment of the DNA template is amplified using a plasmid pKK3535 that encodes a *23S rRNA* gene (Yan et al., 2010). The polymerase that we utilize in this PCR step is Phusion High-Fidelity DNA polymerase (New England BioLabs); however, other polymerases can be used as well. The resultant PCR products are purified using the Qiagen PCR purification kit following the manufacturer's recommendations.

In a typical *in vitro* transcription reaction, 20 μg of DNA template is dissolved in 500 μl solution containing T7 buffer (40 m*M* Tris–HCl, pH 7.9, 16 m*M* $MgCl_2$, 2 m*M* spermidine), 20 m*M* DTT, 3.2 m*M* of each of rNTP, 0.2 U/μl RNase inhibitor from human placenta, and 3 U/μl T7 RNA polymerase (New England BioLabs). The reaction mixture is incubated at 37 °C for 3 h, followed by the addition of 40 U of RQ1 RNase Free DNase (Promega) and further incubation at 37 °C for 30 min. The transcribed rRNA is purified by RNeasy midi kit (Qiagen) following the manufacturer's recommendation. RNA is then precipitated overnight at −20 °C by addition of 1/10th volume of 3 *M* NaOAc, pH 5 and 3 volumes of ethanol. The sample is subsequently pelleted, washed with 70% aqueous ethanol, air-dried, resuspended in RNase-free water, and stored at −20 °C. The size and the purity of the resulting rRNA fragments are verified by gel electrophoresis on denaturing 4.5% TBE, 7 *M* urea-PAGE gel.

5.3 Protocol for Testing the Activities of RlmN and Cfr Using Total Radioactivity Incorporation

A typical activity assay for RlmN and Cfr is performed in 100 μl of HEPES buffer (100 m*M*, pH 8.0) containing 100 m*M* KCl, 10 m*M* $MgCl_2$, and

2 m*M* DTT to which 20 μ*M* Flavodoxin (Flv), 2 μ*M* Flavodoxin reductase (Flx), 0.14 μCi [^{14}C-methyl]-SAM (58 mCi/mmol), 1 m*M* NADPH, 1 μ*M* RlmN (or 2 μ*M* Cfr), and 2 μ*M* rRNA fragment are added. Alternatively, [^{3}H-methyl]-SAM can be used as a source of radiolabeled SAM. All of the components except the radiolabeled SAM are made anaerobic by purging with argon prior to mixing in an MBraun glove box. The reaction mixture containing rRNA, enzyme, Flv, Flx, and the buffer is prepared, and subsequently removed from the anaerobic chamber in a gas-tight glass vials (Supelco). The content of the vial is kept under a positive pressure of argon by attachment of an argon balloon through a septum. The vials are kept on ice until the commencement of the reaction. [^{14}C-methyl]-SAM (or [^{3}H-methyl]-SAM) is supplied in a 9:1 mixture of 10 m*M* H_2SO_4:EtOH, which needs to be removed prior to reaction. The radiolabeled SAM is aliquoted into an empty glass vial (Supelco) and dried by an argon stream. The vial is then capped with a septum and maintained under a positive pressure of argon. The dried SAM is dissolved in 4× buffer (400 m*M* HEPES pH 8, 400 m*M* KCl) added by a gas-tight syringe. To the content of this vial, the reaction mixture lacking NADPH is added by a gas-tight syringe, and the resulting mixture is left to equilibrate at 37 °C for 10 min under a positive pressure of argon. The reaction is then initiated by the addition of an anaerobic solution of NADPH (5 μl of 20 m*M* stock solution, final concentration of 1 m*M*). An initial sample (5 μl, $t=0$ min) is immediately removed. This sample is used to determine the total amount of the radioactivity in the vial to account/correct for a possible variability between different samples. Typically, reaction mixture is incubated for 90 min at 37 °C, after which the reaction is quenched via addition of cold SAM to a final concentration of 120 m*M*. Samples are frozen on dry ice and stored at −20 °C until further analysis.

In order to quantify the amount of total radioactivity incorporated into the product, RNA is purified from the rest of the reaction mixture. We commonly separate the RNA by using the RNA Clean&Concentrate kit (Zymo Research), a method that we find more consistent than phenol/chloroform extractions. Following purification, RNA is eluted in 100 μl of water and added directly into a scintillation vial containing 5 ml of Ultima Gold Scintillation Cocktail (Perkin-Elmer). Samples are mixed by vigorously shaking and left to rest in the dark overnight. Each scintillation vial is then counted on LSC (Beckman Coulter LS6500 multipurpose scintillation counter) for 5 min.

5.4 Protocol for Determining the Regiochemistry of Methylation by RlmN and Cfr

In order to establish the regioselectivity of the modification on the adenosine ring by RlmN and Cfr, the radiolabeled RNA from the *in vitro* assay is purified and enzymatically digested to single nucleosides. The resulting nucleosides are analyzed by HPLC, and their retention times compared to synthetic standards of predicted methylated products, 2-methyladenosine (RlmN), or 8-methyladenosine and 2,8-dimethyladenosine (Cfr).

Following the purification of the product RNA as described above (Section 5.3), RNA is enzymatically digested using nuclease P1 (Sigma–Aldrich), snake venom phosphodiesterase (Sigma–Aldrich), and Antarctic phosphatase (New England BioLabs). First, 1/20th volume of 1 *M* NH_4OAc pH 5.3 is added to purified rRNA, followed by nuclease P1 (4 U/50 μg of RNA) and incubated at 45 °C for 3 h. 1/40th volume of 1 *M* NH_4HCO_3 and snake venom phosphodiesterase (4 mU/50 μg of RNA) are then added and incubation is continued for an additional 2 h at 37 °C. Lastly, 1/10th volume of 10× Antarctic phosphatase reaction buffer (500 m*M* bis–Tris–propane HCl pH 6, 10 m*M* $MgCl_2$, 1 m*M* $ZnCl_2$) is added together with Antarctic phosphatase (1 U/50 μg of RNA) and the digestion mixture is incubated for 1 h at 37 °C. Upon completion of the reaction, samples are pelleted at 10,000 rpm for 5 min at room temperature. Supernatant is loaded onto a Luna analytical C18 column (10 μm, 4.6 mm × 250 mm), in a solvent system consisting of 40 m*M* ammonium acetate, pH 6.0 (A) and 40% aqueous acetonitrile (B). The nucleotides are eluted of the column at a flow rate of 1 ml/min with a step gradient of 0% B (0–2 min), 0–25% B (2–27 min), and 25–60% B (27–37 min). The retention times for the products are: 8-methyladenosine (29.5 min), 2-methyladenosine (30 min), and 2,8-dimethyladenosine (35 min). The mononucleosides are detected by a Packard radiomatic 515TR flow scintillation analyzer (Perkin-Elmer) and their retention times are compared to the retention times of the synthetic standards monitored by their UV absorption at 254 nm (Fig. 3). Synthetic standards were synthesized in-house following previously published procedures (Van Aerschot et al., 2002; van Tilburg, Gremmen, von Frijtag Drabbe Kunzel, de Groote, & IJzerman, 2003; Yan et al., 2010).

5.5 Protocol for Testing the Activities of RlmN and Cfr Using Nonradioactive SAM

In this procedure, the methylation reaction is performed at a larger scale (typically 200 μl) in order to assure that sufficient quantities of methylated

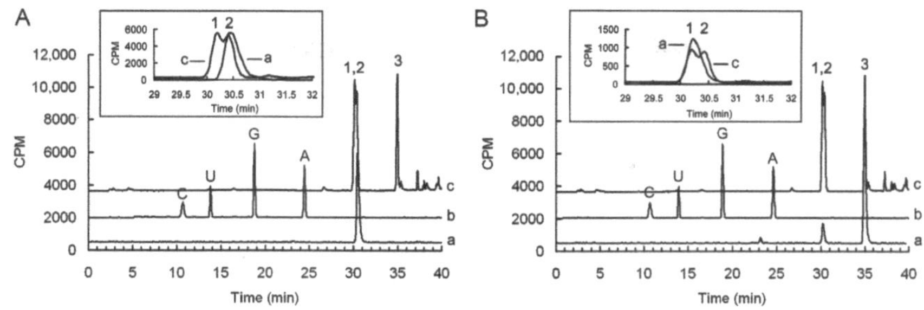

Figure 3 HPLC analysis of the methylation products from RlmN and Cfr reactions. (A) Reaction of RlmN with 23S rRNA isolated from BW2113/Δ*rlmN* strain; (B) reaction of Cfr with 23S rRNA isolated from BW2113/Δ*rlmN* strain. (a) Reverse-phase HPLC flow ^{3}H radiogram of digested RNA; (b) reverse phase HPLC UV–vis chromatogram of digested RNA measured at 254 nm; and (c) reverse phase HPLC UV–vis chromatogram of synthetic methyladenosine standards at 254 nm. (1) m^{8}A; (2) m^{2}A; and (3) m^{2}m^{8}A. *Figure was taken and modified with permission from Yan et al. (2010). Copyright 2010 American Chemistry Society.*

rRNA product are obtained for the detection by LC/MS analysis. A typical reaction is performed in 200 μl of the reaction buffer (100 m*M* HEPES pH 7.5, 100 m*M* KCl) containing 10 m*M* $MgCl_2$, 200 μ*M* SAM, 20 μ*M* RNA, 2 m*M* DTT, 20 μ*M* Flv, 2 μ*M* Flx, 2 m*M* NADPH, and 20 μ*M* either RlmN or Cfr. All the reaction components are prepared anaerobic prior to mixing in an anaerobic chamber. The reaction is initiated by the addition of the enzyme, and incubated at 37 °C for 1 h. Upon completion of the reaction, vials are frozen in liquid nitrogen and stored at −80 °C. The RNA is then digested to nucleosides via the treatment with nuclease P1, snake venom phosphodiesterase and Antarctic phosphatase as described above (Section 5.4), with the exception that no prior purification of methylated rRNA is performed. Upon completion of the digestion, samples are pelleted at 10,000 rpm for 5 min at room temperature, and filtered through Amicon Ultra (3 kDa MWCO) filters (Millipore). Filters are washed twice with 200 μl of water, and eluent is lyophilized overnight. Nucleosides are then separated by reverse-phase HPLC using the method described above (Section 5.4). Fractions containing 2-methyladenosine, 8-methyladenosine, 2,8-methyladenosine, and 5′-deoxyadenosine are collected separately based on their known retention times, and dried by lyophilization prior to LC/MS analysis. The lyophilized material is then dissolved in water, and analyzed by a Waters Alliance HPLC, equipped with a Waters 248y diode array detector and a Waters/Micromass ZQ single-quadruple mass spectrometer (Waters). The solutions are loaded onto an Xterra MS C18 column (3.5 μm, 2.1 mm × 50 mm) (Waters) and eluted at a flow rate

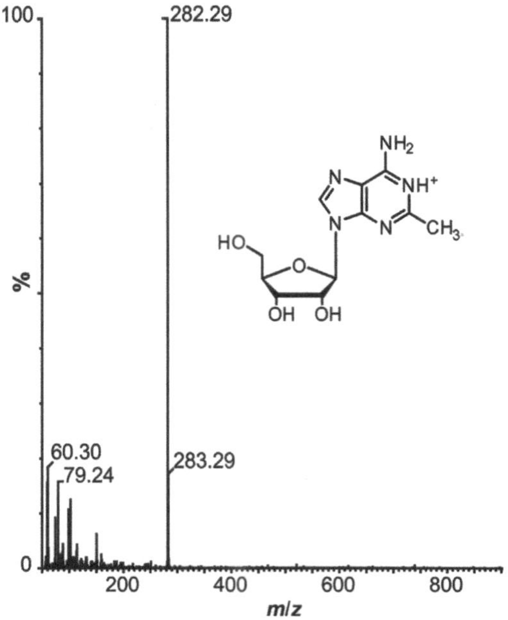

Figure 4 Liquid chromatography/mass spectrometry (LC/MS) analysis of methylated adenosine isolated from the reaction of RlmN with rRNA (2496–2582). This reaction was performed with cold SAM.

of 0.2 ml/min with a 6-min gradient from 100% water to 20% acetonitrile. Data are analyzed by MassLynx software version 4.0 (Fig. 4). While this method is less sensitive than radioactivity incorporation, an advantage of this approach is the ability to monitor the coupling between the product formation (methylated adenosines) and the production of 5′-deoxyadenosine formed in the reaction (Fig. 2).

6. *IN VIVO* ACTIVITY ASSAY FOR RlmN AND Cfr

6.1 General Overview of the Method

A number of experimental approaches have been used in the past to evaluate RNA modifications in cellular contexts. Some of the commonly used techniques are primer extension analysis, thin layer chromatography of digested RNA, and mass spectrometry (Basturea, 2013; Kellner, Burhenne, & Helm, 2010). Since our goal is to detect the extent of methylation (unmodified, mono-, or dimethylated adenosine) at a specific position, A2503, mass spectrometry of an RNA fragment encompassing this nucleotide is a particularly well-suited method. An alternative approach is a reverse transcriptase primer extension analysis, which has been extensively used to detect different RNA modifications (Boorstein & Craig, 1989; Motorin, Muller, Behm-Ansmant, & Branlant, 2007). However, in comparison to the mass spectrometry method, this method requires additional radioactive labeling. Moreover, reverse transcriptase is sensitive to the secondary structure of RNA (Basturea, 2013; Boorstein & Craig, 1989; Lane, Field, Olsen, & Pace, 1988; Motorin et al., 2007), and does not provide a large difference in the intensity of the stop between mono- and dimethylated A2503. As a result, this method is less well suited for the analysis of RlmN- and Cfr-mediated methylation *in vivo*. Another method commonly used to identify new modifications in RNA is thin-layer chromatography (TLC). In comparison to the mass spectroscopy method, this method requires smaller amounts of purified RNA; however, the harsh chemical conditions can lead to degradation of the labile-modified nucleosides, as well as need for radioactive labeling deterred us from utilizing this approach (Basturea, 2013; Motorin et al., 2007).

Matrix-assisted laser desorption ionization-time of flight (MALDI-TOF) mass spectrometry method is an extensively used analytical approach to assess the post-transcriptional modification status in RNA (Andersen, Porse, & Kirpekar, 2004; Douthwaite & Kirpekar, 2007; Kirpekar, Douthwaite, & Roepstorff, 2000). The advantages of MALDI-TOF MS technique include: sensitivity, tolerance to impurities, the ability to handle complex mixtures without any prior fractionation, and relative simple sample preparation (Douthwaite & Kirpekar, 2007). Moreover, MALDI spectra are easy to interpret since they are dominated by single charged ions. Although amenable to analysis of RNAs as long as intact rRNAs (Berhane & Limbach, 2003), in most protocols, RNA is digested with an appropriate RNase (e.g., RNase T1, RNase A, RNase H, or RNase U) prior to MS analysis in order to obtain fragments that are less than 20 nucleotides long.

The activity of an RNA modifying enzyme can be identified *in vivo* by comparing the modified nucleotides in RNAs isolated from the wild-type strain to those of a knockout strain that does not contain the gene encoding for the methylating enzyme. Moreover, it is feasible to test the activity of a putative RNA modifying enzyme from various species using *E. coli* or *S. cerevisiae* as a host. In this approach, the putative methylase gene is cloned into a plasmid suitable for transformation into a bacterial or a yeast strain lacking the homologous methylating enzyme to ensure that the methylation site is not premodified (Fig. 1). Our protocol for analysis of RlmN and Cfr activity *in vivo* is based on a method published by Anderson and coworkers, with several modifications (Andersen et al., 2004). The enzyme of interest is expressed in cells during the mid-log phase, and the total RNA isolated prior to the culture reaching the stationary phase. In order to detect both the RlmN-like C2 methylation and Cfr-like C8 methylation of A2503, both the WT *E. coli* and the BW2113/Δ*rlmN* are used as host strains. In the WT strain, the C2 position of A2503 in 23S rRNA is already premodified by the endogenous copy of *rlmN*, preventing its use as a substrate of transformed RlmN-like enzymes. On the other hand, A2503 in BW2113/Δ*rlmN* strain is unmodified and can be used to evaluate RlmN-like methylation. Since the C8 position of A2503 in unmodified in both strains, both can be used to evaluate Cfr-like activity.

6.2 Protocol for Isolation of Total rRNA from Strains Overexpressing RlmN and Cfr

E. coli BL21(DE3) and BW2113/Δ*rlmN* strains transformed with a plasmid encoded RlmN or Cfr (e.g., pET21a-RlmN, pET15b-Cfr, pZA-RlmN (Cfr)) are grown in the LB medium at 37 °C and 180 rpm to an OD_{600} of 0.1. At this point, production of the protein is induced by the addition of either IPTG (for pET vector—final concentration of 200 μ*M*) or anhydrous tetracycline (ATH; for pZA vector—final concentration 20 ng/ml), and growth of the culture is continued for an additional 3.5 h at 18 °C. Following incubation, the cells are harvested by centrifugation (10 min, 4 °C 3000 rpm). Ten milliliters of the culture is sufficient for isolation of enough RNA for the subsequent analysis. The cell pellets are thawed on ice, and then resuspended in TE buffer (10 m*M* Tris–HCl pH 8.0, 1 m*M* EDTA). Lysozyme is added to a final concentration of 2 mg/ml, and the resuspensions are left at room temperature for 10 min. The total RNA is purified using the RNeasy Midi Kit (Qiagen) following manufacturer's recommendations.

6.3 Protocol for Isolation of a Defined rRNA Fragment

To isolate a region of 23S rRNA that contains A2503, 2000 pmol of a synthetic oligodeoxynucleotide complementary to C2480–C2520 nucleotide region of *E. coli* 23S rRNA is incubated with 200 pmol of total RNA in 0.3 volumes of hybridization buffer (250 m*M* HEPES pH 7, 500 m*M* KCl). The hybridization mixture is incubated at 90 °C for 7 min, and subsequently slowly cooled to 45 °C over 3.5 h in order to allow for hybridization to occur. Single-stranded RNA and DNA are digested with mung bean nuclease (30 units, New England BioLabs) and RNaseA (0.5 μg, Sigma–Aldrich) over 1 h at 37 °C in a buffer solution (50 m*M* NaOAc pH 5, 30 m*M* NaCl, and 1 m*M* $ZnCl_2$). The RNA:DNA hybrid is extracted by the addition of 1 volume of a 1:1 phenol–chloroform solution, and the hybrid is precipitated from the aqueous layer by addition of 3 *M* NaOAc pH 5 (1/10th volume) and ethanol (3 volumes). Following an overnight precipitation at −20 °C, the precipitate is dissolved in a 1:2 solution of H_2O:formamide. The RNA:DNA hybrid is separated by gel electrophoresis (13% polyacrylamide gel containing 7 *M* urea, 80 V), and bands are visualized by ethidium bromide staining. The RNA band is excised and the RNA is eluted overnight at 4 °C in 200 μl of 1 *M* NH_4OAc pH 5.3. The RNA is subsequently precipitated overnight at −20 °C with 2 volumes of a 1:1 mixture of ethanol and isopropanol. The precipitate is pelleted by centrifugation at 14,000 rpm for 40 min at 4 °C. The supernatant is then removed and the pellet washed with 70% aqueous EtOH to completely remove any residual SDS and free salts. Subsequently, the RNA is pelleted, solvent is removed, and sample is left to air dry.

6.4 Protocol for RNase Digestion and MALDI-TOF Analysis

The isolated RNA fragment (Section 6.3) is first dissolved in nuclease-free water (typically 10–15 μl) and the sample concentration is determined by NanoDrop. A solution containing 25–30 pmol of isolated RNA is mixed with 0.5 volumes of freshly prepared 0.5 *M* 3-hydroxypicolinic acid (3-HPA). The RNA is then digested with 500 units of RNase T1 (USBiological) for 1 h at 37 °C. Cyclic phosphates that result from the RNase T1 digestion are linearized by the addition of 0.25 volume of 0.5 *M* HCl (room temperature, 30 min). The samples are lyophilized, and then redissolved in 5 μl of water. One microliter of the sample is spotted onto the target plate and mixed with 1 μl of 0.5 *M* 3-HPA matrix. The sample is allowed to dry, and spectra recorded in a reflector and positive ion mode on a Voyager Elite STR MALDI-TOF mass spectrometer (PE-Biosystems). RNase T1 digestion of RNA results in a three nucleotide fragment that encompasses A2503. The m/z ratio of the monomethylated fragment is 1013.16, while m/z ratio of the dimethylated fragment is 1027.18. A representative MALDI spectrum of both mono and dimethylated fragment is shown in Fig. 5.

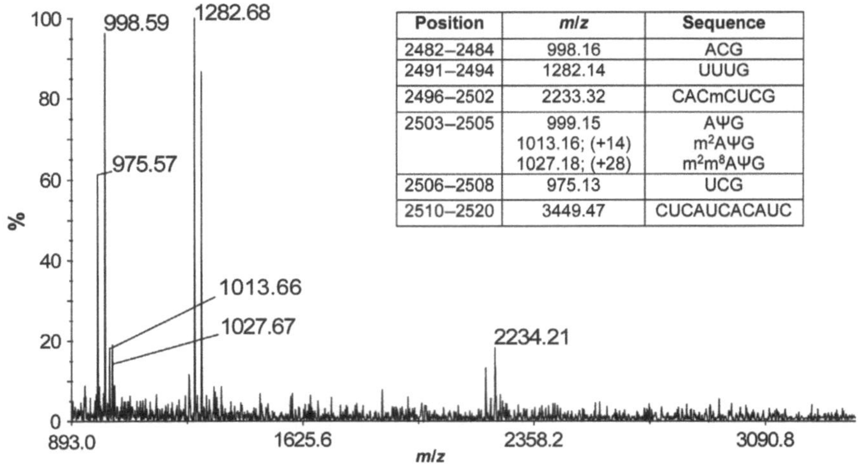

Position	m/z	Sequence
2482–2484	998.16	ACG
2491–2494	1282.14	UUUG
2496–2502	2233.32	CACmCUCG
2503–2505	999.15 1013.16; (+14) 1027.18; (+28)	AΨG m^2AΨG m^2m^8AΨG
2506–2508	975.13	UCG
2510–2520	3449.47	CUCAUCACAUC

Figure 5 MALDI-TOF analysis of Cfr methylation *in vivo*. Fragment C2480–C2520 of *E. coli* 23S rRNA, isolated from BW2113/Δ*rlmN* strain transformed with pZA-Cfr plasmid, was digested with RNase T1 and fragments were analyzed by MALDI-TOF. Masses of the expected fragments from RNase T1 digestion, based on known nucleotide modifications, are shown in the inset. There is a slight shift in observed *m/z* values in comparison to predicted values due to difference in calibration of the instrument. m^2A is 2-methyladenosine, m^8A is 8-methyladenosine, and Ψ is pseudouridine.

7. SUMMARY AND CONCLUDING REMARKS

While RNA methylation occurs in all kingdoms of life, the type and the distribution of different methylated species varies substantially among archaea, bacteria, and eukaryotes. The most prevalent type of RNA methylation is methylation of nucleobases. However, despite recent advances in our knowledge of these marks, the biological roles of such modifications are still incompletely understood (Machnicka et al., 2013; Motorin & Helm, 2011; Sergeeva et al., 2014; Sergiev et al., 2011). A number of mechanisms have evolved to enable RNA methylation, which are tuned to the electronic demands of the substrate. Herein, we provide an overview of methods for expression, purification, and activity analysis of a specific type of RNA methylating enzymes, radical SAM methylsynthases. These enzymes modify the amidine carbon atoms of an adenosine, A2503, in bacterial 23S rRNA. The activities of these enzymes have only been recently reconstituted (Yan et al., 2010), which can be attributed to the complex anaerobic catalysis that they perform. As the substrate A2503 is located at the nascent peptide exit tunnel of the bacterial ribosome, methylations catalyzed by these enzymes have profound impact on the biology of the host strain. RlmN, an endogenous protein found in all bacteria, methylates the C2 amidine carbon and contributes to the translational fidelity (Benitez-Paez et al., 2012; Ramu et al., 2011; Vazquez-Laslop, Ramu, Klepacki, Kannan, & Mankin, 2010). Cfr, found in pathogenic species, methylates the C8 amidine carbon, a modification that confers resistance to various classes of antibiotics (Giessing et al., 2009; Long et al., 2006; Smith & Mankin, 2008). Interestingly, C2 methylated adenosine was recently detected in a subset of tRNAs, raising the question of the physiological role of this modification (Benitez-Paez et al., 2012). With an increase in available whole genome sequences, the development of methods to identify target substrates of RNA methylating enzymes (Khoddami & Cairns, 2013; Meyer et al., 2012; Tim, Katharina, & Matthias, 2010), as well as advances in the characterization of their activities, we anticipate the coming years will unravel novel aspects of mechanisms of the RNA methylation and deepen insight into the function of the resulting modification.

ACKNOWLEDGMENTS

We thank Lindsey Pack and Christina Fitzsimmons for critical comments on the manuscript, Dr. Feng Yan, Dr. Kevin McCusker, and current members of the Fujimori lab for useful discussions, and NIAID (R01AI095393) and NSF (CAREER 1056143) to D.G.F. for funding.

REFERENCES

Agarwalla, S., Stroud, R. M., & Gaffney, B. J. (2004). Redox reactions of the iron-sulfur cluster in a ribosomal RNA methyltransferase, RumA: Optical and EPR studies. *Journal of Biological Chemistry*, *279*, 34123–34129.

Agris, P. F. (1996). The importance of being modified: Roles of modified nucleosides and Mg^{2+} in RNA structure and function. *Progress in Nucleic Acid Research and Molecular Biology*, *53*, 79–129.

Andersen, T. E., Porse, B. T., & Kirpekar, F. (2004). A novel partial modification at C2501 in *Escherichia coli* 23S ribosomal RNA. *RNA*, *10*, 907–913.

Basturea, G. N. (2013). Research methods for detection and quantification of RNA modifications. *Materials and Methods*, *3*, 186.

Benitez-Paez, A., Villarroya, M., & Armengod, M. E. (2012). The *Escherichia coli* RlmN methyltransferase is a dual-specificity enzyme that modifies both rRNA and tRNA and controls translational accuracy. *RNA*, *18*, 1783–1795.

Berhane, B. T., & Limbach, P. A. (2003). Stable isotope labeling for matrix-assisted laser desorption/ionization mass spectrometry and post-source decay analysis of ribonucleic acids. *Journal of Mass Spectrometry*, *38*, 872–878.

Boal, A. K., Grove, T. L., McLaughlin, M. I., Yennawar, N. H., Booker, S. J., & Rosenzweig, A. C. (2011). Structural basis for methyl transfer by a radical SAM enzyme. *Science*, *332*, 1089–1092.

Boorstein, W. R., & Craig, E. A. (1989). Primer extension analysis of RNA. *Methods in Enzymology, 180*, 347–369.
Chervin, S. M., Kittendorf, J. D., & Garcia, G. A. (2007). Probing the intermediacy of covalent RNA enzyme complexes in RNA modification enzymes. *Methods in Enzymology, 425*, 121–137.
Clarke, S., & Banfield, K. (2001). *S*-adenosylmethionine-dependent methyltransferases: Potential targets in homocysteine-linked pathology. In R. Carmel & D. W. Jacobsen (Eds.), *Homocysteine in health and disease* (pp. 63–78). Cambridge, England: Cambridge University Press.
Douthwaite, S., & Kirpekar, F. (2007). Identifying modifications in RNA by MALDI mass spectrometry. *Methods in Enzymology, 425*, 3–20.
Dunkle, J. A., Vinal, K., Desai, P. M., Zelinskaya, N., Savic, M., West, D. M., et al. (2014). Molecular recognition and modification of the 30S ribosome by the aminoglycoside-resistance methyltransferase NpmA. *Proceedings of the National Academy of Sciences of the United States of America, 111*, 6275–6280.
Giessing, A. M., Jensen, S. S., Rasmussen, A., Hansen, L. H., Gondela, A., Long, K., et al. (2009). Identification of 8-methyladenosine as the modification catalyzed by the radical SAM methyltransferase Cfr that confers antibiotic resistance in bacteria. *RNA, 15*, 327–336.
Grove, T. L., Benner, J. S., Radle, M. I., Ahlum, J. H., Landgraf, B. J., Krebs, C., et al. (2011). A radically different mechanism for *S*-adenosylmethionine-dependent methyltransferases. *Science, 332*, 604–607.
Grove, T. L., Livada, J., Schwalm, E. L., Green, M. T., Booker, S. J., & Silakov, A. (2013). A substrate radical intermediate in catalysis by the antibiotic resistance protein Cfr. *Nature Chemical Biology, 9*, 422–427.
Hansen, L. H., Kirpekar, F., & Douthwaite, S. (2001). Recognition of nucleotide G745 in 23S ribosomal RNA by the rrmA methyltransferase. *Journal of Molecular Biology, 310*, 1001–1010.
Kellner, S., Burhenne, J., & Helm, M. (2010). Detection of RNA modifications. *RNA Biology, 7*, 237–247.
Khoddami, V., & Cairns, B. R. (2013). Identification of direct targets and modified bases of RNA cytosine methyltransferases. *Nature Biotechnology, 31*, 458–464.
King, M. Y., & Redman, K. L. (2002). RNA methyltransferases utilize two cysteine residues in the formation of 5-methylcytosine. *Biochemistry, 41*, 11218–11225.
Kirpekar, F., Douthwaite, S., & Roepstorff, P. (2000). Mapping posttranscriptional modifications in 5S ribosomal RNA by MALDI mass spectrometry. *RNA, 6*, 296–306.
Lane, D. J., Field, K. G., Olsen, G. J., & Pace, N. R. (1988). Reverse transcriptase sequencing of ribosomal RNA for phylogenetic analysis. *Methods in Enzymology, 167*, 138–144.
Leclercq, R., & Courvalin, P. (1991). Bacterial resistance to macrolide, lincosamide, and streptogramin antibiotics by target modification. *Antimicrobial Agents and Chemotherapy, 35*, 1267–1272.
Li, S., & Mason, C. E. (2014). The pivotal regulatory landscape of RNA modifications. *Annual Review of Genomics and Human Genetics, 15*, 127–150.
Long, K. S., Poehlsgaard, J., Kehrenberg, C., Schwarz, S., & Vester, B. (2006). The Cfr rRNA methyltransferase confers resistance to phenicols, lincosamides, oxazolidinones, pleuromutilins, and streptogramin A antibiotics. *Antimicrobial Agents and Chemotherapy, 50*, 2500–2505.
Lövgren, J. M., & Wikström, P. M. (2001). The rlmB gene is essential for formation of Gm2251 in 23S rRNA but not for ribosome maturation in *Escherichia coli*. *Journal of Bacteriology, 183*, 6957–6960.
Machnicka, M. A., Milanowska, K., Osman Oglou, O., Purta, E., Kurkowska, M., Olchowik, A., et al. (2013). MODOMICS: A database of RNA modification pathways—2013 update. *Nucleic Acids Research, 41*, D262–D267.

McCusker, K. P., & Fujimori, D. G. (2012). The chemistry of peptidyltransferase center-targeted antibiotics: Enzymatic resistance and approaches to countering resistance. *ACS Chemical Biology, 7*, 64–72.

McCusker, K. P., Medzihradszky, K. F., Shiver, A. L., Nichols, R. J., Yan, F., Maltby, D. A., et al. (2012). Covalent intermediate in the catalytic mechanism of the radical S-adenosyl-L-methionine methyl synthase RlmN trapped by mutagenesis. *Journal of the American Chemical Society, 134*, 18074–18081.

Meyer, K. D., Saletore, Y., Zumbo, P., Elemento, O., Mason, C. E., & Jaffrey, S. R. (2012). Comprehensive analysis of mRNA methylation reveals enrichment in 3′ UTRs and near stop codons. *Cell, 149*, 1635–1646.

Motorin, Y., & Helm, M. (2011). RNA nucleotide methylation. *Wiley Interdisciplinary Reviews: RNA, 2*, 611–631.

Motorin, Y., Muller, S., Behm-Ansmant, I., & Branlant, C. (2007). Identification of modified residues in RNAs by reverse transcription-based methods. *Methods in Enzymology, 425*, 21–53.

Poehlsgaard, J., & Douthwaite, S. (2005). The bacterial ribosome as a target for antibiotics. *Nature Reviews Microbiology, 3*, 870–881.

Ramu, H., Vázquez-Laslop, N., Klepacki, D., Dai, Q., Piccirilli, J., Micura, R., et al. (2011). Nascent peptide in the ribosome exit tunnel affects functional properties of the A-site of the peptidyl transferase center. *Molecular Cell, 41*, 321–330.

Sergeeva, O. V., Bogdanov, A. A., & Sergiev, P. V. (2014). What do we know about ribosomal RNA methylation in *Escherichia coli*? *Biochimie*, 1–9.

Sergiev, P. V., Golovina, A. Y., Prokhorova, I. V., Sergeeva, O. V., Osterman, I. A., Nesterchuk, M. V., et al. (2011). Modification of ribosomal RNA: From enzymes to function. In M. V. Rodnina, W. Wintermeyer, & R. Green (Eds.), *Ribosomes: Structure, function, and dynamics* (pp. 97–110). Vienna: Springer Verlag GmbH.

Silakov, A., Grove, T. L., Radle, M. I., Bauerle, M. R., Green, M. T., Rosenzweig, A. C., et al. (2014). Characterization of a cross-linked protein-nucleic acid substrate radical in the reaction catalyzed by RlmN. *Journal of the American Chemical Society, 136*, 8221–8228.

Smith, L. K., & Mankin, A. S. (2008). Transcriptional and translational control of the mlr operon, which confers resistance to seven classes of protein synthesis inhibitors. *Antimicrobial Agents and Chemotherapy, 52*, 1703–1712.

Sofia, H. J., Chen, G., Hetzler, B. G., Reyes-Spindola, J. F., & Miller, N. E. (2001). Radical SAM, a novel protein superfamily linking unresolved steps in familiar biosynthetic pathways with radical mechanisms: Functional characterization using new analysis and information visualization methods. *Nucleic Acids Research, 29*, 1097–1106.

Tim, P., Katharina, H., & Matthias, S. (2010). Detection of cytosine methylation in RNA using bisulfite sequencing. *Cold Spring Harbor Protocols, 10*, 1–7.

Tscherne, J. S., Nurse, K., Popienick, P., & Ofengand, J. (1999). Purification, cloning, and characterization of the 16S RNA m2G1207 methyltransferase from *Escherichia coli*. *Journal of Biological Chemistry, 274*, 924–929.

Van Aerschot, A. A., Mamos, P., Weyns, N. J., Ikeda, S., De Clercq, E., & Herdewijn, P. A. (2002). Antiviral activity of C-alkylated purine nucleosides obtained by cross-coupling with tetraalkyltin reagents. *Journal of Medicinal Chemistry, 36*, 2938–2942.

van Tilburg, E. W., Gremmen, M., von Frijtag Drabbe Kunzel, J., de Groote, M., & IJzerman, A. P. (2003). 2,8-Disubstituted adenosine derivatives as partial agonists for the adenosine A2A receptor. *Bioorganic and Medicinal Chemistry, 11*, 2183–2192.

Vazquez-Laslop, N., Ramu, H., Klepacki, D., Kannan, K., & Mankin, A. S. (2010). The key function of a conserved and modified rRNA residue in the ribosomal response to the nascent peptide. *EMBO Journal, 29*, 3108–3117.

Vester, B., & Long, K. S. (2009). Antibiotic resistance in bacteria caused by modified nucleosides in 23S ribosomal RNA. In H. Grosjean (Ed.), *DNA and RNA modification enzymes: Structure, mechanism, function and evolution* (pp. 537–549). Austin, TX: Landes Bioscience.
Wilson, D. N. (2014). Ribosome-targeting antibiotics and mechanisms of bacterial resistance. *Nature Reviews Microbiology, 12*, 35–48.
Yan, F., & Fujimori, D. G. (2011). RNA methylation by radical SAM enzymes RlmN and Cfr proceeds via methylene transfer and hydride shift. *Proceedings of the National Academy of Sciences of the United States of America, 108*, 3930–3934.
Yan, F., LaMarre, J. M., Rohrich, R., Wiesner, J., Jomaa, H., Mankin, A. S., et al. (2010). RlmN and Cfr are radical SAM enzymes involved in methylation of ribosomal RNA. *Journal of the American Chemical Society, 132*, 3953–3964.

AUTHOR INDEX

Note: Page numbers followed by "*f*" indicate figures and "*t*" indicate tables.

H

I

J

K

L

M

Q

R

S

T

U

V

SUBJECT INDEX

Note: Page numbers followed by "*f*" indicate figures and "*t*" indicate tables.

Q

R

U

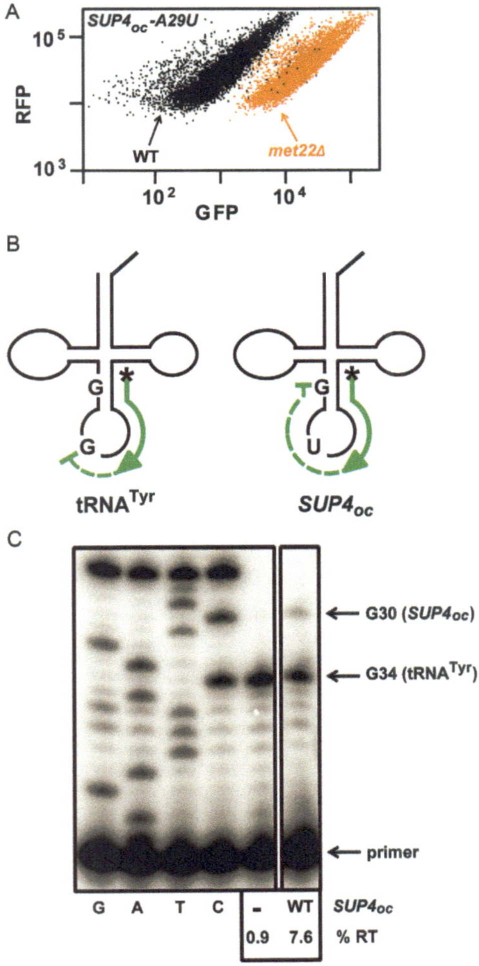

Matthew J. Payea *et al.*, Figure 3 Methods for validating $SUP4_{oc}$ RTD candidates identified through a high-throughput approach. (A) Scatterplot of flow cytometry of the reconstructed $SUP4_{oc}$-*A29U* variant in WT (RTD on, median GFP/RFP = 0.045 ± 0.003) and *met22*Δ cells (RTD off, median GFP/RFP = 0.47 ± 0.06) with the integrated RNA-ID reporter. (B) Schematic representation of the poison primer extension assay with ddCTP used for determining levels of $SUP4_{oc}$ variants. (C) The poison primer extension assay measures the levels of $SUP4_{oc}$ in bulk RNA. Bulk RNA from the indicated strains was analyzed by primer extension in the presence of ddCTP. A sequencing ladder is on the left.

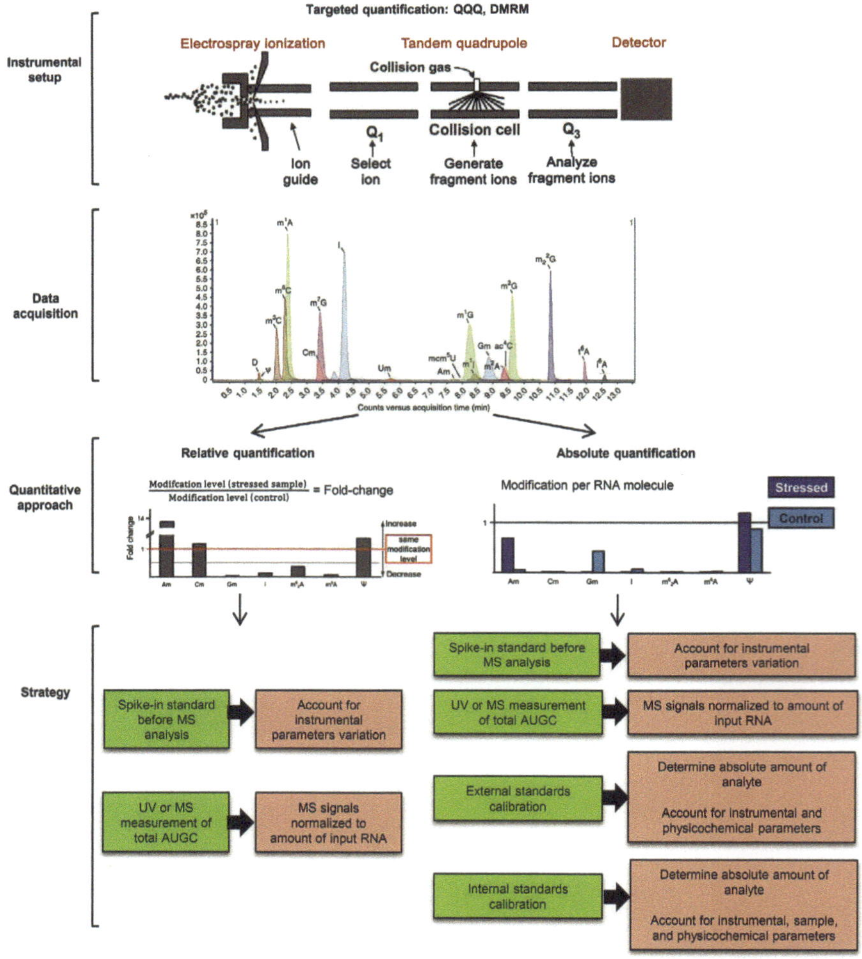

Weiling Maggie Cai *et al.*, Figure 3 Conceptual workflow for targeted quantification of modified ribonucleosides by LC–MS/MS. Instrument setup: The MS system is programmed with a work list containing estimated retention times and CID transition information for each ribonucleoside, with dynamic MRM (DMRM) mode for data acquisition. Data acquisition: The signal intensity for each ribonucleoside is acquired in DMRM mode as the ribonucleoside elutes from the HPLC, with the peak area calculated from these data. Quantitative approach: Relative quantification provides fold-change information for stress-induced changes in modification levels (left), while absolute quantification indicates the abundance of the ribonucleoside in measurements per nucleotide or per tRNA molecule (right). *Strategy*: Steps to achieve precision and accuracy in both relative (left) and absolute (right) quantification are illustrated. Refer to main text for detailed description.

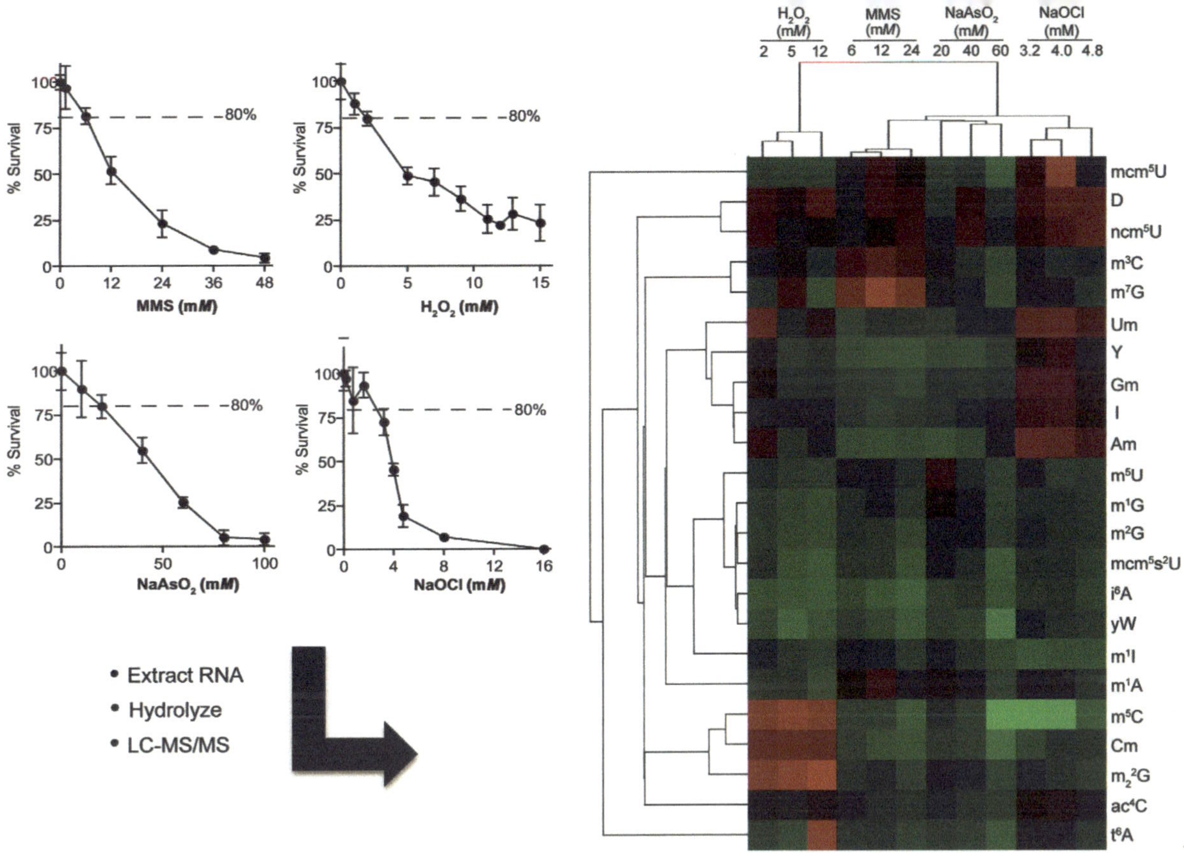

Weiling Maggie Cai *et al.*, Figure 5 Stress-induced reprogramming of tRNA modifications analyzed by quantitative LC–MS/MS. To assess the reprogramming of tRNA modifications in cells exposed to different toxic stresses, cytotoxicity dose–response curves were determined for *S. cerevisiae* exposed H_2O_2, MMS, $NaAsO_2$, and NaOCl (left panel). Following exposure of cells to equitoxic doses of each toxicant (LC_{20}, LD_{50}, LD_{80}), tRNA was isolated and hydrolyzed, and each modified ribonucleoside was subjected to LC–MS/MS analysis. The data were used to calculate fold-change values for exposed cells compared to unexposed controls, with the fold-change data analyzed by hierarchical clustering analysis (right panel) to reveal stress-induced tRNA modification signatures. *Cytotoxicity plots and heatmap image were adapted from Chan et al. (2010).*

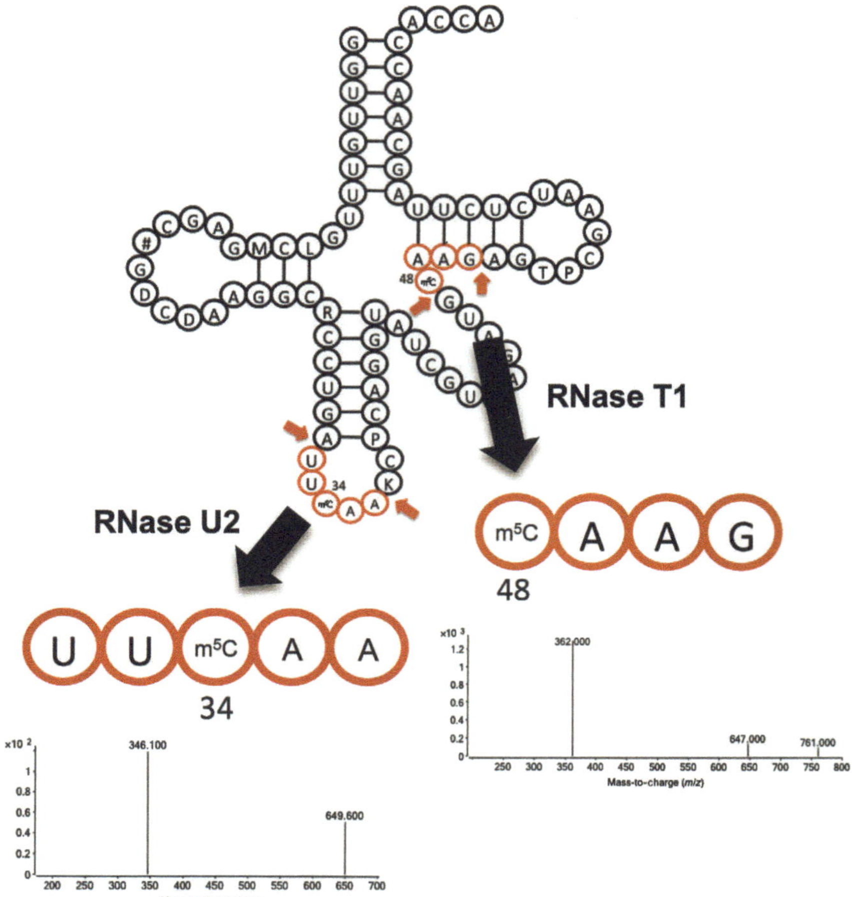

Weiling Maggie Cai *et al.*, Figure 6 Quantitative mapping of modified ribonucleosides in a specific tRNA. To assess stress-induced changes in m^{5}C levels in tRNA$^{\text{Leu}(\underline{\text{C}}\text{AA})}$, two RNase-generated oligonucleotides possessing m^{5}C-modified sites in this tRNA were identified using the Mongo Oligo Mass Calculator. RNase T1 yields a 4-mer oligonucleotide harboring either C or m^{5}C at position 48 ($\underline{\text{C}}$AAG), while RNase U2 produces a 5-mer oligonucleotide with C or m^{5}C at the wobble position 34 of (UU$\underline{\text{C}}$AA). The UUCAA oligonucleotide was sequenced and quantified by LC–MS/MS analysis with affinity-purified tRNA and the CAAG oligonucleotide was analyzed by LC–MS/MS analysis using total tRNA.

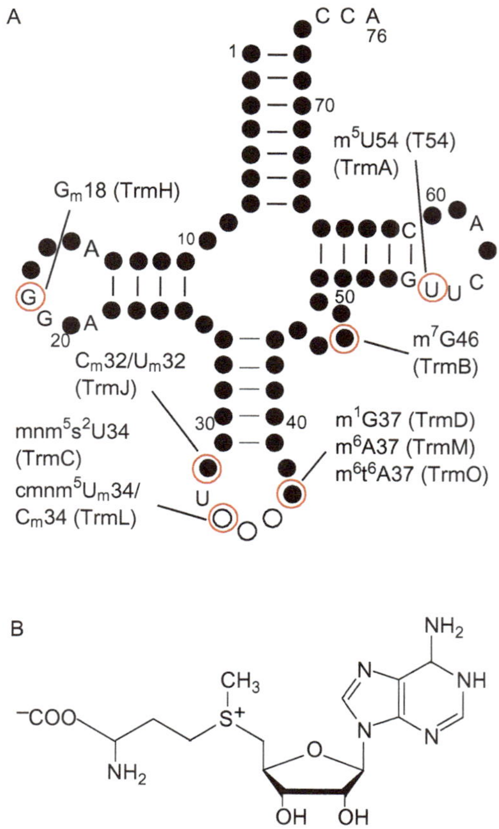

Ya-Ming Hou and Isao Masuda, Figure 1 AdoMet-dependent methyl transfer to *E. coli* tRNA. (A) The cloverleaf structure of tRNA is made up of nucleotides as circles. Conserved nucleotides are shown in letters, nonconserved nucleotides are shown as closed circles, and the anticodon nucleotides are shown as open circles. The position of each AdoMet-dependent methylation (catalyzed by the *E. coli* enzyme in parentheses) is indicated. The nucleotide numbering is based on the standard sequence framework of tRNA (Sprinzl, Horn, Brown, Ioudovitch, & Steinberg, 1998). (B) The chemical structure of AdoMet.

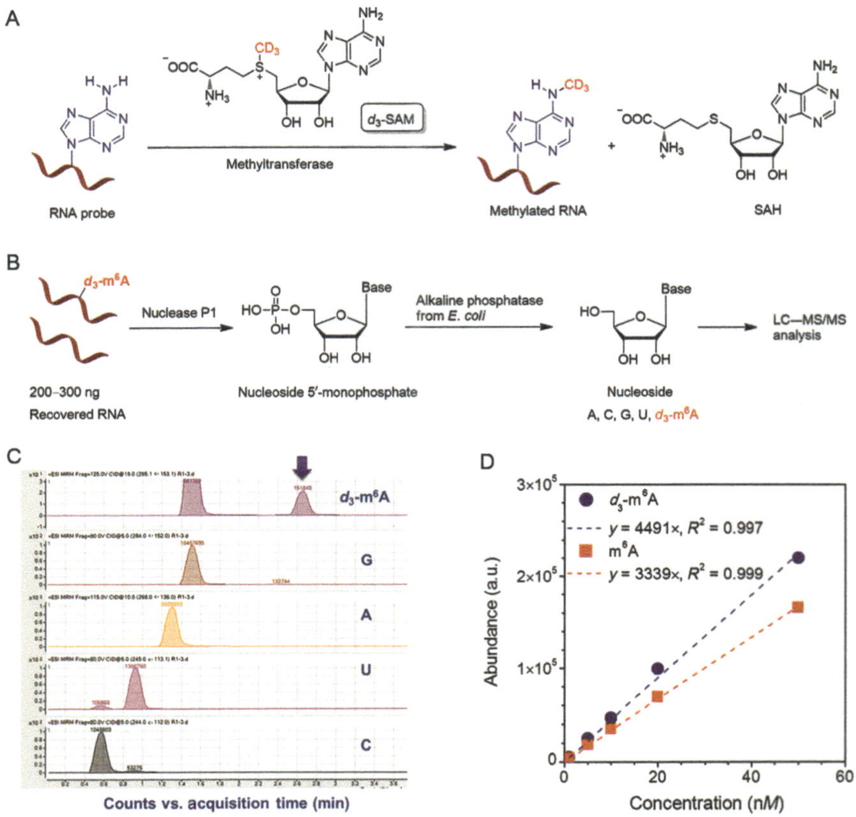

Jianzhao Liu *et al.*, Figure 2 Typical procedure for quantification of m^6A methylation reaction yield by LC–MS/MS. (A) Schematic illustration of m^6A biochemistry reaction. (B) A flowchart of digestion of RNA for quantitative LC–MS/MS analysis. (C) On a LC–MS/MS system, the amount of each nucleoside was quantified by its integration area in the corresponding chromatogram. (D) The m^6A and d_3-m^6A standard calibration curves were obtained from five standard samples with different amounts of pure m^6A and d_3-m^6A, respectively. For *in vitro* biochemistry reaction, the d_3-m^6A calibration curve is applied to calculate the methylation reaction yield, while an m^6A calibration curve could be used to determine the m^6A content from cellular mRNA or other RNAs. *Figure (D) is adapted from supplementary figure 2B in Liu et al. (2014), with permission from Nature Publishing Group.*

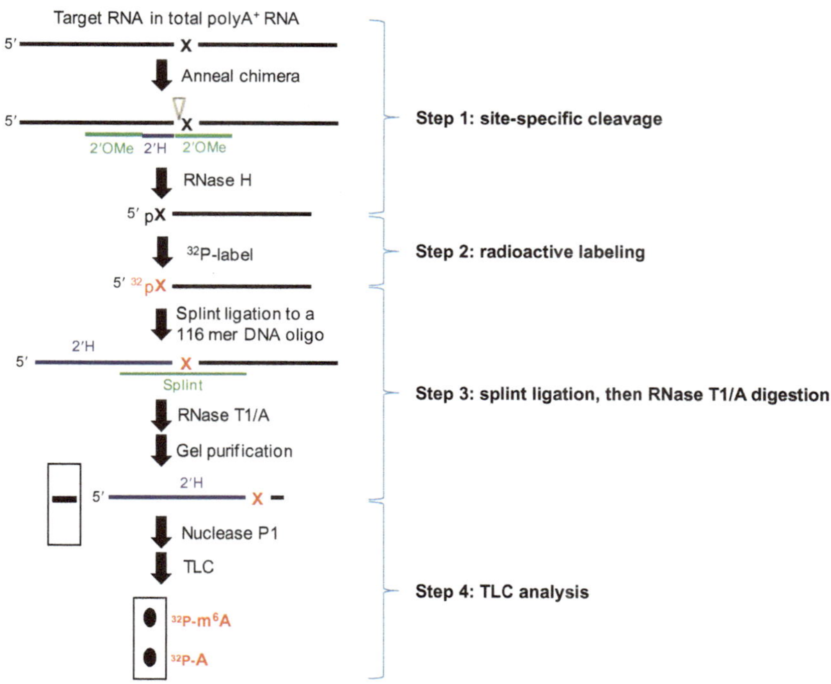

Nian Liu and Tao Pan, Figure 1 Schematic diagram of SCARLET. SCARLET consists of four steps: site-specific cleavage at the target nucleotide site, radioactive labeling of the target nucleotide, splint-assisted ligation followed by RNases T1/A digestion, and thin layer chromatography (TLC).

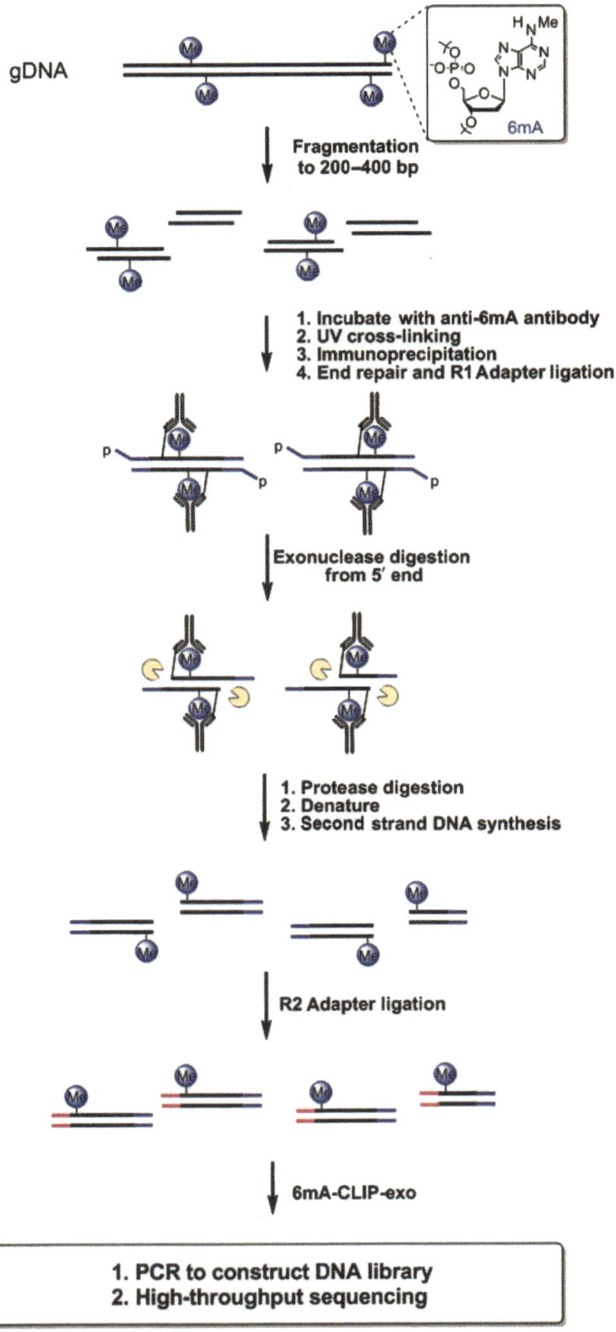

Kai Chen *et al.*, Figure 2 Scheme of 6mA-CLIP-exo. *Adapted from Figure 2 in Fu et al., 2015.*

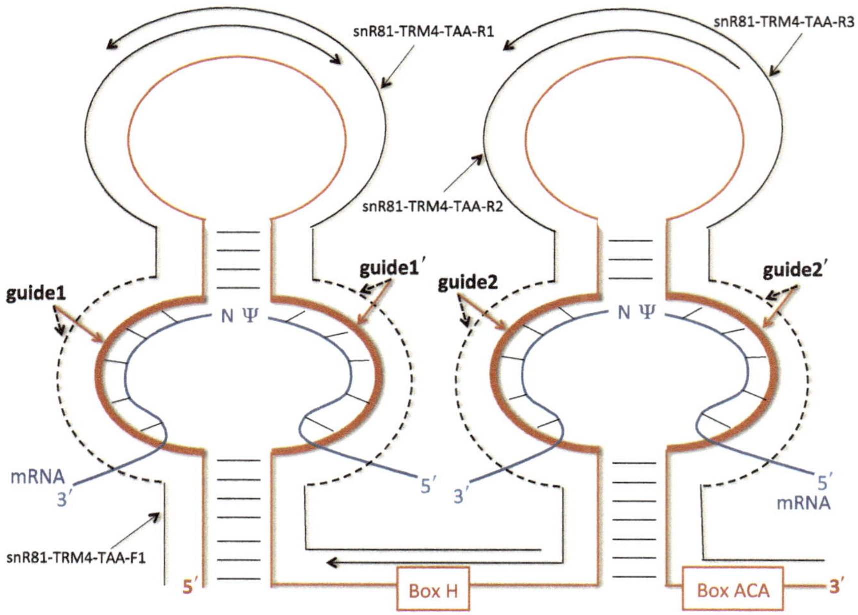

Guowei Wu *et al.*, Figure 2 Schematic representation of box H/ACA guide RNA and its interaction with its substrate RNAs. The guide RNA (red) folds into a hairpin-hinge-hairpin-tail structure. The pseudouridylation pockets responsible for target site specification are the internal loops (thick red lines) within the hairpin structures. The H and ACA boxes and the guide sequences (guide1, guide1′, guide2, and guide2′) are indicated. The mRNA substrates (blue) are paired with the guide sequences of the guide RNA, as indicated by short lines between strands. The pseudouridylated uridine (Ψ) and its downstream residue (N) in the mRNA substrates are left unpaired. The four overlapping primers, used for guide RNA construction (see text), are also shown (black lines). The dotted black lines denote specific guide sequences.

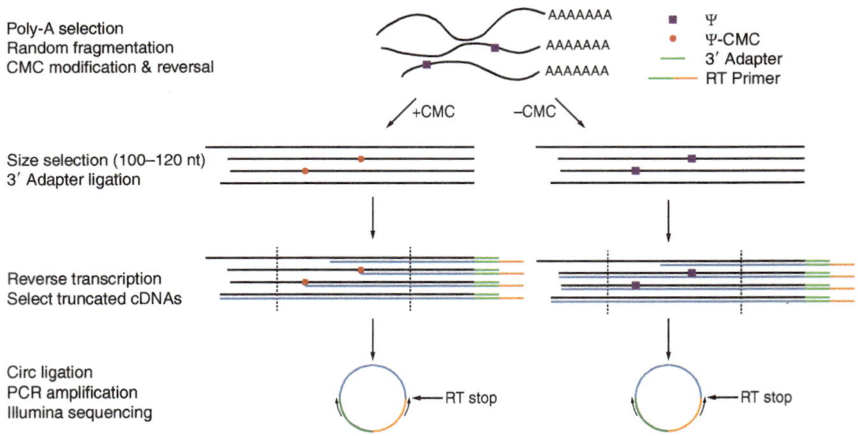

Thomas M. Carlile *et al.*, Figure 2 A schematic of Pseudo-seq library preparation. CMC-dependent RT stops correspond to sites of pseudouridylation. See Section 3 for details.

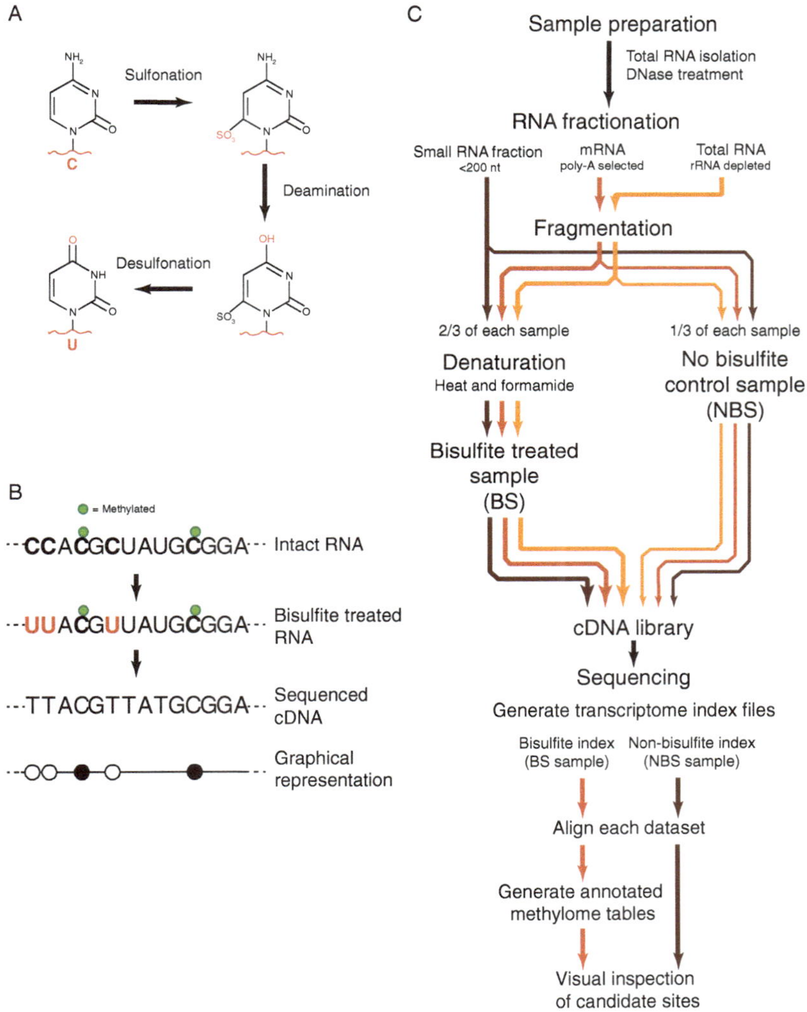

Vahid Khoddami *et al.*, Figure 1 (A) Principle of bisulfite conversion. Unmethylated cytosine is sulfonated by sodium bisulfite and is subsequently deaminated to uridine. Methylated cytosine is refractory, and remains as cytosine. (B) Impact of cDNA synthesis. Upon conversion to cDNA, unmethylated cytosine is read by sequencing as thymine. (C) Bisulfite treatment of RNA. A flowchart of RNA bisulfite sequencing and analysis is depicted. Fractionation of RNA samples is recommended for full coverage of all possible sites in different RNA populations. It is also useful to have a matched nontreated control for each fraction to filter out false positives during the analysis phase. For high-throughput bisulfite sequencing and analysis, it is recommended to align both BS and NBS samples. Once methylome tables are calculated for the BS samples, these can be compared to the NBS samples to filter out false positives arising from incorrect mapping.

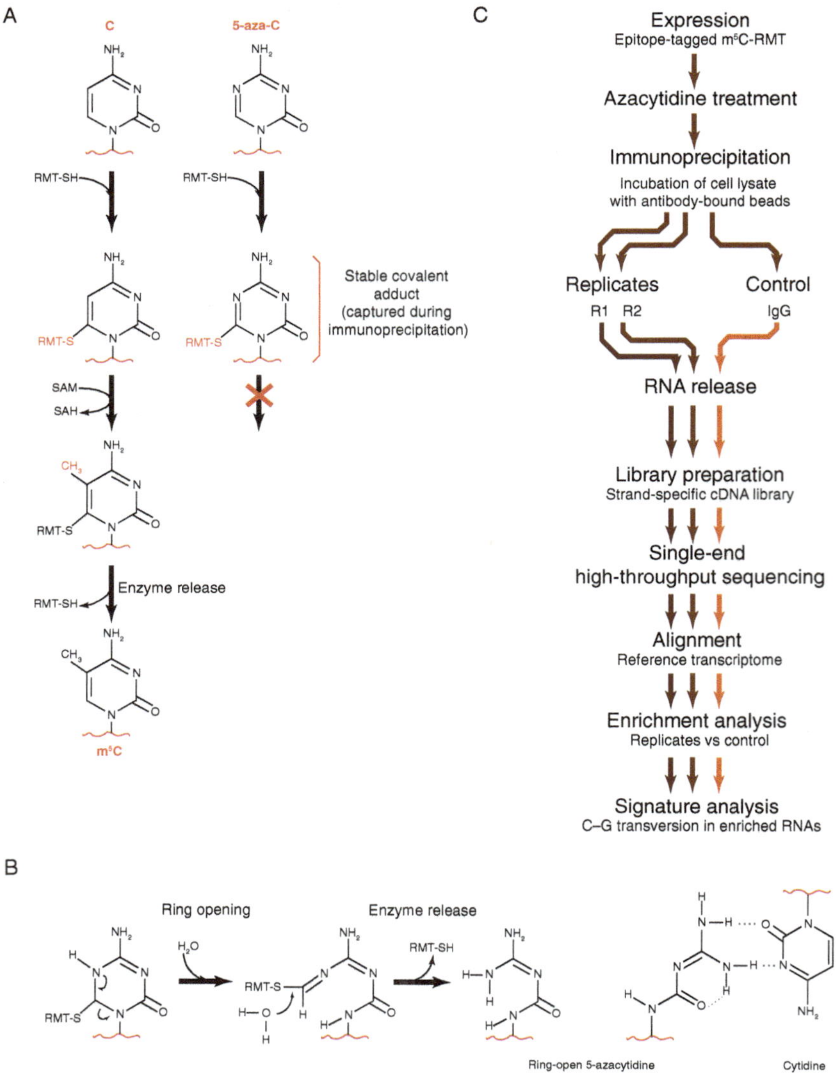

Vahid Khoddami *et al.*, Figure 3 (A) Catalytic mechanism of m^{5}C-RMTs. During catalysis, the m^{5}C-RMT forms a covalent bond with the target cytosine, and is released upon catalyzing the methylation reaction. The nucleotide analog 5-aza-C inhibits the final enzyme release, leaving the enzyme covalently bound to its substrate and unable to participate in further catalysis. (B) C > G transversion. When the enzyme is subsequently released from the RNA, it often leaves the 5-aza-C in a "ring-open" conformation, which can then base-pair with another cytosine, leading to a C > G transversion upon sequencing. (C) Aza-IP. For Aza-IP, it is recommended to set up two replicate samples (R1 and R2) and a matched negative control pulldown using a nonspecific antibody (e.g., IgG), and the same lysate as the pulldown. During analysis, this negative control sample can be used both to calculate true enrichment rates of transcripts, and to filter SNPs for identifying the true C > G transversions.

Kolibri 360 GmbH
Ferdinand-Jühlke-Straße 7
99095 Erfurt, Deutschland
produktsicherheit@kolibri360.de